Lecture Notes in Computer Science 7137

Commenced Publication in 1973
Founding and Former Series Editors:
Gerhard Goos, Juris Hartmanis, and Jan van Leeuwen

Till Mossakowski   Hans-Jörg Kreowski (Eds.)

Recent Trends in Algebraic Development Techniques

20th International Workshop, WADT 2010
Etelsen, Germany, July 1-4, 2010
Revised Selected Papers

 Springer

Volume Editors

Till Mossakowski
DFKI – Deutsches Forschungszentrum für künstliche Intelligenz
Cyber-Physical Systems
28359 Bremen, Germany
E-mail: till.mossakowski@dfki.de

Hans-Jörg Kreowski
Universität Bremen
Fachbereich Mathematik und Informatik
28359 Bremen, Germany
E-mail: kreo@informatik.uni-bremen.de

ISSN 0302-9743 e-ISSN 1611-3349
ISBN 978-3-642-28411-3 ISBN 978-3-642-28412-0 (eBook)
DOI 10.1007/978-3-642-28412-0
Springer Heidelberg Dordrecht London New York

Library of Congress Control Number: 2012931155

CR Subject Classification (1998): F.3, D.2.4, D.3.1, F.4, I.1, C.2.4

LNCS Sublibrary: SL 1 – Theoretical Computer Science and General Issues

Typesetting: Camera-ready by author, data conversion by Scientific Publishing Services, Chennai, India

Printed on acid-free paper

Springer is part of Springer Science+Business Media (www.springer.com)

Preface

This volume contains three invited papers and 15 selected papers from the 20th International Workshop on Algebraic Development Techniques (WADT 2010) which took place at Schloss Etelsen, a castle near Bremen, at the beginning of July 2010 (see also the picture on p. 16).

The algebraic approach to system specification encompasses many aspects of the formal design of software systems. Originally born as a formal method for reasoning about abstract data types, it now covers new specification frameworks and programming paradigms (such as object-oriented, aspect-oriented, agent-oriented, logic and higher-order functional programming) as well as a wide range of application areas (including information systems, concurrent, distributed and mobile systems). The workshop provided an opportunity to present recent and ongoing work, to meet colleagues, and to discuss new ideas and future trends. Typical topics of interest are:

- Foundations of algebraic specification
- Other approaches to formal specification including process calculi and models of concurrent, distributed and mobile computing
- Specification languages, methods, and environments
- Semantics of conceptual modeling methods and techniques
- Model-driven development
- Graph transformations, term rewriting and proof systems
- Integration of formal specification techniques
- Formal testing and quality assurance validation, and verification

The workshop took place under the auspices of IFIP WG 1.3 and was sponsored by the University of Bremen. The event was organized by the Computer Science Department of the University of Bremen and DFKI Bremen group Safe and Secure Cognitive Systems. The local organizers were Mihai Codescu, Hans-Jörg Kreowski (Chair), Christian Maeder, Till Mossakowski (Chair), Sylvie Rauer, and Lutz Schröder.

The scientific program consisted of three invited talks by Hans-Dieter Ehrich, Jan Kofron, and Martin Wirsing and 32 presentations based on selected abstracts. After the workshop, all authors were invited to submit full papers, which underwent a thorough refereeing process, using EasyChair. Each paper was reviewed by three reviewers. We would like to thank both the Program Committee (PC) and the additional reviewers for their work. Special thanks go to José Luiz Fiadeiro, who on behalf of the PC Chairs handeled the PC Chair function of EasyChair, in order to keep the reviewing process strictly anonymous to all authors. Moreover, we are very grateful to Alfred Hofmann and his team at Springer for the excellent cooperation once again.

As this was the 20th ADT Workshop, some reminiscing may be allowed. The workshop series started at Sorpesee (Germany) in 1982, and further events took place in Passau (Germany, 1983), Bremen (Germany, 1984), Braunschweig (Germany, 1986), Gullane (UK, 1987), Berlin (Germany, 1988), Wusterhausen (Germany, 1990), Dourdan (France, 1991), Caldes de Malavella (Spain, 1992), S. Margherita Ligure (Italy, 1994), Oslo (Norway, 1995), Tarquinia (Italy, 1997), Lisbon (Portugal, 1998), Château de Bonas (France, 1999), Genova (Italy, 2001), Frauenchiemsee (Germany, 2002), Barcelona (Spain, 2004), La Roche en Ardenne (Belgium, 2006), and Pisa (Italy, 2008). The 8th to 11th WADT events were held jointly with the COMPASS Workshops, the yearly meetings of the ESPRIT Basic Research Working Group A Compehensive Algebraic Approach to System Specification and Development. Starting with the 12th WADT, the name was changed from Workshop on Abstract Data Types to Workshop on Algebraic Development Techniques while keeping the acronym. While the workshop series started as a regular meeting of the abstract data type community, it soon became clear that this was a too narrow objective. The new name reflects the widening scope and the broadening range of topics of the workshop. It should also be noted that the ADT workshops took place under the auspices of the IFIP Working Group 1.3 (Foundation of System Specifications) for nearly two decades. Since 2005, the CMCS and WADT communities have jointly organized the biannual conference CALCO together in turn with their ordinary workshops.

The first, second, fourth, and sixth proceedings appeared as technical reports, the third proceedings as Informatik-Fachberichte 116 under the title *Recent Trends in Data Type Specification*, the fifth volume and all proceedings from the seventh on were published as Springer Lecture Notes volumes (332, 534, 655, 785,906, 1130, 1376, 1589, 1827, 2267, 2755, 3423, 4409, 5486). With the name of the workshop, the proceedings were renamed as *Recent Trends in Algebraic Development Techniques*.

Altogether, it is quite justified to call WADT an ongoing success story. Therefore we wish it at least 20 further editions.

October 2011

Till Mossakowski
Hans-Jörg Kreowski

Organization

Program Committee

Michel Bidoit	INRIA Saclay-Île-de-France, France
Mihai Codescu	DFKI GmbH, Bremen, Germany
Andrea Corradini	Università di Pisa, Italy
José Luiz Fiadeiro	University of Leicester, UK
Fabio Gadducci	Università di Pisa, Italy
Reiko Heckel	University of Leicester, UK
Rolf Hennicker	Ludwig-Maximilians-Universität München, Germany
Hans-Jörg Kreowski	University of Bremen (Co-chair), Germany
Narciso Marti-Oliet	Universidad Complutense de Madrid, Spain
Till Mossakowski	DFKI Lab Bremen (Co-chair), Germany
Fernando Orejas	Universitat Politècnica de Catalunya, Barcelona, Spain
Francesco Parisi-Presicce	Università di Roma La Sapienza, Italy
Markus Roggenbach	Swansea University, UK
Grigore Rosu	Runtime Verification, Inc., and The University of Illinois at Urbana-Champaign, USA
Donald Sannella	University of Edinburgh, UK
Lutz Schröder	DFKI Bremen and Universität Bremen, Germany
Andrzej Tarlecki	Warsaw University, Poland

Steering Committee

Michel Bidoit	INRIA Saclay-Île-de-France, France
Andrea Corradini	Università di Pisa, Italy
José Luiz Fiadeiro	University of Leicester, UK
Rolf Hennicker	Ludwig-Maximilians-Universität München, Germany
Hans-Jörg Kreowski	University of Bremen, Germany
Narciso Marti-Oliet	Universidad Complutense de Madrid, Spain
Till Mossakowski	DFKI Lab Bremen (Chair), Germany
Fernando Orejas	Universitat Politècnica de Catalunya, Barcelona, Spain
Francesco Parisi-Presicce	Università di Roma La Sapienza, Italy
Grigore Rosu	Runtime Verification, Inc., and The University of Illinois at Urbana-Champaign, USA
Andrzej Tarlecki	Warsaw University, Poland

Additional Reviewers

Bauer, Sebastian
Baumeister, Hubert
Bocchi, Laura
Bruni, Roberto
Caires, Luís
Cîrstea, Corina
Dietrich, Dominik
Ehrig, Hartmut
Gîrlea, Codruţa
Goncharov, Sergey
Heckel, Reiko
Klin, Bartek
Knapp, Alexander

Lluch Lafuente, Alberto
Ölveczky, Peter
Palomino, Miguel
Peña, Ricardo
Popescu, Andrei
Rodríguez-Hortalá, Juan
Serbanuta, Traian
Ulidowski, Irek
van Breugel, Franck
Vandin, Andrea
Wasowski, Andrzej
Wolter, Uwe

Table of Contents

Building a Modal Interface Theory
for Concurrency and Data[*]

Sebastian S. Bauer, Rolf Hennicker, and Martin Wirsing

Ludwig-Maximilians-Universität München, Germany

Abstract. Treating control and data in an integrated way is an important issue in system development. We discuss a compositional approach for specifying concurrent behavior of components with data states on the basis of interface theories. The dynamic aspects of a system are specified by modal I/O-transition systems, whereas changes of data states are specified by pre- and postconditions. In this setting we study refinement and behavioral compatibility of components. We show that refinement is compositional and that compatibility is preserved by refinement; thus the requirements for interface theories are satisfied. As a consequence, our approach supports independent implementability and reusability of concurrently interacting components with data states.

1 Introduction: Basic Principles of System Design

Algebraic development techniques support the rigorous construction of verifiable software systems. Having their origin in the algebraic specification of abstract data types, algebraic techniques have been particularly valuable for the formal development of function-oriented systems; in particular, they provide them with a precise mathematical semantics based on isomorphically closed classes of algebras. In the simplest case an algebraic specification consists just of a pair (Σ, E) where Σ is an algebraic signature, determining sorts and function symbols, and E is a set of equations describing the required properties of a program. For building specifications of complex systems algebraic structuring operators have been introduced, most importantly for combining smaller specifications to larger ones or for parametrized specifications and their instantiations [12,25]. Several variations of structuring operators have been realized in algebraic specification languages like, e.g., OBJ [15], ASL [24], ACT-ONE and ACT TWO [11], Maude [8] and CASL [1], for an overview see [26].

Structuring specifications is important to cope with the complexity of large scale software systems. It is often called the "horizontal" dimension of software development. But for constructing correct implementations from abstract specifications, it is equally important to be able to refine specifications in a stepwise

---

[*] This work has been partially sponsored by the EU project ASCENS, 257414. The first author has been partially supported by the German Academic Exchange Service (DAAD), grant D/10/46169.

T. Mossakowski and H.-J. Kreowski (Eds.): WADT 2010, LNCS 7137, pp. 1–12, 2012.

manner. Moving from abstract to concrete specifications is often called the "vertical" dimension of software development. Algebraic specifications come with formally defined refinement relations and corresponding verification techniques.

Depending on the features of concrete specification languages and on their underlying semantics, e.g. initial, loose or observational, various formal refinement relations have been proposed. For instance, the idea of the loose semantics approach goes back to Hoare 1972 [17] with the meaning that a specification describes all its correct implementations. In this case the semantics of a specification S is given by the class $Mod(S)$ of all models of the specification and refinement is simply defined by model class inclusion.

Refinement:
    S' refines S, written $S' \leq S$, if $Mod(S') \subseteq Mod(S)$.

Of course, for stepwise program development it is crucial that refinements can be vertically composed, i.e. that the refinement relation is transitive. In the loose case this is a trivial consequence of the definition.

For applying structuring and refinement concepts in a methodologically useful way it is essential that these concepts fit properly together, i.e. that refinement is compatible with composition. For instance, assume that $\otimes$ denotes a binary operator for the combination of specifications. Then the principle of horizontal composition expresses the following requirement, for specifications S, S', T and T'.

Horizontal composition:
    If $S' \leq S$ and $T' \leq T$, then $S' \otimes T' \leq S \otimes T$.

Horizontal and vertical composition are indispensable prerequisites for so-called compositional system development where a system is composed of independently developed parts. This "holy grail" of compositionality has been proven to hold for several algebraic specification formalisms including the ACT TWO and the ASL languages (see e.g. [12,24]).

These algebraic specification approaches are highly developed tools for constructing systems with functional behavior; but they are not tailored towards the specification and analysis of systems exhibiting a dynamic and concurrent behavior as it is typical, e.g., for reactive components. For this purpose techniques based on execution traces, automata, Petri nets or (nondeterministic) rewriting are more appropriate. On the other hand, refinement and composition are obviously important principles for the development of dynamic systems as well. The methodological requirements of vertical and horizontal composition remain valid in this context. But these properties are not sufficient in the case of reactive systems which heavily rely on interactions of components with the environment. For such systems, it is essential that no communication errors occur when components interact. This is ensured by a compatibility property which is usually formulated on the level of the interfaces of the components (and not on components directly) in order to abstract from the the particular realizations of the components. For compositionality one has to guarantee that components which are correct w.r.t. their interface specifications are interacting properly with each other [10], i.e. refinement of interface specifications must preserve interface compatibility.

Horizontal compositionality and preservation of interface compatibility are fundamental properties of any formalism that is supposed to support independent implementability of interface specifications for reactive components. In addition we require that compatibility of two interface specifications implies that they can actually be composed. These principles can be formally captured by the following notion of an interface theory (inspired by [10]).

An *interface theory* is a tuple $(\mathcal{A}, \otimes, \leq, \leftrightarrows)$ consisting of a class $\mathcal{A}$ of interface specifications, a partial composition operator $\otimes : \mathcal{A} \times \mathcal{A} \to \mathcal{A}$, a reflexive and transitive refinement relation $\leq \subseteq \mathcal{A} \times \mathcal{A}$, and a symmetric compatibility relation $\leftrightarrows \subseteq \mathcal{A} \times \mathcal{A}$, such that the following conditions are satisfied. Let $S, S', T, T' \in \mathcal{A}$ be specifications.

(1) *Compatibility implies composability*
    If $S \leftrightarrows T$ then $S \otimes T$ is defined.
(2) *Compositional refinement*
    If $S' \leq S$ and $T' \leq T$ and $S \otimes T$ is defined,
    then $S' \otimes T'$ is defined and $S' \otimes T' \leq S \otimes T$.
(3) *Preservation of compatibility*
    If $S \leftrightarrows T$ and $S' \leq S$ and $T' \leq T$, then $S' \leftrightarrows T'$.

In the following we will present a compositional approach - called MIOD - for the specification of interfaces for concurrently running reactive components with encapsulated data states. We model interfaces of reactive components by modal I/O-transition systems (as introduced in [18]) enhanced by data constraints (as in [4]) and show that the MIOD approach forms an interface theory, i.e. it satisfies the vertical and horizontal composition properties and preserves interface compatibility.

2   Modal Input/Output Automata with Data Constraints and Their Refinement

Components interact with the environment by accepting inputs and sending outputs which both are modeled by incoming or outgoing operation calls, for provided and required operations resp. An *I/O-operation signature* $O = O^{prov} \uplus O^{req} \uplus O^{int}$ consists of pairwise disjoint sets of provided, required, and internal operations, resp. An *I/O-state signature* $V = V^{prov} \uplus V^{req} \uplus V^{int}$ consists of pairwise disjoint sets of provided, required and internal state variables, resp. State variables are used to model data states of components. The provided and the internal state variables together form the "local" variables of a component, denoted by $V^{loc} = V^{prov} \uplus V^{int}$; the required state variables are used to access the visible data states of the environment. An *I/O-signature* is a pair $\Sigma = (V, O)$ consisting of an I/O-state signature V and an I/O-operation signature O.

We extend modal I/O-transition systems (MIOs) introduced in [18] to take into account constraints on data states. The resulting transition systems, called

MIODs, provide interface specifications for components with data states. They do not only specify the control flow of behaviors but also the effect on data states in terms of pre- and postconditions. Moreover, the modalities stemming from MIOs allow additionally to distinguish may and must transitions thus supporting a flexible concept for refinement.

For specifying pre- and postconditions we assume given a set $\mathcal{S}(W, X)$ of *state predicates* and a set $\mathcal{T}(W, W', X)$ of *transition predicates*. State predicates, often denoted by φ, refer to single states and transition predicates, often denoted by π, to pairs of states (pre- and poststates). Given an I/O-signature $\Sigma = (V, O)$, the set $\mathcal{L}(\Sigma)$ of Σ-labels consists of the following expressions where operations (of any kind) are surrounded by pre- and postconditions which may contain the operation's formal parameters, denoted by $par(op)$, as logical variables (being disjoint from the state variables).

- $[\varphi] op?[\pi]$ with $\varphi \in \mathcal{S}(V, par(op))$, $op \in O^{prov}$, $\pi \in \mathcal{T}(V, V^{loc}, par(op))$.
- $[\varphi] op![\pi]$ with $\varphi \in \mathcal{S}(V, par(op))$, $op \in O^{req}$, $\pi \in \mathcal{T}(V, V^{req}, par(op))$.
- $[\varphi] op; [\pi]$ with $\varphi \in \mathcal{S}(V, par(op))$, $op \in O^{int}$, $\pi \in \mathcal{T}(V, V^{loc}, par(op))$.

The symbols "?" ("!",";") are just used as decorations to emphasize that op is a provided (required, internal) operation, resp. An *input* label $[\varphi] op?[\pi]$ models that a provided operation op can be invoked under the precondition φ and then the postcondition π will hold after the execution of op. The postcondition π of an input is a transition predicate which must only specify changes of data states for local state variables. An *output* label $[\varphi] op![\pi]$ models that a component issues a call to a required operation op if the precondition φ is satisfied and after execution of the invoked operation the component expects that the postcondition π holds. The postcondition of an output is a transition predicate which must only specify the expected changes of the visible data states in the environment, i.e. for required state variables. Hence, outputs are not expected to alter the data state of the calling component itself. Finally, an *internal* label $[\varphi] op; [\pi]$ stands for the execution of an internal operation op.

Definition 1 (MIOD). *A modal I/O automaton with data constraints (MIOD)*

$$S = (\Sigma, St, init, \varphi^0, \Delta^{\mathsf{may}}, \Delta^{\mathsf{must}})$$

consists of an I/O-signature Σ, a finite set of states St, the initial (control) state $init \in St$, the initial (data) state predicate $\varphi^0 \in \mathcal{S}(V^{loc}, \emptyset)$, a finite may transition relation $\Delta^{\mathsf{may}} \subseteq St \times \mathcal{L}(\Sigma) \times St$, and a finite must transition relation $\Delta^{\mathsf{must}} \subseteq \Delta^{\mathsf{may}}$. The class of all MIODs is denoted by $\mathcal{M}_d$.

Example 1. We exemplify MIODs by specifying a simple protocol of a robot leg, see Fig. 1 (for a more elaborated description of this example, see [3]). Provided and required operations are indicated by incoming arrows on the left border and outgoing arrows on the right border of the frame respectively. The operations $swing(a)$ and $update(x)$ have the parameters a and x respectively which are the logical variables used in predicates. Primed variables refer to the value in the poststate. We assume the initial state predicate *true* (omitted in the figure).

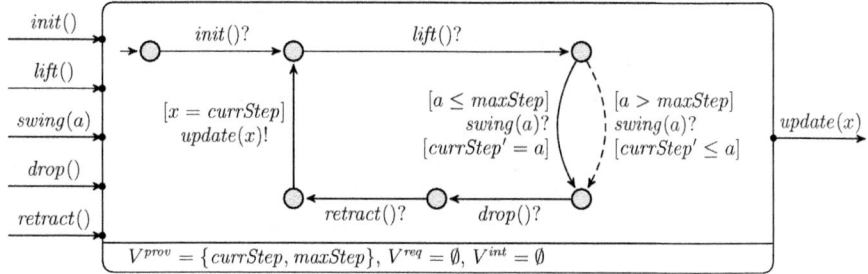

Fig. 1. Specification of a robot leg

Preconditions are written above/in front of and postconditions below/after operation names; conditions of the form [*true*] are omitted. As usual, must (may) transitions are drawn with solid (dashed) arrows, and may transitions originating from must transitions are not drawn.

The leg component has two provided variables *maxStep* (for the leg's maximal step size) and *currStep* (for the current step size). The locomotion of a robot leg usually proceeds in four phases: (1) the leg is lifted, (2) the leg is swung forward, (3) the leg is dropped, and finally (4) the leg is pulling the robot's body forward by retracting the leg. After the first phase (lifting the leg) the leg component must accept all calls to $swing(a)?$ for which $a \leq maxStep$; then the guarantee of the leg component is that in the next state the current step size equals a. The leg component may, however, also accept inputs $swing(a)?$ for which $a > maxStep$. In this case, the guarantee is the weaker condition $maxStep' \leq a$. ∎

In [4] we have provided a semantic foundation of our approach where any MIOD is equipped with a model-theoretic semantics describing the class of all its correct implementations. Implementations are modeled by guarded I/O-transition systems where concrete data states are simply algebras assigning values to state variables. For implementation correctness we have defined in [4] a particular simulation relation taking into account the satisfaction of data constraints before and after a transition has been fired. If a given MIOD has a correct implementation, i.e. its model class is not empty, then the MIOD is consistent.

Let us now turn again to the syntactical aspects of MIODs. For the definition of a (syntactic) refinement relation between MIODs, we follow the basic idea of modal refinement [19] where must transitions of an abstract specification must be respected by the concrete specification and, conversely, may transitions of the concrete specification must be allowed by the abstract one. Concerning the impact of data constraints, every must transitions of an abstract MIOD, say T, with a precondition φ_T must be simulated by a corresponding must transition of a more concrete MIOD, say S, whose precondition does not require more than φ_T does. This condition is formalized by the first item of condition 1 in Def. 2 (by taking into account that it is sufficient if the precondition on a must transition of T is matched by the disjunction of several preconditions distributed over different transitions of S which all maintain the simulation relation). This condition is independent of the kind of the labels. Concerning postconditions the situation is different, because postconditions are not related to the executability of transitions but rather to the specification of admissible poststates after a transition has fired. In this case, if the must transition of T concerns input or internal labels, the corresponding must transition of the refinement S should lead to a postcondition which guarantees the postcondition π_T of T. This is formalized by the second item of condition 1 in Def. 2 (by taking into account the splitting into different transitions as explained above). If a must transition of T concerns an output label, then the postcondition π_T expresses the expectation of T about the next state of the environment. Then, obviously, the postcondition of the refinement should not be stronger than π_T which is formalized, again for the general case of splitting transitions, in the third item of Def. 2(1).

When moving from concrete to abstract specifications concrete may transitions must be allowed by the abstract specification which is formalized in condition 2 of Def. 2. In this case, a similar splitting of transitions is possible just the other way round.

Definition 2 (Modal Refinement). *Let S and T be two MIODs with the same I/O-signature. A binary relation $R \subseteq St_S \times St_T$ is a* modal refinement *between the states of S and T iff for all $(s,t) \in R$,*

1. *from abstract to concrete*

    if $(t, [\varphi_T]op[\pi_T], t') \in \Delta_T^{\mathsf{must}}$ and φ_T is satisfiable
    then there exists $N \geq 0$ and transitions $(s, [\varphi_{S,i}]op[\pi_{S,i}], s_i') \in \Delta_S^{\mathsf{must}}$,
    $0 \leq i \leq N$, such that
       − *$\models \varphi_T \Rightarrow \bigvee_i \varphi_{S,i}$*
       − *for all i, if $op \in O^{prov} \uplus O^{int}$ then $\models \varphi_T \wedge \varphi_{S,i} \wedge \pi_{S,i} \Rightarrow \pi_T$*
       − *for all i, if $op \in O^{req}$ then $\models \varphi_T \wedge \varphi_{S,i} \wedge \pi_T \Rightarrow \pi_{S,i}$*
       − *for all i, $(s_i', t') \in R$*
    are satisfied.

2. *from concrete to abstract*

    if $(s, [\varphi_S]op[\pi_S], s') \in \Delta_S^{\mathsf{may}}$ and φ_S is satisfiable
    then there exists $N \geq 0$ and $(t, [\varphi_{T,i}]op[\pi_{T,i}], t_i') \in \Delta_T^{\mathsf{may}}$, $0 \leq i \leq N$,
    such that
       − *$\models \varphi_S \Rightarrow \bigvee_i \varphi_{T,i}$*
       − *for all i, if $op \in O^{prov} \uplus O^{int}$ then $\models \varphi_S \wedge \varphi_{T,i} \wedge \pi_S \Rightarrow \pi_{T,i}$*
       − *for all i, if $op \in O^{req}$ then $\models \varphi_S \wedge \varphi_{T,i} \wedge \pi_{T,i} \Rightarrow \pi_S$*
       − *for all i, $(s', t_i') \in R$*
    are satisfied.

A state $s \in St_S$ refines a state $t \in St_T$, written $s \leq_{md} t$, iff there exists a modal refinement between the states of S and T containing (s,t). S is a modal refinement of T, written $S \leq_{md} T$, iff $init_S \leq_{md} init_T$ and $\models \varphi_S^0 \Rightarrow \varphi_T^0$.

It can be easily verified that $\leq_{md}$ is a reflexive and transitive relation on the class of all MIODs. Moreover, we have shown in [4], that modal refinement implies inclusion of model classes of two MIODs in the sense of the refinement relation

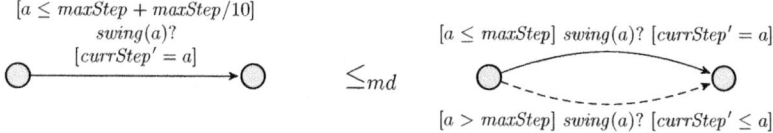

Fig. 2. Refinement of the swing phase of the robot leg

for algebraic specifications with loose semantics discussed in Sect. 1. This is in general not true the other way round such that it remains still a challenge to investigate in syntactic approximations of semantic refinement.

Example 2. In the following we will rather focus on the treatment of data constraints than on the treatment of the control flow. Fig. 2 shows two excerpts of two MIODs specifying the swing phase of the robot leg; the abstract one on the right hand side is like in Fig. 1.

The concrete MIOD on the left hand side refines the abstract MIOD on the right hand side by requiring that the leg component must receive calls to *swing(a)?* for values of the parameter a exceeding the maximal step size at most by ten percent; for all other values of the parameter a the reception is not allowed. Note that the postcondition is the same for the precondition $a \leq maxStep$; for the precondition $maxStep < a \leq maxStep + maxStep/10$ the postcondition is strengthened to $currStep' = a$. ■

3   Compositionality and Compatibility Results

MIODs can be composed to specify the behavior of concurrent systems of interacting components with data states. The composition operator extends the synchronous composition of MIOs [18,5]. The composition is only defined if syntactic restrictions requiring composability of I/O-signatures are satisfied. We require that overlapping of operations only happens on complementary types and that the same holds for state variables. Two composable I/O-signatures $\Sigma_S = (V_S, O_S)$ and $\Sigma_T = (V_T, O_T)$ can be composed to the I/O-signature $\Sigma_S \otimes \Sigma_T = (O_S \otimes O_T, V_S \otimes V_T)$ where shared variables as well as shared operations are internalized.

The synchronous composition $S \otimes_d T$ of two (composable) MIODs S and T synchronizes transitions whose labels refer to shared operations. For instance, a transition with label $[\varphi_S]op![\pi_S]$ of S is synchronized with a transition with label $[\varphi_T]op?[\pi_T]$ of T which results in a transition with label $[\varphi_S \wedge \varphi_T]op[\pi_T]$ where the original preconditions are combined by logical conjunction and only the postcondition π_T of the input is kept. Since the postcondition π_S of the output expresses an assumption on the environment and since input and output actions synchronize to an internal action, π_S is irrelevant for the composition. Transitions whose labels concern shared operations which cannot be synchronized are dropped (as usual) while all other transitions are interleaved in the composition. Concerning modalities we follow the usual modal composition operator [18] which yields a must transition if two must transitions are synchronized and a may transition otherwise. For the precise formal definition of MIOD composition and for the proof of the next theorem see [4].

The following theorem shows that modal refinement is a precongruence with respect to the composition of MIODs which provides our first compositionality result.

Theorem 1. *Let* S, S', T, T' *be MIODs and let* S *and* T *be composable. Then* $S' \leq_{md} S$ *and* $T' \leq_{md} T$ *imply* $S' \otimes_d T' \leq_{md} S \otimes_d T$.

When we want to compose two MIODs we have seen that it is first necessary to check composability which is a purely syntactic condition. But then it is of course important that the two components work properly together without communication errors, i.e. are behaviorally compatible. The following compatibility notion builds upon (strong) modal compatibility as defined in [5]. From the control point of view (strong) compatibility requires that in any reachable state of the product $S \otimes_d T$ of two MIODs S and T, if one MIOD *may* issue an output (in its current control state) then the other MIOD is in a control state where it *must* be able to take the corresponding input.[1] In the context of data states we have the additional requirement that the data constraints of the two MIODs S and T must be compatible. This is respected in condition 1(a) of Def. 3 which requires that the operation call to *op* issued by S under the condition that φ_S holds, must be accepted by T, hence there must exist accepting transitions in T such that the disjunction of their preconditions is not stronger than φ_S. Condition 1(b) of Def. 3 requires that the postcondition π_S of the caller S is respected: for any may transition with a corresponding input label the assumption π_S is not stronger than the guarantee π_T.

For practical verification of compatibility of MIODs, we go through all syntactically reachable states of $S \otimes_d T$ and check whether the pre- and postconditions of synchronizing transitions match. The set of the syntactically reachable states of S is given by $\mathcal{R}(S) = \bigcup_{n=0}^{\infty} \mathcal{R}_n$ where $\mathcal{R}_0(S) = \{init_S\}$ and $\mathcal{R}_{n+1}(S) = \{s' \mid \exists s \in \mathcal{R}_n(S), \exists \ell \in \mathcal{L}(\Sigma) : (s, \ell, s') \in \Delta_S^{may}\}$. Note that taking the syntactically reachable states is, of course, an over-approximation of the (semantically) reachable states in the composition of implementation models.

Definition 3 (Modal Compatibility of MIODs). *Let* S *and* T *be two composable MIODs.* S *and* T *are modally compatible, denoted by* $S \leftrightarrows_d T$, *iff for all reachable states* $(s, t) \in \mathcal{R}(S \otimes_d T)$,

1. *for all* $op \in O_S^{req} \cap O_T^{prov}$, *whenever* $(s, [\varphi_S]op![\pi_S], s') \in \Delta_S^{may}$ *and* φ_S *is satisfiable then*
   (a) *there exists* $(t, [\varphi_{T,i}]op?[\pi_{T,i}], t_i') \in \Delta_T^{must}$, $0 \leq i \leq N$, *such that* $\models \varphi_S \Rightarrow \bigvee_i \varphi_{T,i}$, *and*

---

[1] We follow the "pessimistic" approach to compatibility where two components should be compatible in any environment, in contrast to the "optimistic" approach pursued in [10,18] which relies on the existence of a "helpful" environment.

(b) for all $(t, [\varphi_T]op?[\pi_T], t') \in \Delta_T^{\mathsf{may}}$, it holds that $\models \varphi_S \wedge \varphi_T \wedge \pi_T \Rightarrow \pi_S$;

2. *symmetrically for all $op \in O_T^{req} \cap O_S^{prov}$.*

We can now state that compatibility of MIODs is preserved by refinement, which in combination with Theorem 1 shows, that MIODs together with their synchronous composition, modal refinement and modal compatibility form an interface theory as defined in Sect. 1.

Theorem 2. *Let S, S', T, T' be MIODs such that S and T are composable. Then $S \leftrightarroweq_d T$, $S' \leq_{md} S$ and $T' \leq_{md} T$ imply $S' \leftrightarroweq_d T'$.*

Corollary 1. *The algebra of all MIODs $(\mathcal{M}_d, \otimes_d, \leq_{md}, \leftrightarroweq_d)$ forms an interface theory.*

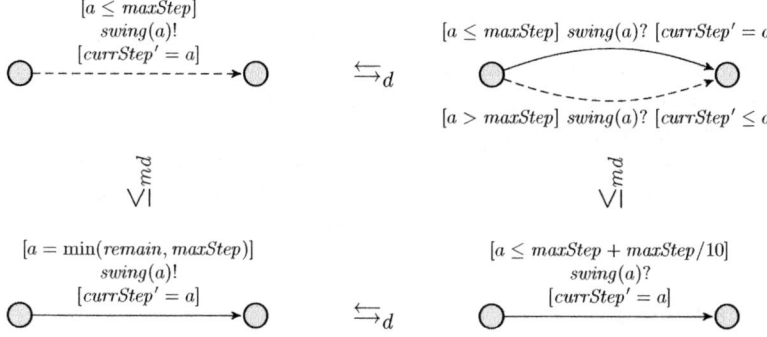

Fig. 3. Independent implementability: leg controller and leg specification

Example 3. In Fig. 3, the principle of independent implementability is illustrated in terms of our running example showing small excerpts of four MIODs. We start from the two abstract specifications in the first row of the figure where the left hand side specifies a possible leg controller and the right hand side stems from the already known specification of the leg component (see Fig. 1). We can easily verify their compatibility: for the output $swing(a)!$ of the leg controller, there exists an input in the leg specification which is required to appear in any implementation, and the expected change of $currStep$ is respected. Note that in case of $a > maxStep$ the postcondition $currStep' \leq a$ would not meet the caller's expectation $currStep' = a$; however the leg controller guarantees to send the output only for $a \leq maxStep$. Then we refine both independently of each other by the MIODs shown in the second row. The variable $remain$ in the precondition in the refined specification of the leg controller stores the remaining distance to be gone. First, Thm. 1 ensures that the composition of the refined specifications refine the composition of the abstract specifications, and second, Thm. 2 guarantees that the refined specifications are compatible again. ∎

4   Related Work

Specifications of interaction behavior and of changing data states are often considered separately from each other. Complex interaction behavior can be well specified by process algebraic approaches [6,20]; sequence diagrams (see e.g. [7]) or basic message sequence charts (see e.g. [16]) are popular formalisms to specify the temporal ordering of messages, and pre/postconditions are commonly used to specify the effects of operations w.r.t. data states. Though approaches like CSP-OZ [14] or Circus [23,27] offer means to specify interaction and data aspects, they do, however, not support modalities expressing allowed and required behavior and compatibility. Other related approaches are based on symbolic transition systems (STS) [13,2], but STS are mainly focusing on model checking and not on (top down) development of concurrent systems by refinement. Closely related to our work is the study of Mouelhi et al. [21] who consider an extension of the theory of interface automata [10] to data states. However, they do neither consider modal refinements nor the contract principle between interface specifications regarding data which, in our case, is based on a careful and methodologically important separation of provided, internal and required state variables. Sociable interfaces [9] are another data-oriented extension of interface automata which support n-ary communication but do not consider modalities and modal refinement. On the other hand, existing work on modal transition systems and their use as specification formalism for component interfaces [18,22] does not take into account explicit data states.

5   Conclusion

We have described an interface theory for concurrently running components with encapsulated, local data states. Interface specifications are formalized by modal input/output automata with data constraints; the refinement and compatibility notions extend those for modal I/O-transition systems to take into account data states. We have shown that our approach satisfies the "holy grail" of stepwise compositional development: refinement of MIODs is transitive, MIOD composition is monotone w.r.t. refinement, and MIOD compatibility implies composability and is preserved under refinement. Thus MIODs possess the properties of vertical and horizontal composition which are required for compositional algebraic specification approaches; moreover, MIODs satisfy also the additional compatibility requirements for ensuring safe interaction of reactive components.

  Currently our approach assumes atomic execution of transitions but, apart from this, we have allowed arbitrary interference of behaviors. We believe that more powerful refinement relations utilizing postconditions of previous computation steps can be obtained if we assume certain interference-freeness constraints; but this is left for further investigation. Of course, modal refinement should be extended to take into account data refinement in order to establish refinements between MIODs with different I/O-state signatures. Also, more elaborated verification techniques for refinement and compatibility of MIODs are envisaged, so that they get closer to semantic refinement and semantic compatibility. More-over, we plan to study more operators on MIODs, like conjunction and quotient, and to implement our framework, for a particular assertion language, in the MIO Workbench [5], a verification tool for modal input/output automata.

References

1. Astesiano, E., Bidoit, M., Kirchner, H., Krieg-Brückner, B., Mosses, P.D., Sannella, D., Tarlecki, A.: CASL: the Common Algebraic Specification Language. Theor. Comput. Sci. 286(2), 153–196 (2002)
2. Barros, T., Ameur-Boulifa, R., Cansado, A., Henrio, L., Madelaine, E.: Behavioural models for distributed fractal components. Annales des Télécommunications 64(1-2), 25–43 (2009)
3. Bauer, S.S., Hennicker, R., Bidoit, M.: A modal interface theory with data con-straints. In: Davies, J., Davies, J., Silva, L., Simão, A. (eds.) SBMF 2010. LNCS, vol. 6527, pp. 80–95. Springer, Heidelberg (2011)
4. Bauer, S.S., Hennicker, R., Wirsing, M.: Interface theories for concurrency and data. Theor. Comput. Sci. 412(28), 3101–3121 (2011)
5. Bauer, S.S., Mayer, P., Schroeder, A., Hennicker, R.: On Weak Modal Compati-bility, Refinement, and the MIO Workbench. In: Esparza, J., Majumdar, R. (eds.) TACAS 2010. LNCS, vol. 6015, pp. 175–189. Springer, Heidelberg (2010)
6. Bergstra, J.A., Middelburg, C.A.: An interface group for process components. Fun-dam. Inform. 99(4), 355–382 (2010)
7. Cengarle, M.V., Knapp, A., Mühlberger, H.: Interactions. In: Lano, K. (ed.) UML 2 Semantics and Applications, pp. 205–248 (2009)
8. Clavel, M., Durán, F., Eker, S., Lincoln, P., Martí-Oliet, N., Meseguer, J., Talcott, C.: All About Maude - A High-Performance Logical Framework, How to Spec-ify, Program and Verify Systems in Rewriting Logic. LNCS, vol. 4350. Springer, Heidelberg (2007)
9. de Alfaro, L., da Silva, L.D., Faella, M., Legay, A., Roy, P., Sorea, M.: Sociable Interfaces. In: Gramlich, B. (ed.) FroCos 2005. LNCS (LNAI), vol. 3717, pp. 81–105. Springer, Heidelberg (2005)
10. de Alfaro, L., Henzinger, T.A.: Interface Theories for Component-Based Design. In: Henzinger, T.A., Kirsch, C.M. (eds.) EMSOFT 2001. LNCS, vol. 2211, pp. 148–165. Springer, Heidelberg (2001)
11. Ehrig, H., Claßen, I., Boehm, P., Fey, W., Korff, M., Löwe, M.: Algebraic Concepts for Software Development in ACT ONE, ACT TWO and LOTOS (eingeladener Vortrag). In: Lippe, W.-M. (ed.) Software-Entwicklung. Informatik-Fachberichte, vol. 212, pp. 201–224. Springer, Heidelberg (1989)
12. Ehrig, H., Mahr, B.: Fundamentals of Algebraic Specification 2. EATCS Mono-graphs of Theoretical Computer Science, vol. 21. Springer, Berlin (1990)
13. Fernandes, F., Royer, J.-C.: The STSLib project: Towards a formal component model based on STS. Electr. Notes Theor. Comput. Sci. 215, 131–149 (2008)
14. Fischer, C.: CSP-OZ: a combination of Object-Z and CSP. In: Bowman, H., Derrick, J. (eds.) Proc. FMOODS, Canterbury, UK, pp. 423–438. Chapman and Hall, London (1997)
15. Goguen, J.A., Kirchner, C., Kirchner, H., Mégrelis, A., Meseguer, J., Winkler, T.C.: An Introduction to OBJ 3. In: Kaplan, S., Jouannaud, J.-P. (eds.) CTRS 1987. LNCS, vol. 308, pp. 258–263. Springer, Heidelberg (1988)

16. Harel, D., Thiagarajan, P.S.: Message sequence charts. In: Lavagno, L., Martin, G., Selic, B. (eds.) UML for Real: Design of Embedded Real-time Systems. Kluwer Academic Publishers (2003)

17. Hoare, C.A.R.: Proof of correctness of data representations. Acta Inf. 1, 271–281 (1972)

18. Larsen, K.G., Nyman, U., Wąsowski, A.: Modal I/O Automata for Interface and Product Line Theories. In: De Nicola, R. (ed.) ESOP 2007. LNCS, vol. 4421, pp. 64–79. Springer, Heidelberg (2007)

19. Larsen, K.G., Thomsen, B.: A Modal Process Logic. In: 3rd Annual Symp. Logic in Computer Science, LICS 1988, pp. 203–210. IEEE Computer Society (1988)

20. Montesi, F., Sangiorgi, D.: A Model of Evolvable Components. In: Wirsing, M., Hofmann, M., Rauschmayer, A. (eds.) TGC 2010, LNCS, vol. 6084, pp. 153–171. Springer, Heidelberg (2010)

21. Mouelhi, S., Chouali, S., Mountassir, H.: Refinement of interface automata strengthened by action semantics. Electr. Notes Theor. Comput. Sci. 253(1), 111–126 (2009)

22. Raclet, J.-B., Badouel, E., Benveniste, A., Caillaud, B., Passerone, R.: Why Are Modalities Good for Interface Theories? In: 9th Int. Conf. Application of Concurrency to System Design, ACSD 2009, pp. 119–127. IEEE Computer Society, Los Alamitos (2009)

23. Sampaio, A., Woodcock, J., Cavalcanti, A.: Refinement in Circus. In: Eriksson, L.-H., Lindsay, P.A. (eds.) FME 2002. LNCS, vol. 2391, pp. 451–470. Springer, Heidelberg (2002)

24. Sannella, D., Wirsing, M.: A Kernel Language for Algebraic Specification and Implementation. In: Karpinski, M. (ed.) FCT 1983. LNCS, vol. 158, pp. 413–427. Springer, Heidelberg (1983)

25. Wirsing, M.: Algebraic Specification. In: van Leeuwen, J. (ed.) Handbook of Theoretical Computer Science, Volume B: Formal Models and Sematics (B), pp. 675–788 (1990)

26. Wirsing, M.: Algebraic Specification Languages: An Overview. In: Astesiano, E., Reggio, G., Tarlecki, A. (eds.) Abstract Data Types 1994 and COMPASS 1994. LNCS, vol. 906, pp. 81–115. Springer, Heidelberg (1995)

27. Woodcock, J., Cavalcanti, A.: The semantics of Circus. In: Bert, D., Bowen, J.P., Henson, M.C., Robinson, K. (eds.) ZB 2002. LNCS, vol. 2272, pp. 184–203. Springer, Heidelberg (2002)

My ADT Shrine

Hans-Dieter Ehrich

Institut für Informationssysteme
Technische Universität Braunschweig
D-38092 Braunschweig
HD.Ehrich@tu-bs.de

Abstract. The 20th WADT 2010 is put into perspective by giving afterglows of the 1st WADT 1982 in Langscheid near Dortmund, and the 10th WADT 1994 in Santa Margherita near Genova. First encounters with pioneers in the field are recalled, in particular with the ADJ group who initiated the initial-algebra approach. The author's contributions at that time are put in this context. Around 1982, the emphasis of his work moved to databases and information systems, in particular conceptual modeling. His group used a triple of layers to model information systems, data—objects—systems, where the focus of interest now was on objects and systems. The interest in data issues paled in comparison. There were cases, however, where benefits could be drawn from the early work on ADTs and the foundations established in this field.

1 Opening the Shrine

In March 1982, Udo Lipeck and the author organized a workshop on Algebraic Specification in Langscheid, Sorpesee, near Dortmund. It took place in the *Heimvolkshochschule Sorpesee*, an institution of adult education. Figure 1 shows its logo at that time.

There were 29 participants coming from 9 universities, 8 German and 1 Dutch. Table 1 shows the list grouped by universities and ordered by alphabet.

As for the Dortmund group: at the time of the workshop, Volker Lohberger had left for Essen, Gregor Engels for Osnabrück, and Udo Pletat for Stuttgart. Klaus Drosten, the author, Martin Gogolla, and Udo Lipeck were about to leave for Braunschweig. Many of the other participants moved as well more or less shortly after the workshop. So the group felt that the event should be repeated in order to remain in contact.

This way, the Langscheid workshop became the 1st WADT. It was followed by WADTs in Passau (2nd:1983), Bremen (3rd:1984), Warberg Castle near Braunschweig (4th:1986), Gullane near Edinburgh (5th:1987), Berlin (6th:1988), Wusterhausen near Berlin (7th 1990), Dourdan near Paris (8th:1991), and Caldes de Malavella near Barcelona (9th:1992).

The 10th WADT was held from May 30 to June 6, 1994, in Santa Margherita Ligure near Genova, together with the 5th COMPASS Workshop. Figure 2 gives a view of the beautiful location (left) and the organizer, Egidio Astesiano (right), on an excursion to Genova.

T. Mossakowski and H.-J. Kreowski (Eds.): WADT 2010, LNCS 7137, pp. 13–24, 2012.
© Springer-Verlag Berlin Heidelberg 2012

Fig. 1. Site of the 1st WADT

Table 1. Participants of the 1st WADT

Aachen	Herbert Klaeren, Heiko Petzsch
Berlin	Hartmut Ehrig, Werner Fey, Horst Hansen, Klaus-Peter Hasler, Hans-Jörg Kreowski, Michael Löwe, Peter Padawitz, Michaela Reisin
Bonn	Christoph Beierle, Peter Raulefs, Angelika Voß
Bremen	Herbert Weber
Dortmund	Klaus Drosten, Hans-Dieter Ehrich, Gregor Engels, Martin Gogolla, Udo Lipeck, Volker Lohberger, Udo Pletat, Axel Poigné
Karlsruhe	Heinrich C. Mayr
Leiden	Jan Bergstra
München	Harald Ganzinger, Peter Pepper, Martin Wirsing
Saarbrücken	Claus-Werner Lermen, Jacques Loeckx

90 participants were registered at the workshop, the author refrains from listing them. The program offered 62 presentations. Selected papers were published after the conference in the LNCS 906 volume entitled *Recent Trends in Data Type Specification*. It was edited by Egidio Astesiano, Gianna Reggio and Andrzej Tarlecki, and it was published in 1995 after the conference.

The author's contribution to this volume, coauthored by Amilcar Sernadas, was a paper entitled *Local Specification of Distributed Families of Sequential Objects* [12]. So he was away from the ADT field by then—in fact, already for nearly ten years.

Fig. 2. Left: Santa Margherita Ligure, the site of the 10th WADT; Right: Egidio Astesiano speaking to participants on an excursion to Genova

It should be mentioned that the 10^{th}WADT/5^{th}COMPASS Workshop was the starting point of the "Common Framework Initiative for algebraic specification and development" (CoFI, the homepage is [5]) to unify and standardize the algebraic specification languages that were around at that time. There were quite a few. At least the main concepts to be incorporated were thought to be clear—although it was realized that it might not be so easy to agree on a common language to express these concepts. And so it was. The result of the efforts are published in two LNCS volumes [1,22].

Actually, WADT proceedings were continuously published in the Springer LNCS series from the 1987 Gullane meeting on, with a precursor in 1984 when the Bremen WADT proceedings were published in the Springer *Informatik Fachberichte*[1].

The author was very pleased when he was invited to give a retrospect lecture at the 20th anniversary WADT in Schloss Etelsen near Bremen. Figure 3 gives a view of that beautiful place. The author hadn't attended any of the WADT's since 1995, the 11th one, in Oslo.

For his presentation, the author was very generously given one full hour. But still, there was a need to concentrate on what was felt most essential and what could be of interest for the audience. So the author had to select and to simplify more than he had wished to. Also, his selection of topics was quite personal, guided by the material that he had readily available or could easily retrieve, and it was certainly also guided by vanity. So the selection is neither complete nor fairly balanced.

Since this paper is an elaboration of that presentation, it naturally suffers from the same problems: oversimplification, overselectivity, vanity bias. The author is aware that much more work along the ideas outlined here was going on at that time, all over the world. Historical completeness and balance is not among his intentions, though.

---

[1] Informatics Technical Reports.

Fig. 3. Schloss Etelsen near Bremen, site of the 20th WADT

2   Early Treasures

2.1   ADT Roots

It was in the mid to late nineteen hundred and seventies when the author became aware of the seminal ADJ[2] papers [20] and [19].

In [20], ADJ address programming language semantics, not abstract data types yet: the syntax of a programming language $\mathcal{L}$ is represented as an algebraic signature $L = (S, \Sigma)$ where S is a set of sorts and $\Sigma = \{\Sigma_{x,s}\}_{x \in S^*, s \in S}$ is an $S^* \times S$-sorted operator signature, i.e., a collection of operators $\omega \in \Sigma_{x,s}$ also written $\omega : s_1 \times \ldots \times s_n \rightarrow s$, where $x = s_1 \ldots s_n \in S^*$ and $s \in S$.[3] Interpretation is given in the category $\mathbf{Alg}_{\Sigma}^c$ of continuous Σ-algebras where fixpoint equations can be solved. This category has initial algebras which form an isomorphism class, with the term algebra T_{Σ} as a natural representative. T_{Σ} may be seen as the abstract syntax of the programming language. "Abstract" here means concentrating on the structure of the syntax and disregarding the concrete symbols written by the programmer. Mathematical semantics is given implicitly

---

[2] The acronym ADJ denoted a group of authors consisting of varying subsets of {Jim Thatcher, Eric Wagner, Jesse Wright, Joseph Goguen}; the group aimed at establishing an adjunction between category theory and computer science.

[3] Throughout this paper, a coherent notation is used which may deviate from the notations in the papers referred to.

by writing equations over Σ that are supposed to hold. A set of Σ-equations E determines the subcategory $\mathbf{Alg}^c_{\Sigma,E}$ of algebras satisfying these equations. This subcategory also has initial algebras, with a natural representative $T_{\Sigma,E}$, the quotient term algebra $T_\Sigma / \sim_E$ where $\sim_E$ is the congruence relation induced by E. By initiality, there is a unique morphism $\mu : T_\Sigma \to T_{\Sigma,E}$. This initial morphism gives meaning to each syntactic construct by interpreting it in the semantic algebra. This way, it establishes the semantics $[\![\mathcal{L}]\!]$ of the programming language $\mathcal{L}$. This approach has been termed *algebraic* semantics[4].

In [19], ADJ carried this elegant approach over to the specification of abstract data types, concentrating on finitary algebras. The essentials are well-known.

- a *data type* is an algebra
- an *abstract* data type is an isomorphism class of algebras
- the *syntax* of an abstract data type is given by a *specification* $D = (S, \Sigma, E)$ where S is a set of sorts, Σ is an $S^* \times S$-sorted set family of operators, and E is a set of Σ-equations

- the semantics of such a specification is defined in two steps

    1. the *Σ-equations* E determine a category of Σ-algebras satisfying these axioms
    2. the isomorphism class of *initial (Σ, E)-algebras* (or any representative in it) is suggested as *the* abstract data type specified "up to isomorphism" by the equational specification D.

- under natural conditions, there is an *operational* semantics in the form of a term rewriting system that operates precisely in an initial algebra (the term normal form algebra).

This is the basis. There are extensions and ramifications in many respects.

Fig. 4. Left: IBM Research Yorktown Heights in 1981; right, from left: Jim Thatcher, Eric Wagner, NN, the author, Wolfgang Wechler at MFCS'81 in Štrbské Pleso in the High Tatras (then Czechoslovakia, now Slowakia)

---

[4] As opposed to denotational, operational, and axiomatic sematics.

Figure 4 shows photographs of early encounters with members of the ADJ group. Their ideas spread over to Europe, Germany in particular, and gave rise to the 1982 Langscheid workshop, to become the 1st WADT, in order to give the emerging discussions about abstract data types a forum.

Early cooperation of ADJ with the Berlin group is witnessed by [14]. ADJ's influence on the author's own work is best shown in what he considers to be a special treasure in his shrine, namely [6]—not the first but probably the most important of his early publications in prestigious journals.

In this paper, the concepts of implementation and parameterization of abstract data types are explored on the syntactic level of specifications $D = (S, \Sigma, E)$. A *specification morphism* $f : D_1 \rightarrow D_2$ is a pair (f_S, f_Σ) where $f_S : S_1 \rightarrow S_2$ is a map, and $f_\Sigma : \Sigma_1 \rightarrow \Sigma_2$ is an $S^* \times S$-indexed set family of maps $\{f_{\Sigma,x,s} : \Sigma_{1,x,s} \rightarrow \Sigma_{2,x,s}\}_{x \in S^*, s \in S}$ such that, for every equation $l = r$ in E_1, $f(l) = f(r)$ is in E_2^*, the closure of E_2. We trust that the reader has an idea how f is applied to the left and right hand side terms of an equation (to every term operator recursively), and what the "closure" of an equation system is[5].

Referring to Figure 5, specification D_1 *implements* specification D_0 if there is a specification D_2 with two specification morphisms $f : D_1 \rightarrow D_2$ and $t : D_0 \rightarrow D_2$, such that D_0 is embedded "truely" into D_2 wrt t, and D_1 is embedded "fully" into D_2 wrt f. As for the precise definitions and their variants, we refer to the paper.

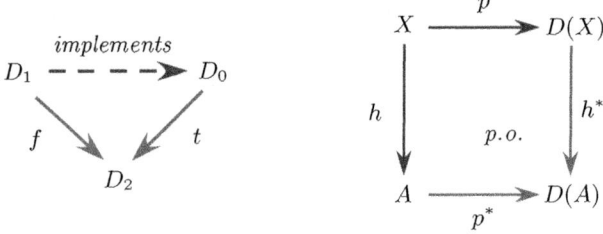

Fig. 5. Implementation (left) and parameterization (right)

A parameterized abstract data type $p : X \hookrightarrow D$ is given by an inclusion morphism p embedding a formal parameter part X into a specification D, written $D(X)$. A parameter assignment is given by a morphism $h : X \rightarrow A$ where A is an actual parameter specification. h says which formal parameter sorts and operators are to be replaced by which actual ones, while equations in X act as constraints that have to be obeyed by the actual parameter A. The result of substituting X by A is given by a pushout in the category of specifications and specification morphisms: the specification $D(A)$ where X is replaced by A, with p^* showing the embedding of the actual parameter in the result, and h^* showing the extension of h to the entire specification.

The author's approach to parameter replacement was an adaptation and simplification of Ehrig, Pfender and Schneider's categorial treatment of graph

---

[5] The expert reader may wonder what is meant here: the deductive closure or the entailment closure. It does not matter, they coincide in equational theories.

replacements [15]. (Co)limits as a mathematical tool for describing system integration had been used at about the same time by Joseph Goguen [17]. The semantics of CLEAR [4] makes extensive use of colimit constructions, also pushouts for handling parameter replacement, albeit on the level of semantic algebras.

There were also other approaches to abstract data type specification, some using categories and some not, some using initial semantics and some not. [2] gives a nice overview of abstract data type models at that time.

2.2   Algebraic Domain Equations

The 1983 paper [11], coauthored by the author and Udo Lipeck, sadly didn't find due attention, but the author always kept a warm remembrance of it. The idea was stimulated by Dana Scott's work on data types as lattices where data types (domains) are characterized implicitly as solutions of domain equations [23]. Was there an analogue in the realm of algebraic data types? What could the term "domain equation" possibly mean here? The authors' basic idea was to remodel a domain equation as a parametric specification $p : X \hookrightarrow D(X)$ with a further morphism $e : X \to D(X)$ with the idea that the targets of p and e had to be merged.

$$X = D(X) \quad \text{vs.} \quad X \overset{p}{\underset{e}{\rightrightarrows}} D(X)$$

In order to give a flavour of the approach, here are a couple of examples. With appropriate definitions of X, $\times$, $+1$, p and e (the details of which can be found in [11]), we have, for example, natural numbers specified by $X \rightrightarrows X + 1$, stacks with data of sort S specified by $X \rightrightarrows X \times S + 1$, binary trees with data of sort S at each node specified by $X \rightrightarrows X \times X \times S + 1$, binary trees with data of sort S at the leaves specified by $X \rightrightarrows X \times X + S$, etc.[6].

In order to give an idea of how it works, we expand the first example, natural numbers, specified by $X \overset{p}{\underset{e}{\rightrightarrows}} X + 1$.[7] X has one sort which we also denote by X, and one constant $x_0 :\to X$. "$+1$" adds one new sort N, one new constant $n_0 :\to N$, and a "construction" operator $\sigma : X \to N$ which connects the sorts. Let $p : X \to X + 1$ be the embedding sending all items in X to the ones in $X + 1$ with the same denotaton. Let $e : X \to X + 1$ send sort X to sort N and constant x_0 to constant n_0. Then the coequalizer of p and e identifies the two sorts, let us denote the resulting sort by $I\!N$, and also the two constants, let us denote the resulting constant by 0. If we suggestively rename the operator σ by $succ$, then its signature is $succ : I\!N \to I\!N$. So we have the basic specification of the natural numbers with one sort, constant 0 and the successor operator.[8]

---

[6] However, there are no interesting counterparts of Scott's reflexive domain $X = X^X$ and his powerset domain $X = 2^X$.

[7] This example is a simplification of example 7.1 in [11].

[8] A slightly more complex version of this example adds to "$+1$" a "projection" operator $\pi : N \to X$ and the (not too far fetched) equation $\pi(\sigma(x)) = x$. In the coequalizer, these then become the predecessor operator $pred : I\!N \to I\!N$ and the equation $pred(succ(n)) = n$. There is no equation for $pred(0)$, so this value is left undefined.

The semantics of algebraic domain equations roughly works as follows. As is well-known, each specification morhism $f : D_1 \rightarrow D_2$ gives rise to adjoint functors $F : D_1\text{-alg} \rightarrow D_2\text{-alg}$, the free functor, and $\bar{F} : D_2\text{-alg} \rightarrow D_1\text{-alg}$, the forgetful functor. An Algebraic Domain Equation $X \underset{e}{\overset{p}{\rightrightarrows}} D(X)$ then defines the pair $X\text{-alg} \underset{\bar{E}}{\overset{P}{\rightleftarrows}} D\text{-alg}$ of functors.

Let (c, Q) be the coequalizer of p and e, $X \underset{e}{\overset{p}{\rightrightarrows}} D \overset{c}{\longrightarrow} Q$. Referring to the functors $X\text{-alg} \underset{\bar{E}}{\overset{P}{\rightleftarrows}} D\text{-alg} \overset{\bar{C}}{\longleftarrow} Q\text{-alg}$, the *fixpoints* of $\bar{E}P$ are precisely the algebras $A\bar{C}$ for algebras $A \in Q\text{-alg}$. These are the *solutions* of an Algebraic Domain Equation $X \underset{e}{\overset{p}{\rightrightarrows}} D(X)$. Apparently, any initial Q-algebra is an *initial* solution of the Algebraic Domain Equation.

Fascinating work must be mentioned that has been influential in one or the other way, although there is no room for giving detailed references. Smyth's and Plotkin's work on a categorial solution for recursive domain equations [24] was of great interest when working on the above. Burstall's and Goguen's work on the ADT specification language CLEAR [3,4], and Goguen's work on OBJ beginning with [18] were continuing sources of inspiration, also later on when the author worked on database conceptual modeling languages.

Initial semantics was a great idea that guided the author quite a bit in the early ADT days. It soon became apparent, though, that initiality was not enough. In 1984, Martin Gogolla, Klaus Drosten, Udo Lipeck, and the author published a paper [16] supporting all forms of error handling: error introduction, error propagation, and error recovery; there is an initial semantics which coincides with the term-rewriting semantics if the latter is finite Church-Rosser; a specification-correctness criterion allows for non-initial models—loosely speaking, they are "initial on the ok values" and "loose on the error values". This means that carrier sets are partitioned into an ok part and an error part, and the initial morphisms must be injective on the former while there are no constraints for the latter. In specification, operators are classified into those which introduce errors in normal situations and those which preserve ok states. Two types of variables are introduced, one for non-error situations only and the others for ok states and exceptional states.

3   Drifting Away

When moving from Dortmund to Braunschweig where the author took the chair of databases and information systems, conceptual modeling became his major interest. ADTs were largely unknown in this community, and there was disappointment in the beginning that they were of little help. The author expressed that in his contribution to the 3rd WADT 1984 in Bremen [7].

This marked the beginning of the author's drifting away from the ADT area. Rather, *object* concepts caught his interest. Corresponding concepts have been dealt with in the ADT community as well, probably first by Joseph Goguen [17] who made objects fundamental in his OBJ series of languages [18]. Although OBJ has found practical applications in interesting projects, the author thinks it is fair to say that the ADT-related object concepts did not find their way into everyday practice. Practice in conceptual modeling, and in software engineering at large, came to be heavily dominated by UML which did not succeed in raising much enthusiasm among ADT theorists—so far.

But of course, it is not easy to abandon an Old Love; the author made attempts to use ADT theory for database aspects. An example is [8]. In brief, the ideas are as follows:

- the *data* in a database are organized in ADTs—but they are usually standard, not open to user specification
- database objects are identified by *keys* specified, for instance, in the context of relational tables; keys are usually immutable
- database keys are organized in ADTs which are *extensions* of the data-level ADTs
- for their specification, a *final* algebra approach is appropriate, relative to the initial-algebra data layer.

Ironically, the term *abstract data type* had always been used in the programming area in the sense that later would become standard as *object*, namely a stateful unit of state-based operations reading and writing data.

So the author's migration *from data types to object types* (this is the title of [13]) was, in a way, a turn back to the programming roots. Although substantial agreement has been achieved in practice, there is still no coherent "theory of objects" which copes with all aspects, including interaction, aggregation, inheritance, types, classes, specification, implementation, correctness, verification, ..., and which can provide a sufficiently rich and reliable basis for designing, implementing, and using object-oriented languages and systems.

The problem is not that there is no theory. The problem is that there are so many. And that they are so diverse.

4 Visits Home

Writing textbooks takes time. So the book [9][9] appeared long after initiating the project, when all three authors had abandoned the ADT area for years. They were relieved when the book was finished, and they did not find the push to work on an English version. Unfortunately.

So the author was quite glad when Jacques Loeckx asked him years later to coauthor an English textbook on abstract data types where the basic material in [9] could be reused [21][10]. In order to find wider distribution, this textbook does not use categories. Figure 6 shows the covers of the textbooks.

---

[9] Authored by the author, Martin Gogolla and Udo Lipeck.
[10] Authored by Jacques Loeckx, the author and Markus Wolf.

Fig. 6. Textbooks [9] and [21]

Another visit home—in the sense of using categories and cooperating with an ADT pioneer, not exactly in the sense of working on an ADT subject—was [10], coauthored with Joseph Goguen and Amílcar Sernadas. It is an amalgamation of two approaches, Joseph's *objects–as–sheaves* approach and Amilcar's and the author's *objects as observed processes* approach. So it put two of the many theories of objects into relationship.

A moment of commemoration is in order. Joseph Amadee Goguen passed away on July 3, 2006. We owe him a lot.

5 Closing the Shrine

After having been away from the WADT community for most of the last 20 years, the author felt during WADT 2010 that the subjects covered were not that far away from what he had been doing in the more theoretical parts of his work. Partly, at least. So he could as well have stayed...

For the future, the author feels inspired by the following citation from a tragedy of German classic literature: Johann Wolfgang von Goethe, *Die natürliche Tochter*. Trauerspiel, 1. Akt, Herzog.[11]

> "...
> Und heute noch, verwahrt im edlen Schrein,
> Erhältst du Gaben, die du nicht erwartet.
> ..."

> "... / And still today, coffered in a noble shrine /
> You receive gifts that you did not expect. / ..." [12]

Indeed—that happens all the time. For instance when the author was invited to the 20th WADT. He was surprised. And pleased that the WADT community remembered him. So he now feels encouraged to keep in closer touch.

---

[11] The Natural Daughter, Tragedy, 1st Act, Duke.
[12] Translation: the author.

And it goes on:

> "...
>
> Hier ist der Schlüssel! Den verwahre wohl!
> Bezähme deine Neugier! Öffne nicht,
> Eh' ich dich wieder sehe, jenen Schatz.
> Vertraue niemand, sei es, wer es sei.
> Die Klugheit rät's, der König selbst gebeut's.
> ..."

> "... / Here is the key! Keep it well under lock! / Restrain your curiosity! Do not open / That shrine before we meet again / Trust nobody, whoever it may be / Judiciousness suggests it, the king himself demands it. / ..."

The conclusion is that the author shall not reopen the shrine. Never!   Well, ... certainly not before the 30th WADT!

Acknowledgements. Warmest thanks are due to the organizers of the 20th anniversary WADT and the editors of this volume. In the final reviewing process, the three anonymous referees were very friendly and helped a lot to improve the paper. Many thanks to them! It goes without saying that the author admits responsibility for all remaining deficiencies.

References

1. Bidoit, M., Mosses, P.D. (eds.): CASL User Manual. LNCS, vol. 2900. Springer, Heidelberg (2004)
2. Broy, M., Wirsing, M., Pair, C.: A Systematic Study of Models of Abstract Data Types. Theoretical Computer Science 33, 139–174 (1984)
3. Burstall, R.M., Goguen, J.A.: Putting Theories Together to Make Specifications. In: Proc. 5th IJCAI, pp. 1045–1058. MIT, Cambridge (1977)
4. Burstall, R.M., Goguen, J.A.: The Semantics of CLEAR, a Specification Language. In: Bjorner, D. (ed.) Abstract Software Specifications. LNCS, vol. 86, pp. 292–331. Springer, Heidelberg (1980)
5. Common Framework Initiative (CoFI),
   http://www.informatik.uni-bremen.de/cofi/wiki/index.php/CoFI
6. Ehrich, H.-D.: On the Theory of Specification, Implementation, and Parametrization of Abstract Data Types. Journal of the ACM 29, 206–277 (1982)
7. Ehrich, H.-D.: Algebraic (?) Specification of Conceptual Database Schemata (extended abstract). In: Kreowski, H.-J. (ed.) Recent Trends in Data Type Specification. Informatik-Fachberichte, vol. 116. Springer, Berlin (1985)
8. Ehrich, H.-D.: Key Extensions of Abstract Data Types, Final Algebras, and Database Semantics. In: Pitt, D., Abramsky, S., Poigné, A., Rydeheard, D. (eds.) Category Theory and Computer Programming. LNCS, vol. 240, pp. 412–433. Springer, Heidelberg (1986)
9. Ehrich, H.-D., Gogolla, M., Lipeck, U.W.: Algebraische Spezifikation Abstrakter Datentypen. Teubner, Stuttgart (1989)
10. Ehrich, H.-D., Goguen, J.A., Sernadas, A.: A Categorial Theory of Objects as Observed Processes. In: de Bakker, J.W., de Roever, W.-P., Rozenberg, G. (eds.) REX 1990. LNCS, vol. 489, pp. 203–228. Springer, Heidelberg (1991)

11. Ehrich, H.-D., Lipeck, U.: Algebraic Domain Equations. Theoretical Computer Science 27, 167–196 (1983)
12. Ehrich, H.-D., Sernadas, A.: Local Specification of Distributed Families of Sequential Objects. In: Astesiano, E., Reggio, G., Tarlecki, A. (eds.) Abstract Data Types 1994 and COMPASS 1994. LNCS, vol. 906, pp. 219–235. Springer, Heidelberg (1995)
13. Ehrich, H.-D., Sernadas, A., Sernadas, C.: From Data Types to Object Types. J. Inf. Process. Cybern. EIK 26(1-2), 33–48 (1990)
14. Ehrig, H., Kreowski, H.-J., Thatcher, J.W., Wagner, E.G., Wright, J.B.: Parameterized Data Types in Algebraic Specification Languages, pp. 157–168. Springer, Berlin (1980)
15. Ehrig, H., Pfender, M., Schneider, H.J.: Graph-grammars: An algebraic approach. In: Proceedings of the 14th Annual Symposium on Switching and Automata Theory (SWAT 1973), pp. 167–180. IEEE Computer Society, Washington, DC, USA (1973)
16. Gogolla, M., Drosten, K., Lipeck, U., Ehrich, H.-D.: Algebraic and Operational Semantics of Specifications Allowing Exceptions and Errors. Theoretical Computer Science 34, 289–313 (1984)
17. Goguen, J.A.: Objects. International Journal of General Systems, 1563-5104 1, 237–243 (1974)
18. Goguen, J.A.: Some Design Principles and Theory for OBJ-0. In: Yeh, R. (ed.) LNCS, vol. 75, pp. 425–475. Prentice-Hall (1979)
19. Goguen, J.A., Thatcher, J.W., Wagner, E.G.: An Initial Algebra Approach to the Specification, Correctness and Implementation of Abstract Data Types. In: Yeh, R. (ed.) Current Trends in Programming Methodology IV, pp. 80–149. Prentice-Hall (1978)
20. Goguen, J.A., Thatcher, J.W., Wagner, E.G., Wright, J.B.: Initial Algebra Semantics and Continuous Algebras. Journal of the ACM 24, 68–95 (1977)
21. Loeckx, J., Ehrich, H.-D., Wolf, M.: Specification of Abstract Data Types. J. Wiley & Sons and B.G.Teubner Publishers (1996)
22. Mosses, P.D. (ed.): CASL Reference Manual. LNCS, vol. 2960. Springer, Heidelberg (2004)
23. Scott, D.S.: Data Types as Lattices. SIAM J. Comp. 5, 522–587 (1976)
24. Smyth, M.B., Plotkin, G.D.: The Category-Theoretic Solution of Recursive Domain Equations. In: Proc. 18th IEEE FOCS, pp. 13–17 (1977)

Evolving SOA in the Q-ImPrESS Project*

Jan Kofroň and František Plášil

Department of Distributed and Dependable Systems, Charles University in Prague
Malostranské náměstí 25, 118 00 Prague 1, Czech Republic
{kofron,plasil}@d3s.mff.cuni.cz
http://d3s.mff.cuni.cz

Model-driven development has become a popular approach to designing modern application. On the other hand, there are a vast number of legacy applications spread over the enterprise systems that are poorly documented and for which no models exist in order to help understand particular aspects of the system, including its architecture and behaviour of individual parts. If a need for an extension of such a system arises, it is difficult to envision the impact of necessary modifications, in particular in terms of performance, cost, and the man power needed to implement the desired changes.

The Q-ImPrESS project [1] is a medium-sized research project (STREP) funded under the European Union's Seventh Framework Programme (FP7), within the ICT Service and Software Architectures, Infrastructures and Engineering priority. Its goal is to provide a platform for reasoning about (extra-functional) properties of the system under modification (SOA), comparing different alternatives of introducing a modification and implementing just the best one, while knowing its properties in advance, i.e., before the system is deployed and run.

In the project, the system is viewed as a set of interacting services forming service architecture. The envisioned workflow supports multiple scenarios and consists of several steps. First, a model of the system needs to be created; the model is an instance of a newly proposed Service Architecture Meta-Model (SAMM) capturing information on the static structure (architecture), behaviour of particular services (in the sense of both extra-functional properties, e.g., response time, and functional ones, e.g., requiring other services to complete the request), deployment (i.e., assignment to particular hardware nodes), and properties of the hardware on which the services are run (e.g., processor speed, connection bandwidth). In general, such models are created either using a reverse engineering tool chain, where Java and C/C++ languages are supported, or manually in the case of a new system. Once a model is available, particular modifications of it can be examined in order to explore different ways in which the new desired functionality can be achieved. Thus, the system architect is encouraged to create at the model level several alternatives realizing the goal, perform their

---

* This work was funded in the context of the Q-ImPrESS research project (http://www.q-impress.eu) by the European Union under the ICT priority of the 7th Research Framework Programme.

T. Mossakowski and H.-J. Kreowski (Eds.): WADT 2010, LNCS 7137, pp. 25–26, 2012.

comparison, and select just the most suitable one for implementation. After performing a modification, a dedicated tool for verification of consistency between the code (in Java) and behaviour models is to be used to check whether it is necessary to further modify either the model or the code.

The proposed method is supported by the Q-ImPrESS IDE, which is a development environment based on Eclipse. Inside the IDE, all the phases of the aforementioned method are supported, mostly in a fully automatized way. Graphical and textual editors exist for entire SAMM. For reasoning about properties of the service architecture, existing tools have been extended and integrated [2,3,4,5] and new ones (e.g., the tool for trade-off analysis) have been proposed and implemented.

Currently, this three-year project is in its final year; the tools have reached the beta stage and are freely available for download from the project page.

References

1. The Q-ImPrESS project, http://www.q-impress.eu/
2. Krogmann, K., Reussner, R.: Palladio — Prediction of Performance Properties. In: Rausch, A., Reussner, R., Mirandola, R., Plášil, F. (eds.) The Common Component Modeling Example. LNCS, vol. 5153, pp. 297–326. Springer, Heidelberg (2008)
3. Bures, T., Hnetynka, P., Plasil, F.: SOFA 2.0: Balancing Advanced Features in a Hierarchical Component Model. In: Proceedings of SERA 2006, Seattle, USA, pp. 40–48. IEEE CS (August 2006) ISBN 0-7695-2656-X
4. Grassi, V., Mirandola, R., Randazzo, E., Sabetta, A.: KLAPER: an Intermediate Language for Model-Driven Predictive Analysis of Performance and Reliability. In: Rausch, A., Reussner, R., Mirandola, R., Plášil, F. (eds.) The Common Component Modeling Example. LNCS, vol. 5153, pp. 327–356. Springer, Heidelberg (2008)
5. Parizek, P., Plasil, F., Kofron, J.: Model Checking of Software Components: Combining Java PathFinder and Behavior Protocol Model Checker. In: Proceedings of the 30th Annual IEEE/NASA Software Engineering Workshop, SEW, April 24-28, pp. 133–141. IEEE Computer Society, Washington, DC (2006)

Sharing in the Graph Rewriting Calculus[*]

Paolo Baldan[1] and Clara Bertolissi[2]

[1] Dipartimento di Matematica Pura e Applicata, Università di Padova, Italy
[2] LIF-Université de Provence, Marseille, France

Abstract. The graph rewriting calculus is an extension of the ρ-calculus, handling graph like structures, with explicit sharing and cycles, rather than simple terms. We study a reduction strategy for the graph rewriting calculus which is intended to maintain the sharing in the terms as long as possible. We show that the corresponding reduction relation is adequate w.r.t. the original semantics of the graph rewriting calculus, formalising the intuition that the strategy avoids useless unsharing.

1 Introduction

The lambda-calculus [6], with its solid mathematical theory, has been classically taken as a foundation for functional languages (e.g., it inspired the development of Lisp). On the other hand, term rewriting [3] is a well known formal framework for analysing the behaviour of functional and rewrite-based languages. Along the years, in order to get closer to the practice of functional programming language design, the two formalisms have enriched one another by combining their features (see, e.g., the recent [17] which considers a lambda calculus with patterns and references therein).

In particular, the *rewriting calculus* (ρ-calculus) [11] has been introduced in the late nineties as a generalisation of term rewriting and of the λ-calculus. The notion of ρ-reduction of the ρ-calculus generalises β-reduction by considering matching on patterns which can be more elaborated than simple variables.

For improving the efficiency of implementations, terms, which would naturally correspond to trees, are often seen as graphs [7]. As an example, consider a rewrite system for multiplication $\mathcal{R} = \{x * 0 \rightarrow 0, \quad x * s(y) \rightarrow (x * y) + x\}$. By using term graphs, the second rule can duplicate the reference to x instead of duplicating x itself (see Fig. 1 (a)).

The use of graphical structures with explicit sharing is useful for the optimization of functional and declarative language implementation [19]. For example, graph rewriting is explicitly used in order to get an efficient implementation of the functional language Clean [22]. The theoretical bases are provided by suitable results proving that term-graph rewriting is adequate for term rewriting (see, e.g., [20,15,13]). Additionally, the use of graphs naturally leads to consider cyclic structures. This brings an increased expressive power that allows one to represent easily regular (i.e., with a finite number of different substructures) infinite data structures [7,1,14,21]. For example, the circular list *ones* = 1 : *ones*,

[*] Supported by the projects SiSteR (MIUR) and AVIAMO (University of Padova).

T. Mossakowski and H.-J. Kreowski (Eds.): WADT 2010, LNCS 7137, pp. 27–41, 2012.

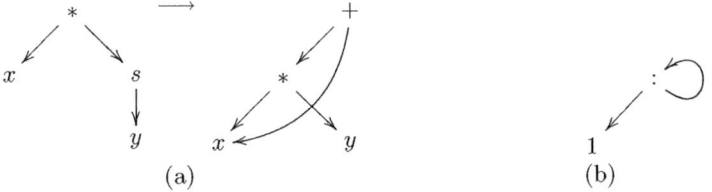

Fig. 1. (a) Rule for multiplication (b) Infinite list of ones

where ":" denotes the concatenation operator, can be represented by the cyclic graph of Fig. 1 (b).

The *graph rewriting calculus* (ρ_{g}-calculus), introduced in [8], combines the features of the cyclic λ-calculus [2] and the ρ-calculus [11], providing a framework where pattern matching, graphical structures and higher-order capabilities are primitive. In the ρ_{g}-calculus matching problems arising from rule applications are solved at the level of the calculus. The substitution arising from a successful match is computed, step by step, in the form of a set of equality constraints of the kind $x = G$, where x is a variable and G a ρ_{g}-term. The "application" of a substitution is then captured by explicit substitution rules. Matching equalities naturally allow the ρ_{g}-calculus to express various forms of sharing and cycles (as equalities can be recursive). The calculus, under suitable linearity constraints for patterns, has been shown to be confluent [4] and expressive enough for simulating the cyclic λ-calculus and term graph rewriting.

In the ρ_{g}-calculus, the loss of sharing is caused by the application of the substitution rules, which create copies of (sub)terms of a ρ_{g}-term. Indeed, during the computation, some unsharing can be unavoidable, for example for making a rule application explicit or for solving a matching constraint. However, in the basic formulation of the ρ_{g}-calculus, substitution rules can be used freely, so that, at any time, terms could be unnecessarily duplicated.

With the aim of improving the efficiency of the ρ_{g}-calculus, building on some preliminary work in [5], we study a strategy which is intended to maintain the sharing information as long as possible. Intuitively, the strategy prevents the application of substitution rules (responsible for unsharing) when they are not useful for activating new redexes. The strategy is obviously sound *w.r.t.* the original semantics of the ρ_{g}-calculus, i.e., any reduction in the strategy is a valid reduction in the original calculus. Moreover, we prove an *adequacy* result: for any reduction in the original calculus we can find a reduction in the strategy which produces essentially the same term, up to unsharing and possibly additional reductions for shared terms. As a consequence, we can show that whenever a term is normalising in the strategy, in the produced normal form sharing is maximal (in a suitably defined sense). The paper generalises and complements [5], where the focus was on *confluence* and a form of completeness for the strategy was proved only for the subclass of terms normalising in the ρ_{g}-calculus, which excludes, in particular, all those terms representing cyclic graphs.

Terms

$\mathcal{G}, \mathcal{P} ::= \mathcal{X}$ (Variables)
     $\mid \mathcal{K}$ (Constants)
     $\mid \mathcal{P} \to \mathcal{G}$ (Abstraction)
     $\mid \mathcal{G}\,\mathcal{G}$ (Functional application)
     $\mid \mathcal{G} \wr \mathcal{G}$ (Structure)
     $\mid \mathcal{G}[\mathcal{C}]$ (Constraint application)

Constraints

$\mathcal{C} ::= \epsilon$ (Empty constraint)
     $\mid \mathcal{X} = \mathcal{G}$ (Recursion equation)
     $\mid \mathcal{P} \ll \mathcal{G}$ (Match equation)
     $\mid \mathcal{C}, \mathcal{C}$ (Conjunction)

Fig. 2. Syntax of the ρ_{g}-calculus

The paper is organized as follows. In Section 2 we review the ρ_{g}-calculus. In Section 3 we present the reduction strategy ∇ proposed for preserving sharing in ρ_{g}-calculus terms. In Section 4 we show the adequacy of strategy ∇ for ρ_{g}-calculus and a maximal sharing property. We conclude in Section 5 by presenting some perspectives of future work. Some proofs are omitted for space limitations.

2 The Graph Rewriting Calculus

The syntax of the ρ_{g}-calculus is presented in Fig. 2. The symbols $G, H, P \ldots$ range over the set $\mathcal{G}$ of terms, $x, y, \ldots$ range over the set $\mathcal{X}$ of variables, $a, b, \ldots$ range over the set $\mathcal{K}$ of constants and $E, F, \ldots$ range over the set $\mathcal{C}$ of constraints.

As in the ρ-calculus, λ-abstraction is generalized by a rule abstraction $P \to G$. There are two different application operators: functional application, denoted simply by concatenation, and constraint application, denoted by "$\_[\_]$". As formalised later in the semantics, a functional application $(P \to G)H$ evaluates to a constraint application $G[H \ll P]$, the intuition being that solving the match will produce a substitution to be applied to G.

Terms can be grouped together into *structures* built using the operator "$\_ \wr \_$". This is useful for representing the (non-deterministic) application of a set of rewrite rules and consequently, the non-deterministic result.

In the ρ_{g}-calculus constraints are conjunctions (built using the operator "$\_, \_$") of match equations of the form $P \ll G$ and recursion equations of the form $x = G$. The empty constraint is denoted by ϵ. The operator "$\_, \_$" is supposed to be associative, commutative, with ϵ as neutral element.

We assume that the application operator associates to the left, while the other operators associate to the right. To simplify the syntax, operators have different priorities. Here are the operators ordered from higher to lower priority: "$\_ \_$", "$\_ \to \_$", "$\_ \wr \_$", "$\_[\_]$", "$\_ \ll \_$", "$\_ = \_$" and "$\_, \_$".

Following [12,1], we denote by $\bullet$ (black hole) a constant representing "undefined" terms that correspond to the expression $x[x = x]$ (self-loop). The notation $x =_\circ x$ is an abbreviation for the sequence $x = x_1, \ldots, x_n = x$. We write $C(\_)$ for a context with exactly one hole $\_$. A ρ_{g}-term is called *acyclic* if it contains no sequence of constraints of the form $C_0(x_0) \lll C_1(x_1), C_2(x_1) \lll C_3(x_2), \ldots,$ $C_m(x_n) \lll C_{m+1}(x_0)$, with $n, m \in \mathbb{N}$ and $\lll \in \{=, \ll\}$.

A term is called *algebraic* if it is of the form $(((f\ G_1)\ G_2)\ldots)\ G_n$, with $f \in \mathcal{K}$ and $G_i \in \mathcal{X} \cup \mathcal{K}$ or G_i algebraic for $i \in \{1,\ldots,n\}$, and we usually write it as $f(G_1, G_2, \ldots, G_n)$. It is called *linear* if each variable occurs free at most once.

Interesting properties for the calculus, like confluence, are obtained by imposing suitable requirements on the patterns P occurring as left-hand sides of abstractions and match equations. Following [4], we require such patterns to be *acyclic* and *linear algebraic terms* in *normal form*. E.g., the ρ_g-term $f(y)[y = g(y)] \to a$ is not allowed since it is an abstraction with a cyclic left-hand side.

The notions of free and bound variables of ρ_g-terms take into account the three binders of the calculus: abstraction, recursion and match. Intuitively, variables on the left hand-side of any of these operators are bound by the operator. As usual, we work modulo α-conversion. The set of free variables of a ρ_g-term G is denoted by $\mathcal{FV}(G)$. Moreover, given a constraint $\mathcal{C}$ we will refer to the set $\mathcal{DV}(\mathcal{C})$, of variables "defined" in $\mathcal{C}$. This set includes, for any recursion equation $x = G$ in $\mathcal{C}$, the variable x and for any match $P \ll G$ in $\mathcal{C}$, the set of free variables of P.

We define next an order over variables bound by a match or an equation. This order will be later used in the definition of the substitution rule of the calculus, which will allow one only "upward" substitutions, a constraint which is essential for the confluence of the calculus (see [4]). We denote by $\leq$ the least pre-order on recursion variables such that $x \geq y$ if $x = \mathsf{C}(y)$, for some context $\mathsf{C}(\_)$. The equivalence induced by the pre-order is denoted $\equiv$ and we say that x and y are cyclically equivalent $(x \equiv y)$ if $x \geq y \geq x$ (they lie on a common cycle). We write $x > y$ if $x \geq y$ and $x \not\equiv y$.

Example 1 (some ρ_g-terms).

1. In the rule $(2 * f(x)) \to ((y + y)[y = f(x)])$ the sharing in the right-hand side avoids the copying of the object instantiating $f(x)$, when the rule is applied to a ρ_g-term.
2. The ρ_g-term $x[x = cons(0, x)]$ represents an infinite list of zeros.
3. The ρ_g-term $f(x, y)[x = g(y), y = g(x)]$ is an example of twisted sharing that can be expressed using mutually recursive constraints (to be read as a `letrec` construct). We have that $x \geq y$ and $y \geq x$, hence $x \equiv y$.

The set of reduction rules of the ρ_g-calculus is presented in Fig. 3. As in the plain ρ-calculus, in the ρ_g-calculus the application of a rewrite rule to a term is represented as the application of an abstraction. A redex can be activated using the ρ rule in the BASIC RULES, which creates the corresponding matching constraint. The computation of the substitution which solves the matching is then performed explicitly by the MATCHING RULES and, if the computation is successful, the result is a recursion equation added to the list of constraints of the term. This means that the substitution is not applied immediately to the term but it is kept in the environment for a delayed application or for deletion if useless, as expressed by the GRAPH RULES.

More in detail, in the GRAPH RULES, the substitution rules copy a ρ_g-term associated to a recursion variable into a variable in the scope of the corresponding constraint. This is needed to make a redex explicit (e.g., in $x\ a[x = a \to b]$) or

BASIC RULES:

(ρ) $(P \rightarrow G_2)[E]\, G_3 \rightarrow_\rho G_2[P \ll G_3, E]$

(δ) $(G_1 \wr G_2)[E]\, G_3 \rightarrow_\delta (G_1\, G_3 \wr G_2\, G_3)[E]$

MATCHING RULES:

$(propagate)$ $P \ll (G[E])$ $\qquad\qquad\qquad \rightarrow_p P \ll G, E \quad if\ P \neq x$

$(decompose)$ $K(G_1,\ldots,G_n) \ll K(G'_1,\ldots,G'_n) \rightarrow_{dk} G_1 \ll G'_1,\ldots,G_n \ll G'_n$
$\qquad\qquad\qquad\qquad\qquad\qquad\qquad\qquad\quad with\ n \geq 0$

$(solved)$ $\qquad x \ll G, E$ $\qquad\qquad\qquad \rightarrow_s x = G, E \quad if\ x \notin \mathcal{DV}(E)$

GRAPH RULES:

$(external\ sub)\ \mathsf{C}(y)[y = G, E]$ $\qquad\quad \rightarrow_{es} \mathsf{C}(G)[y = G, E]$

$(acyclic\ sub)$ $\quad G[P \lll \mathsf{C}(y), y = G_1, E] \rightarrow_{ac} G[P \lll \mathsf{C}(G_1), y = G_1, E]$
$\qquad\qquad\qquad\qquad\qquad\quad if\ G_1\ is\ a\ variable\ or\ (x > y, \forall x \in \mathcal{FV}(P))$
$\qquad\qquad\qquad\qquad\qquad\quad where\ \lll \in \{=, \ll\}$

$(black\ hole)$ $\quad \mathsf{C}(x)[x =_\circ x, E]$ $\qquad\quad \rightarrow_{bh} \mathsf{C}(\bullet)[x =_\circ x, E]$

$\qquad\qquad\quad G[P \lll \mathsf{C}(y), y =_\circ y, E] \rightarrow_{bh} G[P \lll \mathsf{C}(\bullet), y =_\circ y, E]$
$\qquad\qquad\qquad\qquad\qquad\quad if\ x > y,\ \forall x \in \mathcal{FV}(P)$

Fig. 3. Semantics of the ρ_g-calculus

or to solve a match equation (e.g., in $a[a \ll x, x = a]$). As already mentioned, substitutions are allowed only upwards with respect to the order defined on the variables of ρ_g-terms. The *black hole* rules replace the undefined ρ_g-terms, intuitively corresponding to self-loop graphs, with the constant $\bullet$.

Differently from [4], in order to simplify the presentation, we do not have explicit garbage collection rules for getting rid of recursion equations that represent non-connected parts of a term. Instead, we assume that a term of the kind $G[E, x = G']$, with $x \notin \mathcal{FV}(E) \cup \mathcal{FV}(G)$ is automatically simplified to $G[E]$, and we identify the terms $G[\epsilon]$ and G.

We denote by $\mapsto$ the relation induced by the set of rules of Fig. 3 and by $\mapsto\!\!\!\twoheadrightarrow$ its reflexive and transitive closure. For any two rules r and s in this set, we will write $\mapsto\!\!\!\twoheadrightarrow_{r,s}$ to denote a sequence $\mapsto\!\!\!\twoheadrightarrow_r \mapsto_s$.

As mentioned above, the ρ_g-calculus, with linear algebraic patterns, has been shown to be confluent [4]. A term G is in *normal form* if no of the rules in Fig. 3 is applicable to G. A reduction of a term H into its normal form G, when it exists, is denoted by $H \mapsto\!\!\!\twoheadrightarrow^! G$.

Example 2 (simple reduction). An example of reduction in the ρ_g-calculus is reported below. The equality in the last line arise from garbage collection.

$$(f(a, a) \rightarrow a)\, (f(y, y)[y = a])$$
$$\mapsto_\rho \ a[f(a, a) \ll f(y, y)[y = a]]$$
$$\mapsto_p \ a[f(a, a) \ll f(y, y), y = a]$$
$$\mapsto_{dk} \ a[a \ll y, y = a]$$
$$\mapsto_{ac} \ a[a \ll a, y = a]$$
$$\mapsto_{dk} \ a[y = a] \ = \ a$$

Example 3 (reduction to normal form). Consider the term $G = f(y,y)[y = z$ $f(a), z = f(x) \rightarrow x]$. We show one of the possible reductions of G to its normal form (remind that garbage collection is implicit).

$$
\begin{array}{rl}
& f(y,y)[y = z\ f(a), z = f(x) \rightarrow x] \\
\mapsto_{ac} & f(y,y)[y = (f(x) \rightarrow x)\ f(a), z = f(x) \rightarrow x] \\
\mapsto_{\rho} & f(y,y)[y = x[f(x) \ll f(a)]] \\
\mapsto_{dk,s} & f(y,y)[y = x[x = a]] \\
\mapsto_{es} & f(y,y)[y = a[x = a]] \\
\mapsto_{es} & f(a,y)[y = a] \\
\mapsto_{es} & f(a,a)
\end{array}
$$

3   A Sharing Strategy for the ρ_{g}-calculus

In the ρ_{g}-calculus substitutions can be used freely and this can lead to useless and expensive (both in terms of time and space) duplications of terms. For instance, consider the reduction

$$
\begin{array}{rl}
& f(x,x)[x = (a \rightarrow g(b))a] \\
\mapsto_{es} & f((a \rightarrow g(b))a, x)[x = (a \rightarrow g(b))a] \\
\mapsto_{es} & f((a \rightarrow g(b))a, (a \rightarrow g(b))a)[x = (a \rightarrow g(b))a] \\
\mapsto_{\rho} & f(g(b)[a \ll a], g(b)[a \ll a])[x = (a \rightarrow g(b))a] \\
\mapsto_{dk} & f(g(b), g(b))
\end{array}
$$

The same result could be obtained more efficiently with the following reduction

$$
\begin{array}{rl}
& f(x,x)[x = (a \rightarrow g(b))a] \\
\mapsto_{\rho} f(x,x)[x = g(b)[a \ll a]] & \mapsto_{dk}\ f(x,x)[x = g(b)]
\end{array}
$$

For this reason, in this section we study a reduction strategy that aims at keeping the sharing as long as possible in ρ_{g}-terms. Intuitively, the strategy delays as much as possible the application of the substitution rules (*external sub*) and (*acyclic sub*).

The underlying idea is quite simple: we constrain substitution rules to be applied only if they are needed for generating new redexes for the basic or matching rules. For instance, we allow the application of the (*external sub*) rule to the terms $x\ a[x = f(x) \rightarrow x]$ or $x\ a[x = a \wr (a \rightarrow b)]$, since this is useful for creating, respectively, a new (ρ) redex and a new (δ) redex. Instead, (*external sub*) cannot be applied to the terms $f(x,x)[x = g(x)]$ or $x[x = f(x)]$ which are actually considered in normal form in the strategy. As observed in [5], capturing the notion of "substitution needed for generating a new redex" is not straightforward since more than one substitution step can be needed to generate a new redex for the basic or matching rules. This happens, e.g., in the reduction below, where the generated redex is underlined:

$$
\begin{array}{rl}
& y[y = x\ f(a), x = f(z) \rightarrow y] \mapsto_{es} x\ f(a)[y = x\ f(a), x = f(z) \rightarrow y] \\
& \mapsto_{es} \underline{(f(z) \rightarrow y)}\ f(a)[y = x\ f(a), x = f(z) \rightarrow z]
\end{array}
$$

A single step would suffice to generate the redex if we removed the constraint that *(acyclic sub)* can be performed only upward, thus allowing the reduction

$$y[y = x \; f(a), x = f(z) \to y] \mapsto_y [y = (f(z) \to y) \; f(a), x = f(z) \to y]$$

The definition of the strategy for the ρ_g-calculus will rely on the fact, formally proved in [5], that this is a completely general situation: at most two substitutions are needed to generate a new redex (one would suffice removing the constraint on *(acyclic sub)*).

An additional case in which we want to apply the substitution rules is when there are trivial recursion equations of the kind $x = y$ where both sides are single variables, like in $x * y + x[x = z, y = z, z = 1]$. In this situation, the term is simplified to $(z * z + z)[z = 1]$ in which useless names have been eliminated by garbage collection.

Hereafter, we call *basic redex* any term of the shape $(P \to G_2)[E] \; G_3$ or $(G_1 \wr G_2)[E] \; G_3$, which is reducible using the BASIC RULES in Fig. 3. Similarly, a *matching redex* is any term which can be reduced using the MATCHING RULES.

Definition 1 (sharing strategy). *The reduction strategy ∇ is defined by the following clauses:*

1. *All the reduction rules but (external sub) and (acyclic sub) are applicable without any restriction.*
2. *The rules (external sub) and (acyclic sub) are applied to a term G if*
   (a) *they replace a variable by a variable (variable renaming), or*
   (b) *they create (in one step) a basic or a matching redex, or*
   (c) *the term G has the form $C''(x)[x = C(y), y = C'(z), E]$, where $x \equiv y$ and $C(C'(x))$ includes a basic or a matching redex, and the reduction is $G \mapsto_{es} C''(C(y))[x = C(y), y = C'(z), E]$.*

A reduction obeying the strategy ∇ is called a ∇-reduction and denoted by $\mapsto\!\!\!\!\twoheadrightarrow_\nabla$.

In other words the rules *(external sub)* and *(acyclic sub)* are thus applied when their application leads to

- the instantiation of a variable by a variable (condition *2a*);
- the instantiation of an active variable, i.e., a variable which appears free in the left-hand side of an application, by an abstraction or a structure, which produces a *basic redex* (condition *2b*);
- the instantiation of a variable in a match equation, which produces a *matching redex*, i.e., which enables a decomposition or constraint propagation *w.r.t.* the match equation (condition *2b*).

Additionally, condition *2c* captures the fact that, given $G[E]$ if a cyclic substitution in E would generate a redex, then one is allowed to apply external substitutions in order to reproduce the same redex in G, in two steps.

The strategy ∇ is a mild variation of the one in [5], the difference being that here non-substitution rules have no priority over the substitution rules. As a consequence, we do not need to completely reduce a basic or matching redex (using e.g. the matching rules) before activating a new redex by means of a substitution rule. This choice allows us to simplify the presentation, retaining the adequacy and maximal sharing results of Section 4.

Example 4 (multiplication). Let us use an infix notation for the constant "$*$". The following ρ_g-term corresponds to the application of the rewrite rule $\mathcal{R} = x * s(y) \to (x * y + x)$ to the term $1 * s(1)$ where the constant 1 is shared.

$$
\begin{aligned}
&\quad (x * s(y) \to (x * y + x))\,(z * s(z)[z = 1]) \\
&\mapsto_\rho \quad x * y + x[x * s(y) \ll (z * s(z)[z = 1])] \\
&\mapsto_p \quad x * y + x[x * s(y) \ll z * s(z), z = 1] \\
&\mapsto_{dk,s} x * y + x[x = z, y = z, z = 1] \\
&\mapsto_{es} \quad (z * z + z)[x = z, y = z, z = 1] \quad \text{(allowed by Definition 1(2a))} \\
&= \quad\quad (z * z + z)[z = 1] \quad \text{(garbage collection)}
\end{aligned}
$$

Notice that the term $(z * z + z)[z = 1]$ is in normal form *w.r.t.* the strategy ∇ but it can be further reduced to $(1 * 1 + 1)$ in the plain ρ_g-calculus.

Example 5 (reduction to normal form). We consider the term G of Example 3 and we reduce it following the strategy ∇. We obtain:

$$
\begin{aligned}
&\quad\quad f(y, y)[y = z\, f(a), z = f(x) \to x] \\
&\mapsto_{ac} \quad f(y, y)[y = (f(x) \to x)\, f(a), z = f(x) \to x] \quad \text{(by Definition 1(2b))} \\
&\mapsto_\rho \quad f(y, y)[y = x[f(x) \ll f(a)]] \\
&\mapsto_{dk,s} f(y, y)[y = x[x = a]] \\
&\mapsto_p \quad f(y, y)[y = x, x = a] \\
&\mapsto_{ac} \quad f(y, y)[y = a] \quad \text{(by Definition 1(2a))}
\end{aligned}
$$

Note that the normal form *w.r.t.* ∇, i.e., the term $f(y, y)[y = a]$, represents a graph where the arguments of f are shared. Instead, as shown in Example 3, the reduction in the ρ_g-calculus with no reduction strategy leads to the term $f(a, a)$ where the arguments of f are duplicated.

Example 6. Consider the ρ_g-term $G = f(y)[y = x\, a, x = y \wr b]$. Notice that $x \equiv y$, thus the *(acyclic sub)* rule cannot be applied. We have instead the reduction:

$$
f(y)[y = x\, a, x = y \wr b] \mapsto_{es} f(x\, a)[y = x\, a, x = y \wr b] \mapsto_{es}
$$

$$
f((y \wr b)\, a)[y = x\, a, x = y \wr b] \mapsto_\delta f((y\, a \wr b\, a))[y = x\, a, x = y \wr b]
$$

This is a valid ∇-reduction, since there exists a cyclic substitution step which transforms $x\, a$ into a basic redex $(y \wr b)\, a$. Hence, the first *(external sub)* rule step can be performed according to Definition 1(2c). The second *(external sub)* rule step creates the basic redex $(y \wr b)\, a$, thus it is allowed for Definition 1(2b).

4   Properties of the Sharing Strategy

In this section we will show some basic properties of the ρ_g-calculus with the reduction strategy ∇. After observing its confluence, we prove that the strategy ∇ is adequate for the ρ_g-calculus. Relying on adequacy, we can prove that ∇-reductions maximize (in a suitable sense) the sharing.

4.1   Adequacy

In this section we prove that the reduction strategy ∇ is adequate *w.r.t.* the semantics of the ρ_g-calculus presented in Section 2. This means that given a reduction $G \longmapsto H$ there exists a reduction in the strategy $G \longmapsto_\nabla G'$ which performs essentially the same steps, but avoiding useless unsharing. More precisely, G' unshares to a term H', reachable from H in the strategy.

In the following we will denote by $\longmapsto_s$ a step using a substitution rule. Substitutions which violate the strategy will be denoted by $\longmapsto_{s_u}$ (the subscript "u" stand for "useless"), while those obeying the strategy will be denoted $\longmapsto_{s_\nabla}$. Finally, non-substitution steps will be denoted by $\longmapsto_n$.

A first observation is that it is not possible to create a *basic* or *matching* redex by further reducing a term that is in normal form *w.r.t.* the reduction strategy. This is proved in [5] and using this property we immediately obtain the result below.

Proposition 1. *If a ρ_g-term G is in normal form w.r.t. the strategy ∇, then for any reduction $G \longmapsto_s G'$, the term G' is in normal form w.r.t. ∇.*

We next observe that the calculus with the strategy ∇ is confluent. Additionally, we give some easy commutation results for reductions constrained by the strategy and involving only subsets of rules. Recall that two sets of rules A and B are called *commuting* if for any pair of reductions $G_1 \longmapsto_A G_2$ and $G_1 \longmapsto_B G_3$ there exists a diagram:

$$
\begin{array}{ccc}
G_1 & \overset{A}{\longmapsto\!\!\!\!\!\twoheadrightarrow} & G_2 \\
{\scriptstyle B}\Big\downarrow & & \Big\downarrow{\scriptstyle B} \\
G_3 & \underset{A}{\cdots\!\!\!\!\!\twoheadrightarrow} & G_4
\end{array}
$$

Lemma 1 (confluence and commutation). *The reduction relations $\longmapsto_\nabla$ and $\longmapsto_{s_\nabla}$ are confluent. Moreover the following pairs of reduction relations are commuting:*

1. $\longmapsto_\nabla$ *and* $\longmapsto_n$
2. $\longmapsto_\nabla$ *and* $\longmapsto_s$
3. $\longmapsto_{s_u}$ *and* $\longmapsto_{s_\nabla}$
4. $\longmapsto_n$ *and* $\longmapsto_s$

The next result shows that substitutions can be reordered, by putting those which obey the strategy first.

Lemma 2 (reordering substitutions). *Given a reduction $G_1 \mapsto\!\!\!\twoheadrightarrow_s G_2$ there exists*

$$
\begin{array}{ccc}
G_1 & \overset{s}{\mapsto\!\!\!\twoheadrightarrow} & G_2 \\
{\scriptstyle s\nabla}\big\downarrow & & \big\downarrow{\scriptstyle s\nabla} \\
G'_1 & \underset{s_u}{\mapsto\!\!\!\twoheadrightarrow} & G'_2
\end{array}
$$

Next we show that if an n-reduction step follows some substitutions which violate the strategy, then the n-reduction step can be performed first. This is quite intuitive as substitutions which violate the strategy cannot create new n-redexes.

Lemma 3 (anticipating n-reductions). *Given a reduction $G_1 \mapsto\!\!\!\twoheadrightarrow_{s_u} G_2 \mapsto_n G_3$ there exists:*

$$
\begin{array}{ccccc}
G_1 & \overset{s_u}{\mapsto\!\!\!\twoheadrightarrow} & G_2 & \overset{n}{\mapsto} & G_3 \\
{\scriptstyle n}\big\downarrow & & & & \big\downarrow{\scriptstyle n} \\
G'_1 & & \overset{s}{\mapsto\!\!\!\twoheadrightarrow} & & G'_3
\end{array}
$$

Note that, in general, the lemma above does not hold for a reduction involving many n-redexes. For instance let $H = (a \to (a \to b))\, a$. Then the reduction

$$
x\, a[x = H] \quad \mapsto_{s_u} \quad H\, a[x = H] \quad \mapsto\!\!\!\twoheadrightarrow_n \quad (a \to b)\, a[x = H] \quad \mapsto\!\!\!\twoheadrightarrow_n \quad b[x = H]
$$

cannot be factorised as desired. This is due to the fact that the first substitution, when exchanged with the first n-reduction step becomes a substitution in the strategy, which creates the n-redex reduced immediately after.

The next lemma identifies a situation in which the generalisation of Lemma 3 to multiple n-reduction steps holds. This is the case when the multiple n-redexes are residuals of a single one.

Lemma 4 (anticipating n-reductions - generalisation). *Given a diagram of reductions as follows:*

$$
\begin{array}{ccc}
G_2 & \overset{n}{\mapsto} & G_3 \\
{\scriptstyle \nabla}\big\downarrow & & \big\downarrow{\scriptstyle \nabla} \\
G'_1 \overset{s_u}{\mapsto\!\!\!\twoheadrightarrow} G'_2 & \overset{n}{\mapsto\!\!\!\twoheadrightarrow} & G'_3
\end{array}
$$

there exists:

$$
\begin{array}{ccc}
G_2 & \overset{n}{\mapsto} & G_3 \\
{\scriptstyle \nabla}\big\downarrow & & \big\downarrow{\scriptstyle \nabla} \\
G'_1 \overset{s_u}{\mapsto\!\!\!\twoheadrightarrow} G'_2 & \overset{n}{\mapsto\!\!\!\twoheadrightarrow} & G'_3 \\
{\scriptstyle n}\big\downarrow & & \big\downarrow{\scriptstyle n} \\
G''_1 & \overset{s}{\mapsto\!\!\!\twoheadrightarrow} & G''_3
\end{array}
$$

The next lemma provides a way of "exchanging" a reduction violating the strategy with one which obeys the strategy.

Lemma 5 (exchanging reductions). *Given a reduction* $G_1 \mapsto\!\!\!\twoheadrightarrow_{s_u} G_2 \mapsto\!\!\!\twoheadrightarrow_\nabla G_3$ *there exists:*

$$
\begin{array}{ccccc}
G_1 & \xrightarrow{\ s_u\ } & G_2 & \xrightarrow{\ \nabla\ } & G_3 \\
\nabla \downarrow & & & & \downarrow \nabla \\
G_1' & & \xrightarrow{\quad\quad s_u\quad\quad} & & G_3'
\end{array}
$$

We are now ready to show that for any reduction $G_1 \mapsto\!\!\!\twoheadrightarrow G_2$ in the $\rho_{\mathbf{g}}$-calculus, we can find a reduction obeying the strategy which performs "essentially" the same reduction steps, up to unsharing and reductions of shared terms. More precisely, we can find a reduction $G_1 \mapsto\!\!\!\twoheadrightarrow_\nabla G_1'$ such that the unsharing of G_1' is reachable from G_2 in the strategy (intuitively, since the strategy forces the reduction of shared subterms, from G_2 additional reduction steps may be needed on duplicated terms).

Theorem 1 (adequacy). *Given a $\rho_{\mathbf{g}}$-term G_1, if $G_1 \mapsto\!\!\!\twoheadrightarrow G_2$ in the $\rho_{\mathbf{g}}$-calculus, then there exists $\rho_{\mathbf{g}}$-terms G_1' and G_2' such that $G_1 \mapsto\!\!\!\twoheadrightarrow_\nabla G_1'$ in the $\rho_{\mathbf{g}}$-calculus with the strategy ∇, with $G_1' \mapsto\!\!\!\twoheadrightarrow_{s_u} G_2'$ and $G_2 \mapsto\!\!\!\twoheadrightarrow_\nabla G_2'$.*

$$
\begin{array}{ccc}
G_1 & \xrightarrow{\ \nabla\ } & G_1' \\
\downarrow & & \downarrow s_u \\
G_2 & \xrightarrow{\ \nabla\ } & G_2'
\end{array}
$$

Proof. We proceed by induction on the length of the reduction $G_1 \mapsto\!\!\!\twoheadrightarrow G_2$. The base case is trivial. For the inductive case, if the reduction is of the shape $G_1 \mapsto\!\!\!\twoheadrightarrow G_3 \mapsto G_2$, by inductive hypothesis we obtain the diagram:

$$
\begin{array}{ccc}
G & \xrightarrow{\ \nabla\ } & G_1' \\
\downarrow & & \downarrow s_u \\
G_3 & \xrightarrow{\ \nabla\ } & G_3' \\
\downarrow & & \\
G_2 & &
\end{array}
$$

Then we distinguish two cases according to the nature of the last step.

– $G_3 \mapsto_n G_2$ (the last step is not a substitution)

In this case we can construct the following diagram:

$$G_1 \xrightarrow{\ \nabla\ } G_1' \xrightarrow{\ n\ } G_1'' \xrightarrow{\ \nabla\ } G_1'''$$

$$G_3 \xrightarrow{\ \nabla\ } G_3' \quad (II) \quad s \quad (III) \quad s_u$$

$$G_2 \xrightarrow{\ \nabla\ } G_2' \xrightarrow{\ n\ } G_2' \xrightarrow{\ \nabla\ } G_2'''$$

with s_u (top), (I) with n (left), n (center left)

where square (I) is justified by Lemma 1(1), square (II) is given by Lemma 4 and square (III) is constructed using Lemma 2.

– $G_3 \mapsto_s G_2$ (the last step is a substitution)

In this case we can construct the following diagram:

$$G_1 \xrightarrow{\ \nabla\ } G_1' \xrightarrow{\ \nabla\ } G_1''$$

$$G_3 \xrightarrow{\ \nabla\ } G_3' \quad (II) \quad s_u$$

$$G_2 \xrightarrow{\ \nabla\ } G_2' \xrightarrow{\ \nabla\ } G_2''$$

with s_u (top), (I) with s (center)

where squares (I) and (II) are justified by Lemma 1(2) and Lemma 2, respectively.   □

For instance, take $G_1 = f(x,x)[x = (a \to b)\ a]$ and consider the following reduction: $G_1 \mapsto_{es} f((a \to b)\ a, x)[x = (a \to b)\ a] \mapsto_{\rho,dk} f(b,x)[x = (a \to b)\ a]$. Then the reductions provided by the adequacy theorem are as follows:

$$f(x,x)[x = (a \to b)\ a] \xrightarrow{\ \nabla\ } f(x,x)[x = b]$$

$$f(b,x)[x = (a \to b)\ a] \xrightarrow{\ \nabla\ } f(b,b)$$

with s_u on the right side.

4.2   Maximal Sharing

We can now prove that the strategy guarantees a maximal sharing property. More specifically, we start by showing that if a term G_1 of the ρ_g-calculus reduces to a (possibly non-normalising) term G_2 where only substitutions can be performed, then, according to the strategy, G_1 is normalising and, if its ∇-normal form is G_2', then G_2 is an "unsharing" of G_2'.

Theorem 2 (maximal sharing/1). *If $G_1 \mapsto\!\!\!\twoheadrightarrow G_2$ and the only possible reduction steps from G_2 are substitutions, then G_1 is normalising in the ρ_g-calculus with the strategy ∇, i.e., $G_1 \mapsto\!\!\!\twoheadrightarrow^!_\nabla G'_2$ and $G'_2 \mapsto\!\!\!\twoheadrightarrow_{s_u} G_2$.*

$$
\begin{array}{ccc}
G_1 & \longtwoheadrightarrow & G_2 \longtwoheadrightarrow \cdots \\[-2pt]
& {}^! \searrow \quad & \big\uparrow {\scriptstyle s_u} \\[-2pt]
& {\scriptstyle \nabla} \searrow & \vdots \\[-2pt]
& & G'_2
\end{array}
$$

Proof. This is a consequence of Theorem 1. In fact, by such theorem there exists:

$$
\begin{array}{ccc}
G_1 & \longtwoheadrightarrow & G_2 \\
{\scriptstyle \nabla}\Big\downarrow & & \Big\downarrow{\scriptstyle \nabla} \\
G'_1 & \underset{s_u}{\longtwoheadrightarrow} & G'_2
\end{array}
$$

Since starting from G_2 only substitutions can be performed, these will all be s_u-substitutions, not obeying the strategy. Hence the reduction $G_2 \mapsto\!\!\!\twoheadrightarrow_\nabla G'_2$ must be empty, i.e., G_2 and G'_2 are the same. This concludes the proof. □

As an example, consider the term $G_1 = x[x = (f(a) \rightarrow f(x))\ f(a)]$ and its infinite reduction: $G_1 \mapsto f(x)[x = f(x)] \mapsto\!\!\!\twoheadrightarrow_s \dots$.

Then the reductions provided by the maximal sharing theorem are as follows:

$$
\begin{array}{ccc}
x[x = (f(a) \rightarrow f(x))\ f(a)] & \longtwoheadrightarrow & f(x)[x = f(x)] \longtwoheadrightarrow \cdots \\[-2pt]
{}^! \searrow \qquad\qquad & & \big\uparrow {\scriptstyle s} \\[-2pt]
{\scriptstyle \nabla} \searrow & & \vdots \\[-2pt]
& x[x = f(x)] &
\end{array}
$$

As a simple consequence, for any term G_1 normalising in the ρ_g-calculus, we deduce that G_1 is normalising in the ρ_g-calculus with the strategy ∇, and its normal form is as "shared" as possible. This is expressed by saying that for any way of factorising the normalising reduction of G_1 as $G_1 \mapsto\!\!\!\twoheadrightarrow G_2 \mapsto\!\!\!\twoheadrightarrow^!_s G_3$, we have that G_2 is an unsharing of the ∇-normal form of G_1.

Corollary 1 (maximal sharing/2). *Given a normalising ρ_g-term G_1 and a reduction $G_1 \mapsto\!\!\!\twoheadrightarrow G_2 \mapsto\!\!\!\twoheadrightarrow^!_s G_3$ in the ρ_g-calculus, there exists a ρ_g-term G'_2 such that $G_1 \mapsto\!\!\!\twoheadrightarrow^!_\nabla G'_2$ in the ρ_g-calculus with the strategy ∇ and $G'_2 \mapsto\!\!\!\twoheadrightarrow_{s_u} G_2$.*

$$
\begin{array}{ccccc}
G_1 & \longtwoheadrightarrow & G_2 & \overset{!}{\underset{s}{\longtwoheadrightarrow}} & G_3 \\[-2pt]
& {}^! \searrow \quad & \big\uparrow {\scriptstyle s_u} & & \\[-2pt]
& {\scriptstyle \nabla} \searrow & \vdots & & \\[-2pt]
& & G'_2 & &
\end{array}
$$

For instance, take the ρ_g-term $G_1 = f(y, y)[y = z\ f(a), z = f(x) \rightarrow x]$ and its reductions of Example 3 and 5. The maximal sharing theorem diagram is the following

$$G_1 \longmapsto f(a,y)[y=a] \xmapsto{\;!\;}_s f(a,a)$$

$$\nabla \searrow^{\;!\;} \qquad \Big\uparrow s_u$$

$$f(y,y)[y=a]$$

5   Conclusions

We have studied a reduction strategy ∇ for the ρ_g-calculus, an extension of the ρ-calculus able to express shared and cyclic terms. The strategy aims at maintaining the sharing information as long as possible in the ρ_g-terms. We have presented an adequacy result w.r.t. the standard ρ_g-calculus, and, relying on this, we formalised a maximal sharing property for the strategy.

In the literature, a notion of adequacy and corresponding results have been developed for formalising the relation between (first order) term-graph rewriting and term rewriting [20,15,13]. This gave a solid basis for the use of term-graph rewriting as an "efficient implementation" of term rewriting. Although the formal framework is different, our adequacy result is conceptually similar to these results, with substitution rules playing the role of the unravelling function.

The ρ-calculus has been used for giving an operational semantics to rewrite-based languages like Elan [10] or XSLT [16], thus providing a conceptual guide for their implementation. The ρ_g-calculus, with the adequacy result for strategy ∇, can be seen as an "efficient version" of the ρ-calculus and as such it can can serve as a basis for improving the implementation of rewrite-based languages.

There are several interesting directions for future research. The ρ-calculus can be seen as a higher-order rewriting system. This is proved e.g. in [9] and used to deduce some relevant properties for the ρ-calculus, like a standardisation theorem. In the same way, it would be interesting to formally view the ρ_g-calculus as a higher-order term graph rewriting system, trying to import results already available in this general setting (the relation with first-order term rewriting and the cyclic λ-calculus is preliminarly investigated in [4,8]).

We also intend to investigate the issue of optimality for the reduction strategy, where the notion of "optimal" has to be formally defined, for example in terms of time or space. In this case a natural reference to compare with would be the work on optimal reduction for lambda calculus [18].

Acknowledgements. We are grateful to Andrea Corradini, Fabio Gadducci and Stef Joosten for insightful suggestions on a preliminary version of the work.

References

1. Ariola, Z.M., Klop, J.W.: Equational term graph rewriting. Fundamenta Informaticae 26(3-4), 207–240 (1996)
2. Ariola, Z.M., Klop, J.W.: Lambda calculus with explicit recursion. Information and Computation 139(2), 154–233 (1997)

3. Baader, F., Nipkow, T.: Term rewriting and all that. Cambridge University Press, New York (1998)
4. Baldan, P., Bertolissi, C., Cirstea, H., Kirchner, C.: A rewriting calculus for cyclic higher-order term graphs. Mathematical Structures in Computer Science 17(3), 363–406 (2007)
5. Baldan, P., Bertolissi, C., Cirstea, H., Kirchner, C.: Towards a sharing strategy for the graph rewriting calculus. In: Proceedings of WRS 2007. Electr. Notes Theor. Comput. Sci., vol. 204, pp. 111–127. Elsevier (2008)
6. Barendregt, H.: The Lambda-Calculus, its syntax and semantics, 2nd edn. Studies in Logic and the Foundation of Mathematics. North Holland, Amsterdam (1984)
7. Barendregt, H.P., van Eekelen, M.C.J.D., Glauert, J.R.W., Kennaway, J.R., Plasmeijer, M.J., Sleep, M.R.: Term Graph Rewriting. In: de Bakker, J.W., Nijman, A.J., Treleaven, P.C. (eds.) PARLE 1987. LNCS, vol. 259, pp. 141–158. Springer, Heidelberg (1987)
8. Bertolissi, C.: The graph rewriting calculus: properties and expressive capabilities. Thèse de Doctorat d'Université, INPL, Nancy, France (2005)
9. Bertolissi, C., Kirchner, C.: The Rewriting Calculus as a Combinatory Reduction System. In: Seidl, H. (ed.) FOSSACS 2007. LNCS, vol. 4423, pp. 78–92. Springer, Heidelberg (2007)
10. Borovansky, P., Kirchner, C., Kirchner, H., Moreau, P.E., Ringeissen, C.: An overview of ELAN. In: Kirchner, C., Kirchner, H. (eds.) Proc. of the WRLA 1998. Electr. Notes Theor. Comput. Sci., vol. 15, pp. 55–70 (1998)
11. Cirstea, H., Kirchner, C.: The rewriting calculus — Part I *and* II. Logic Journal of the Interest Group in Pure and Applied Logics 9(3), 427–498 (2001)
12. Corradini, A.: Term rewriting in CT_Σ. In: Gaudel, M.-C., Jouannaud, J.-P. (eds.) TAPSOFT 1993. LNCS, vol. 668, pp. 468–484. Springer, Heidelberg (1993)
13. Corradini, A., Drewes, F.: (Cyclic) term graph rewriting is adequate for rational parallel term rewriting. Tech. Rep. TR-97-14, Dipartimento di Informatica, Pisa (1997)
14. Corradini, A., Gadducci, F.: Rewriting on cyclic structures: Equivalence of operational and categorical descriptions. Theoretical Informatics and Applications 33, 467–493 (1999)
15. Kennaway, J.R., Klop, J.W., Sleep, M.R., de Vries, F.J.: On the adequacy of graph rewriting for simulating term rewriting. ACM Transactions on Programming Languages and Systems 16(3), 493–523 (1994)
16. Kirchner, C., Qian, Z., Singh, P.K., Stuber, J.: Xemantics: a rewriting calculus-based semantics of XSLT. Tech. Rep. A01-R-386, Loria Inria (2002)
17. Klop, J.W., van Oostrom, V., de Vrijer, R.: Lambda calculus with patterns. Theor. Comput. Sci. 398(1-3), 16–31 (2008)
18. Lamping, J.: An algorithm for optimal lambda calculus reduction. In: Proc. of POPL 1990, pp. 16–30. ACM (1990)
19. Peyton-Jones, S.: The implementation of functional programming languages. Prentice Hall, Inc. (1987)
20. Plump, D.: Term graph rewriting. In: Handbook of Graph Grammars and Computing by Graph Transformation, vol. 2, pp. 3–61 (1999)
21. Sleep, M.R., Plasmeijer, M.J., van Eekelen, M.C.J.D. (eds.): Term graph rewriting: theory and practice. Wiley, London (1993)
22. Van Eekelen, M., Plasmeijer, R.: Functional Programming and Parallel Graph Rewriting. Addison-Wesley (1993)

A New Strategy for Distributed Compensations with Interruption in Long-Running Transactions*

Roberto Bruni[1], Anne Kersten[2], Ivan Lanese[3], and Giorgio Spagnolo[1]

[1] Department of Computer Science, University of Pisa, Italy
[2] IMT Lucca, Institute for Andvanced Studies, Italy
[3] FOCUS Team, University of Bologna/INRIA, Italy

Abstract. We propose new denotational (trace-based) and operational semantics for parallel Sagas with interruption, prove the correspondence between the two and assess their merits w.r.t. existing proposals. The new semantics is *realistic*, in the sense that it guarantees that distributed compensations may only be observed after a fault actually occurred. Moreover, the operational semantics is defined in terms of (1-safe) Petri nets and hence retains causality and concurrency information about the events that can occur, not evident in the standard trace semantics.

Keywords: cCSP, Sagas, long-running transactions, compensation, Maude.

1 Introduction

Compensations are a well-known and widely used mechanism to ensure the consistency and correctness of long-running transactions in the area of databases, e.g. when locks cannot be enforced for too long or when perfect roll-back is unrealistic. More recently, several compensable workflow languages and calculi emerged in the area of business process modeling, service-oriented and global computing to provide the necessary formal ground for compensation primitives like those exploited in orchestration languages like WS-BPEL [18]. The focus of this work is on the semantics of the workflow-based calculus Sagas [4] and its compensation policy for parallel branches. The choice of the right strategy allows the user to prevent unnecessary actions in case of an abort. Fixing a fully satisfactory formal account of this policy is the main problem we address.

In the past different policies have emerged. A thorough analysis is presented in [3] by comparing the Sagas calculus with compensating CSP [7] (cCSP). Both Sagas and cCSP focus on a core set of operations, namely compensation pairs $A \div B$ for two basic activities A and B, sequential composition of (compensable) processes $PP; QQ$ as well as the parallel composition $PP|QQ$ and a transaction scope $\{[PP]\}$ (called saga). Basic (compensable) activities include, e.g. *skipp*

---

* Research supported by EU Project FP7-231620 HATS, Italian MIUR Project IPODS, and by EU Project ASCENS.

T. Mossakowski and H.-J. Kreowski (Eds.): WADT 2010, LNCS 7137, pp. 42–60, 2012.

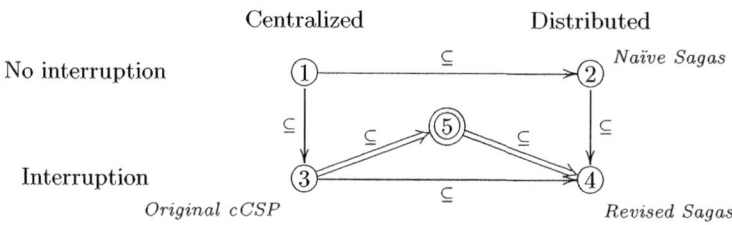

Fig. 1. Compensation policies (arrows stand for trace inclusion)

(vacuous activity) and *throww* (fault issuing). The key idea is that for a pair
$A \div B$ the successful execution of A installs the compensation B, to be executed
for "undoing" A in case the transaction is aborted later on. In case of sequential
composition compensations are unfolded in the reverse order of installation, i.e.
the activity executed last is the first to be compensated. Intuitively, this corresponds to having some sort of *compensation stack* where activities are pushed in
as the normal flow of execution progresses and popped out to be executed when
a fault occurs. The stack is cleared when the whole normal flow enclosed in a
transaction scope is executed without faults. Different strategies emerge when
handling compensations in a concurrent setting, because of two main aspects.
One aspect concerns the interruption of siblings in case of an abort (*interruption*
or *no interruption*). The other depends on whether compensations are started at
the same time (*centralized*) or siblings can start their compensation on their own
(*distributed*). The relation between the different policies is displayed in Fig. 1,
where double lines mark the contribution in this paper.

To clarify the differences between the above policies, we present the different
sets of traces obtained for the process $\{[(A \div A'; B \div B')|(C \div C'; throww)]\}$.
It may stand for example for a workflow for ordering products: A stands for
choosing the product, B for filling in an address form while C is the credit card
check, and each compensation is the sending of failure message by email. A fault
is issued after the credit card check. The example also shows that all semantic
inclusions in Fig. 1 are strict for the particular process under consideration.

For case one, centralized compensation without interruption, the resulting set
of traces is $S_1 \equiv (AB|||C); (B'A'|||C')$, where actions are serialized by juxta-
position, the symbol $|||$ stands for the interleaving of (sets of) traces and ; for
sequential composition of (sets of) traces. Actual traces also have a final $\checkmark$, denoting success (see Section 2), omitted here for the sake of readability. Roughly,
all branches are fully executed forward and only then their (interleaved) compensation is started.

For case two, distributed compensation without interruption, the set of traces
is $S_2 \equiv ABB'A'|||CC'$. Here C may be compensated without waiting for the
completion of the first branch. Moreover, the interleaving of $ABB'A'$ and CC'
includes traces where compensation B' is observed before the *throww* is issued.

For case three, we have $S_3 \equiv CC' \cup (A|||C); (A'|||C') \cup (AB|||C); (B'A'|||C')$,
i.e., the branch for the activities A and B may be interrupted, however

compensation is started only when each branch is ready, i.e., all forward activities precede all compensating activities.

The fourth strategy, distributed interruption, is the most liberal. It allows the set of traces $S_4 \equiv CC' \cup AA'|||CC' \cup ABB'A'|||CC'$. Note that, like in case two, compensation A' may be executed before C, i.e., before the error occurred.

As it turns out, none of the four originally defined semantics is entirely satisfactory. Strategies one to three are too restrictive: it is important to have the possibility to stop a sibling branch and to activate compensations as soon as possible, because typically activities and compensations have a cost. Without interruption (cases one and two) sibling branches finish their execution anyway, even though they will have to compensate. In case three, branches might have to wait until they are allowed to continue together with their siblings. The second and fourth strategies on the other hand are unrealistic: they allow a guessing mechanism where a branch may start its compensation even though the error has not occurred yet. An optimal, realistic semantics should be more "permissive" (i.e., allowing more traces) than strategy three but less than four.

In this paper, we propose new operational and denotational semantics for parallel Sagas with interruption and prove the correspondence between the two. The new semantics is "optimal", in the sense that it guarantees that distributed compensations may only be started after an error actually occurred, but compensations can start as soon as possible. In our new semantics, the traces of the above example are $S \equiv S_3 \cup (CC'AA') \cup (AB|||CC'); B'A'$.

Structure of the paper. In Section 2 we fix the notation and recall the formalization of the four different semantics from [3], in denotational style. In Section 3 we define the denotational semantics for the new policy we propose, we compare it with the semantics in Section 2 and we describe some tool support that has been useful to develop the new semantics. The content of Section 3 is taken from the Master Thesis of the last author [20]. In Section 4 we give an operational and logic account of the new semantics. In particular, we define a concurrent operational model based on Petri nets, we prove that it satisfies the properties we expect to hold for workflow components, and we give a correspondence theorem w.r.t. the denotational semantics. Note that the operational semantics is best suited for traceability of faults for which denotational semantics alone is not satisfactory. In Section 5 we discuss related work and in Section 6 we draw some concluding remarks. Proofs are collected in Appendix A.

2   Background on Parallel Sagas

The compensating CSP (cCSP) [7] and the Sagas calculus [4] were developed independently to study flow composition in long-running transactions with compensations. The two calculi had similar syntax but were presented under different semantic styles: denotational semantics for cCSP and big-step operational semantics for Sagas. Moreover, for Sagas two different semantics for parallel composition were already considered in [4], called *naïve* and *revised*. The comparison in [3] considered a common fragment, that of *parallel Sagas*, and provided a classification scheme for the three existing semantics according to the possibility of interrupting parallel branches or not and of handling compensation in a distributed manner or not (see Fig. 1). A fourth semantics emerged from the scheme (centralized compensation and no interruption) and each policy was accounted for in both the denotational and the operational style.

Below we start by presenting parallel Sagas under the denotational semantics of original cCSP (centralized with interruption). Then we show how to modify the semantics for defining the other three policies and report the main theorem from [3]. For brevity, the different policies are numbered from 1 to 4 as illustrated in Fig. 1. We index the equivalence symbol $=$ by the policy number to disambiguate the notation when needed. Note that parallel Sagas leave out several advanced features present in the original presentation of cCSP and Sagas, like exception handling and nesting: their presence is not relevant for the main result of the paper and their inclusion would just compromise the readability.

Definition 1 (Parallel Sagas). *Let Σ be an alphabet of actions, ranged over by $A, B, \dots$. The set of* parallel Sagas *processes is defined by the following grammar:*

(standard) $\qquad P, Q ::= A \mid P; Q \mid P|Q \mid \{[PP]\} \mid skip \mid throw$

(compensable) $\quad PP, QQ ::= A \div B \mid PP; QQ \mid PP|QQ \mid skipp \mid throww$

A *standard* process is either a basic activity A, the sequential composition $P; Q$ of processes, the parallel composition $P|Q$, the empty process *skip*, the raise of an interruption *throw*, or a transaction block $\{[PP]\}$. A basic *compensable* process is a compensation pair $A \div B$ where A is an atomic activity and B is its compensation. Other basic processes are *skipp*, the basic process with no behavior and no compensation, and *throww*, the basic process that always raises an interrupt (and has no compensation). Compensable processes can be composed either in sequence $PP; QQ$ or in parallel $PP|QQ$.

Policy #3: Interruption and centralized compensation (original cCSP). The denotational semantics for policy #3 is in Fig. 2. A trace for a standard process is a string $s\langle\omega\rangle$, where $s \in \Sigma^*$ is said the *observable flow* and $\omega \in \Omega$ is the *final event*, with $\Omega = \{\checkmark, !, ?\}$ and $\Sigma \cap \Omega = \emptyset$ ($\checkmark$ stands for success, ! for fail, and ? for yield to an interrupt). Note that ? appears only in traces of compensable processes, not of standard ones. We let ϵ denote the empty observable flow. Slightly abusing the notation, we let $p, q, \dots$ range over traces and also observable flows.

The definition for the traces of standard processes is straightforward. The most interesting case is the one of a transaction block $\{[PP]\}$. Note that any trace of a compensable process PP is a pair (p, q), where p is the *forward trace* and q is a *compensation trace* for p. Then, $\{[PP]\}$ selects all successful forward traces $s\langle\checkmark\rangle$ of PP, and the traces sq, corresponding to failed forward flows $s\langle!\rangle$ followed by their compensations q.

The sequential composition of standard traces $p; q$ concatenates the observable flows of p and q only when p terminates with success, otherwise it is p. The composition of two concurrent traces $p\langle\omega\rangle||q\langle\omega'\rangle$ corresponds to the set $p|||q$ of

TRACES OF STANDARD PROCESSES

$$A \triangleq_3 \{A\langle\checkmark\rangle\} \quad \text{for } A \in \Sigma$$
$$P; Q \triangleq_3 \{p; q \mid p \in P \wedge q \in Q\}$$
$$P|Q \triangleq_3 \{r \mid r \in (p\|q) \wedge p \in P \wedge q \in Q\}$$
$$\{\![PP]\!\} \triangleq_3 \{s\langle\checkmark\rangle \mid (s\langle\checkmark\rangle, q) \in PP\} \cup \{sq \mid (s\langle !\rangle, q) \in PP\}$$

$$skip \triangleq_3 \{\langle\checkmark\rangle\}$$
$$throw \triangleq_3 \{\langle !\rangle\}$$

COMPOSITION OF STANDARD TRACES

Sequential $\begin{cases} p\langle\checkmark\rangle; q \triangleq_3 pq \\ p\langle\omega\rangle; q \triangleq_3 p\langle\omega\rangle \text{ when } \omega \neq \checkmark \end{cases}$

Parallel $p\langle\omega\rangle\|q\langle\omega'\rangle \triangleq_3 \{r\langle\omega\&\omega'\rangle \mid r \in (p\|\|q)\}$, where ω'

ω	!	!	!	?	?	$\checkmark$
ω'	!	?	$\checkmark$	?	$\checkmark$	$\checkmark$
$\omega\&\omega'$	!	!	!	?	?	$\checkmark$

and $\begin{cases} p\|\|\epsilon \triangleq_3 \epsilon\|\|p \triangleq_3 \{p\} \\ Ap\|\|Bq \triangleq_3 \{Ar \mid r \in (p\|\|Bq)\} \cup \{Br \mid r \in (Ap\|\|q)\} \end{cases}$

TRACES OF COMPENSABLE PROCESSES

$$A \div B \triangleq_3 \{(A\langle\checkmark\rangle, B\langle\checkmark\rangle), (\langle?\rangle, \langle\checkmark\rangle)\}$$
$$PP; QQ \triangleq_3 \{pp; qq \mid pp \in PP \wedge qq \in QQ\}$$
$$PP|QQ \triangleq_3 \{rr \mid rr \in (pp\|qq) \wedge pp \in PP \wedge qq \in QQ\}$$

$$skipp \triangleq_3 \{(\langle\checkmark\rangle, \langle\checkmark\rangle), (\langle?\rangle, \langle\checkmark\rangle)\}$$
$$throww \triangleq_3 \{(\langle !\rangle, \langle\checkmark\rangle), (\langle?\rangle, \langle\checkmark\rangle)\}$$

COMPOSITION OF COMPENSABLE TRACES

Sequential $\begin{cases} (p\langle\checkmark\rangle, p'); (q, q') \triangleq_3 (pq, q'; p') \\ (p\langle\omega\rangle, p'); (q, q') \triangleq_3 (p\langle\omega\rangle, p') \text{ when } \omega \neq \checkmark \end{cases}$

Parallel $(p, p')\|(q, q') \triangleq_3 \{(r, r') \mid r \in (p\|q) \wedge r' \in (p'\|q')\}$

Fig. 2. Denotational semantics #3 of parallel Sagas

all possible interleavings of the observable flows, with final event $\omega\&\omega'$, where $\&$ is associative and commutative.

When composing compensable traces in series instead, the forward trace corresponds to the sequential composition of the original forward traces, while the compensation trace starts by the second compensation followed by the first one. The parallel composition is defined pairwise as all possible interleavings of the forward flows and those of the backward flows.

Our compensations never fail, thus the final event of all the compensation traces is $\checkmark$. Adding failures to compensations requires to add *throw* as a possible compensation, but this is orthogonal to the aspects we want to highlight here.

Policy #1: No interruption and centralized compensation. The simplest compensation policy differs from policy #3 just by excluding the possibility to interrupt sibling branches. Therefore we discard the possibility to yield from basic compensable processes, which is enough to exclude the presence of yielding traces:

$$A \div B \triangleq_1 \{(A\langle\checkmark\rangle, B\langle\checkmark\rangle)\}$$

Policy #4: Interruption and distributed compensation (revised Sagas). Policy #4 handles the compensation of parallel branches in a fully distributed manner. It differs from policy #3 only by the following definition of parallel composition of compensable traces:

$$(p\langle\checkmark\rangle, p')||(q\langle\checkmark\rangle, q') \triangleq_4 \{(r\langle\checkmark\rangle, r') \mid r \in (p|||q) \wedge r' \in (p'||q')\} \cup$$
$$\{(r\langle?\rangle, \langle\omega\rangle)) \mid r\langle\omega\rangle \in (pp'||qq')\}$$

$$(p\langle\omega\rangle, p')||(q\langle\omega'\rangle, q') \triangleq_4 \{(r\langle\omega\&\omega'\rangle, \langle\omega''\rangle)) \mid r\langle\omega''\rangle \in (pp'||qq')\} \quad \text{if } \omega\&\omega' \neq \checkmark$$

Policy #2: No interruption and distributed compensation (naïve Sagas). Policy #2 excludes the possibility to yield from basic compensable processes (like policy #1), but allows the presence of yielding traces within parallel composition (like policy #4). It differs from policy #3 by combining the above changes together.

Theorem 1 (cf. [3]). *Let PP be a compensable process, and let $\{|PP|\}_i$ denote the set of traces of the saga $\{|PP|\}$ according to policy #i. Then, we have $\{|PP|\}_1 \subseteq \{|PP|\}_2 \subseteq \{|PP|\}_4$ and $\{|PP|\}_1 \subseteq \{|PP|\}_3 \subseteq \{|PP|\}_4$. Moreover, if PP is sequential (i.e., it contains no occurrence of $|$), then $\{|PP|\}_1 = \{|PP|\}_4$.*

The process $\{|(A \div A'; B \div B')|(C \div C'; throww)|\}$ considered in the introduction is such that all semantic inclusions between the different policies are strict and it shows that policies #2 and #3 are not comparable in general.

3   New Semantics for Parallel Sagas

The policies #1 and #2 are of little interest for us, because they disregard interruption: we refer to activities and compensations that may have a cost, hence they should be avoided whenever possible. The absence of interruption may cause the unnecessary execution of forward activities after a sibling's fault: they need to be compensated later on. The policy #3 is not entirely satisfactory, because it is not distributed: a faulty process must wait for its siblings before starting the compensation. On the other end, policy #4 is unrealistic, because it considers traces where a process can be interrupted and its compensation can start before any sibling actually fails. This can be justified either by assuming the availability of an extremely powerful guessing mechanism (some sort of oracle) or by taking asynchronous observations, where the execution order of activities run by parallel processes cannot be detected and is therefore irrelevant.[1]

In this Section we show how to fix the denotational semantics of parallel Sagas so to obtain a fully satisfactory account of distributed interruptions. We call this policy *coordinated compensation*.

Policy #5: Coordinated compensation. Policy #5 differs from policy #3 by slightly changing the semantics of compensation pairs, so to allow a successfully completed activity to yield, and the semantics of parallel composition, to allow distributed compensation without any guessing mechanism:

---

[1] For example, if one process runs in Pisa, a second one in Tokyo and the observer is in Lisbon, then it may happen that the process running in Pisa is interrupted and compensates before the observer can log the activities executed in Tokyo.

$$A \div B \triangleq_5 \{(A\langle\checkmark\rangle, B\langle\checkmark\rangle), (\langle?\rangle, \langle\checkmark\rangle), (A\langle?\rangle, B\langle\checkmark\rangle)\}$$

$$(p\langle\checkmark\rangle, p')\|(q\langle\checkmark\rangle, q') \triangleq_5 \{(r\langle\checkmark\rangle, r') \mid r \in (p\|q) \wedge r' \in (p'\|q')\}$$

$$(p\langle\omega\rangle, p')\|(q\langle\omega'\rangle, q') \triangleq_5 \; itp((p\langle\omega\rangle, p'), (q\langle\omega'\rangle, q')) \cup$$
$$itp((q\langle\omega'\rangle, q'), (p\langle\omega\rangle, p')) \qquad \text{when } \omega, \omega' \neq \checkmark$$

$$(p\langle\omega\rangle, p')\|(q\langle\omega'\rangle, q') \triangleq_5 \; \emptyset \qquad \text{otherwise}$$

$$itp((p\langle\omega\rangle, p'), (q\langle\omega'\rangle, q')) \triangleq_5 \{((p\|q_1)\langle\omega\rangle, (p'\|q_2 q')) \mid q = q_1 q_2\}$$

Let $pp = (p\langle\omega\rangle, p')$ and $qq = (q\langle\omega'\rangle, q')$ with $\omega, \omega' \neq \checkmark$. The function $itp(pp, qq)$ returns the set of all compensable traces obtained by interleaving the activities of p with that of any prefix q_1 of q as forward activities, together with the interleaving of p' with the residual $q_2 q'$ of qq (after removing the prefix q_1). Several cases are possible. If $\omega =\,!$ and $\omega' =\,?$, then it means that qq will be interrupted by the fault raised from pp, which is ok, because qq will yield after all activities in p have been observed. The resulting forward traces have ! as final event. If $\omega =\,?$ and $\omega' =\,!$, then it means that pp is yielding to a sibling different from qq, and therefore pp can legitimately start compensating without waiting for qq to raise the fault. In this case the resulting forward traces have ? as final event. If $\omega = \omega' =\,?$ then it means that pp receives the interrupt before qq. If $\omega = \omega' =\,!$ then it just means that pp is the first to raise the fault.

Comparison. Our first result establishes a formal relation with policies #1–4.

Theorem 2. *Let PP be a compensable process, and let $\{\![PP]\!\}_i$ denote the denotational semantics of saga $\{\![PP]\!\}$ (i.e., its set of traces) according to policy #i. Then we have: $\{\![PP]\!\}_3 \subseteq \{\![PP]\!\}_5 \subseteq \{\![PP]\!\}_4$.*

Proof. The inclusion $\{\![PP]\!\}_3 \subseteq \{\![PP]\!\}_5$ follows by proving (by structural induction) the following implications for any p, p' and any $\omega \in \{?, !\}$:

$$(p\langle\checkmark\rangle, p') \in_3 PP \quad \Rightarrow \quad (p\langle\checkmark\rangle, p') \in_5 PP \; \wedge \; (p\langle?\rangle, p') \in_5 PP$$
$$(p\langle\omega\rangle, p') \in_3 PP \quad \Rightarrow \quad (p\langle\omega\rangle, p') \in_5 PP$$

where $\in_i$ denotes membership according to policy #i.

The inclusion $\{\![PP]\!\}_5 \subseteq \{\![PP]\!\}_4$ follows by proving that for any p, p':

$$(p\langle\checkmark\rangle, p') \in_5 PP \quad \Rightarrow \quad (p\langle\checkmark\rangle, p') \in_4 PP$$
$$(p\langle!\rangle, p') \in_5 PP \quad \Rightarrow \quad \exists q, q' \; (pq\langle!\rangle, q') \in_4 PP \text{ with } p' = qq'$$
$$(p\langle?\rangle, p') \in_5 PP \quad \Rightarrow \quad \exists q, q', \omega \; (pq\langle\omega\rangle, q') \in_4 PP \text{ with } p' = qq' \text{ and } \omega \in \{?, !\}$$

The first implication is quite trivial (from the definition of policies #4 and #5). The other two implications are proved by structural induction.   □

The process $\{\![(A \div A'; B \div B')|(C \div C'; throww)]\!\}$ from the introduction witnesses that all semantic inclusions between the different policies can be strict.

Corollary 1. *If PP is a sequential process (i.e., it contains no parallel composition operator), then $\{\![PP]\!\}_5 = \{\![PP]\!\}_4$.*

Proposition 1. *The policies #2 and #5 are not comparable by inclusion.*

Proof. We show that there exists a process PP such that neither $\{[PP]\}_2 \subseteq \{[PP]\}_5$ nor $\{[PP]\}_5 \subseteq \{[PP]\}_2$ hold. Take $PP = \{[A \div A' \mid (B \div B'; throww)]\}$. Then the trace $p = AA'BB'\langle\checkmark\rangle \in \{[PP]\}_2$, but $p \notin \{[PP]\}_5$, because A' is observed before B (and therefore before the fault occurs). Moreover, the trace $q = BB'\langle\checkmark\rangle \in \{[PP]\}_5$, but $q \notin \{[PP]\}_2$, because it involves the interruption of process $A \div A'$, not allowed in policy #2.   □

Maude support. Given the combinatorial explosion in handling the interleaving of parallel processes, we exploited the rewrite engine of Maude [9] to experiment with the different policies and validate our results. The tool has been developed using the Eclipse plugin Maude Development Tools MOMENT2 [2] and it allows, e.g., to derive the semantics of a process under the policies #1–5, to test for membership of a particular trace under a particular semantics, to test the inclusion between sets of traces and to test for the presence of a particular activity in a set of traces. The tool consists of six Maude functional modules: one for the definitions common to all policies, plus one extension for each policy #1–5. The main advantage in using Maude is the reduced representation distance between the mathematical specifications of the policies and their Maude encoding, which makes immediate the correspondence check. As an additional feature, the evaluation of the semantics of a process is parametric to an environment assigning success and failure to each activity. We remark that the tool has been very useful to assess and refine policy #5 before proving the main correspondence theorem (Th. 2). For more details and source code see [20].

4   Operational and Logical Account

In this section we introduce an operational characterization of the coordinated semantics, and we prove a correspondence result between the operational semantics presented here and the denotational semantics of previous section. Furthermore we show that the semantics satisfies some expected high-level properties.

Petri nets basics. We base our operational semantics on a mapping of sagas into Petri nets [19]. On one side, Petri nets are a well-known model of concurrency and allow us to give a simple account of the interactions between the different activities in a saga. On the other side, this mapping allows us to exploit the well-developed theory of Petri nets and the related tools.

Definition 2 (Petri net). *A Petri net graph is a triple (P, T, F), where: P is a finite set of places; T is a finite set of transitions, disjoint from P; and the flow relation $F \subseteq (P \times T) \cup (T \times P)$ is a set of edges.*

Given a Petri net graph, a marking U for the net is a multiset of places: $U : P \rightarrow \mathbb{N}$. We call Petri net any net N equipped with an initial marking U_N.

The preset of a transition t is the set of its input places: $^\bullet t = \{p \mid (p,t) \in F\}$; its postset is the set of its output places: $t^\bullet = \{p \mid (t,p) \in F\}$. We denote the empty

$$\frac{U : P \to \mathbb{N}}{U : U \to U \in \mathcal{T}(N)} \qquad \frac{t \in T_N}{t : {}^\bullet t \to t^\bullet \in \mathcal{T}(N)}$$

$$\frac{r : U \to V, \; r' : U' \to V' \in \mathcal{T}(N)}{r + r' : U + U' \to V + V' \in \mathcal{T}(N)} \qquad \frac{r : U \to V, \; s : V \to W \in \mathcal{T}(N)}{r ; s : U \to W \in \mathcal{T}(N)}$$

Fig. 3. Inference rules for $\mathcal{T}(N)$

multiset by 0, multiset union by $+$, multiset difference by $-$, multiset inclusion by $\subseteq$ and write $a \in U$ if $U(a) > 0$.

Definition 3 (Firing). *Given a Petri net graph (P, T, F) and a marking U a transition t is* enabled *in U if ${}^\bullet t \subseteq U$. A transition t enabled in U can* fire *leading to the marking $V = U - {}^\bullet t + t^\bullet$.*

A multiset of transitions can fire concurrently, if U contains enough tokens to cover all their presets. After [17], we denote computations over a Petri net as terms of the algebra $\mathcal{T}(N)$ freely generated by the inference rules in Fig. 3 modulo the axioms below (whenever both sides are well-defined):[2]

monoid:	$(p + q) + r = p + (q + r)$	$r + r' = r' + r$	$0 + r = r$
category:	$(p; q); r = p; (q; r)$	$r; V = r = U; r$	
functorial:	$(p; p') + (q; q') = (p + q); (p' + q')$		

Each term $r : U \to V \in \mathcal{T}(N)$ defines a concurrent computation over N, from the marking U to the marking V. We write $N \overset{*}{\to} V$ and say that V is *reachable* in N, if there exists a computation $r : U_N \to V \in \mathcal{T}(N)$. A Petri net N is 1-safe if for any place a and for any reachable marking V we have $V(a) \le 1$. We say that $r : U_N \to V \in \mathcal{T}(N)$ is *maximal* if no transition is enabled in V.

From Sagas to Petri nets. We define the Petri net graph associated to a saga by structural induction on the saga syntax. At the high-level view, each compensable process is a black box with six external places to be interfaced with the environment (Fig. 4a). Places F_1 and F_2 are used for propagating the forward flow of execution: a token in F_1 starts the execution and a token in F_2 indicates that the execution has ended successfully. Places R_1 and R_2 control the reverse flow: a token in R_1 starts the compensation, a token in R_2 indicates that the compensation has ended successfully. The place I_1 is used to interrupt the process from the outside while a token in I_2 is used to inform the environment that an error has occurred. The figure highlights that three auxiliary transitions will be present in any process: two of them handle the catching of the interrupt and reversal of the flow, one handles the disposal of the interrupt in case the process already produced a fault (garbage collection). A standard process such as the saga in Fig. 4e instead has just three places to interact with the environment: F_1 starts its flow, F_2 signals successful termination, and E raises a fault.

---

[2] For category-minded theorists, the computations form a freely generated, strictly symmetric, strict monoidal category.

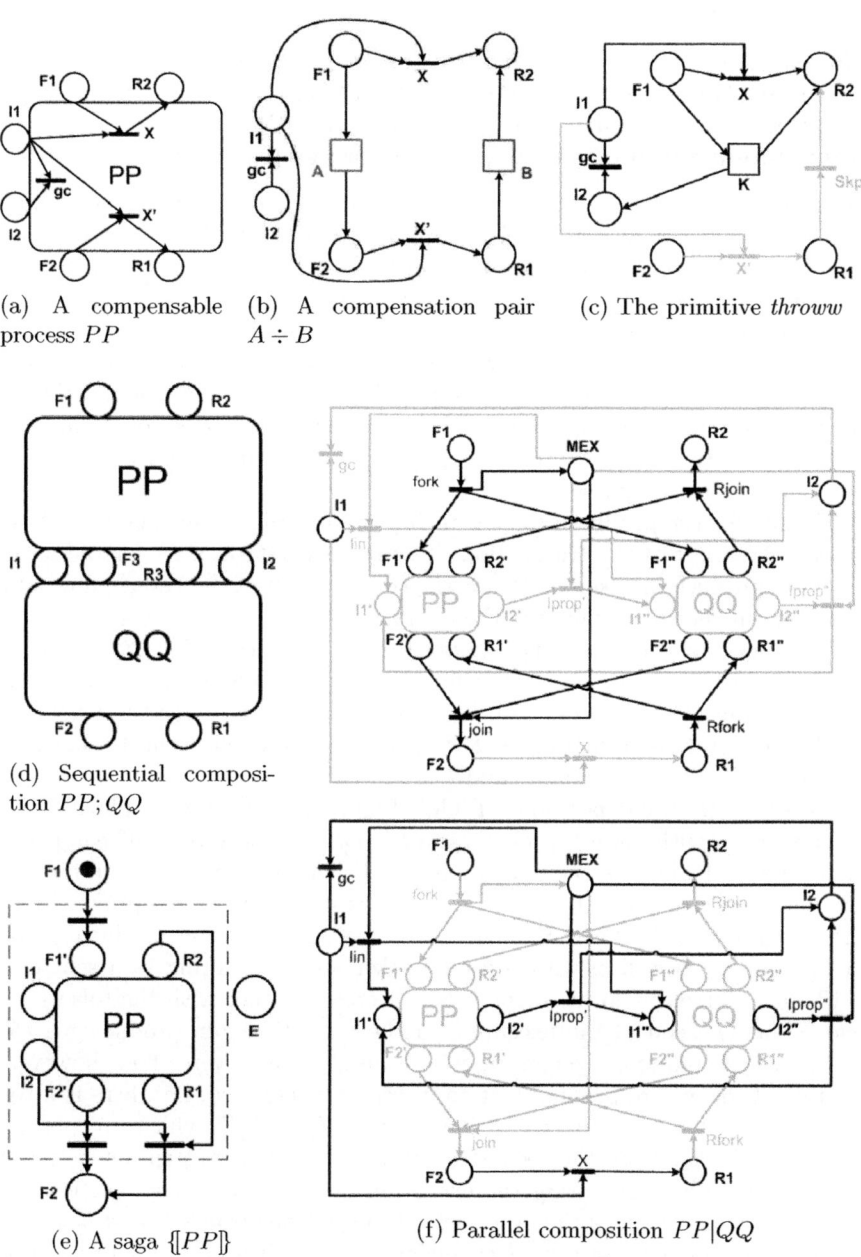

(a) A compensable process PP

(b) A compensation pair $A \div B$

(c) The primitive *throww*

(d) Sequential composition $PP; QQ$

(e) A saga $\{PP\}$

(f) Parallel composition $PP|QQ$

Fig. 4. Petri nets for the encoding of processes

Intuitively, for standard processes, a computation starting in F_1 will lead either to F_2 or to E, while for compensable processes we expect to have the following kinds of computations:

Successful (forward) computation: from marking F_1 the net reaches F_2
Compensating (backward) computation: from R_1 the net reaches R_2.
Aborted computation: from F_1 the net reaches $R_2 + I_2$.
Interrupted computation: from $F_1 + I_1$ the net reaches R_2.

We will formalize this behavior and prove that our model behaves accordingly.

The nets for processes are depicted in Fig. 4. When drawing transitions we use larger boxes for representing sagas activities and thinner, black filled boxes for auxiliary transitions. For a compensation pair $A \div B$ (Fig. 4b) there is a transition called A that consumes a token in F_1 and produces a token in F_2 representing the execution of the forward flow, as well as a transition B that consumes a token in R_1 and produces a token in R_2 corresponding to the reverse flow. Furthermore the process can be interrupted at the beginning with the transition X and at the end with the transition X'. The net for *skipp* is similar (A and B are replaced by vacuous silent activities).

The net for the primitive *throww* is displayed in Figure 4c. The transition K models the abort of the transaction. It consumes the token F_1 for the forward flow and produces a token in R_2 for the continuation of the reverse flow and in I_2 to inform the environment of the abort.

In the net for the sequential composition for a process $PP; QQ$ (Figure 4d) the token in F_3 produced by PP for the forward flow is passed on to start the execution of the process QQ. Equally the token in R_3 for the reverse flow produced by QQ is passed on to PP to start its compensation. Both PP and QQ share also the places for I_1 and I_2.

For the parallel composition $PP|QQ$ (Figure 4f) we use two subnets for the two processes, with places $F_1', F_2', \ldots$ and $F_1'', F_2'', \ldots$ respectively. The upper part of the figure highlights the transitions used in absence of interruptions while the lower part concentrates on transitions exploited by interruption. To start the execution of the forward flow there is a fork transition producing tokens in F_1' and F_1'' as well as an additional token MEX working as a semaphore. Without an error or an interrupt it is collected together with the tokens in F_2' and F_2'' at the end of the execution in the transition join producing a token in F_2. A fork-and-join mechanism is used also for the reverse flow, however no additional tokens are produced. If an interrupt is received, i.e. there is a token in I_1, it is split using the transition Iin into I_1' and I_1'' which are processed by PP and QQ. Here we need the semaphore MEX to guarantee that the interrupt is only split during the execution of the parallel composition. If an error occurs inside one of the processes PP or QQ, the places I_2' and I_2'' are used to inform its sibling and the environment: the transition Iprop$'$ (resp. Iprop$''$) consumes a token from I_2' (resp. I_2'') and produces a token in I_2 and in I_1'' (resp. I_2 and in I_1'). The semaphore MEX is used here to guarantee that only one branch sends the interrupt to the environment, and that no interrupt is sent if an external interrupt has been already received. As usual we have the transition

$X : F_2 + I_1 \rightarrow R_1$, for aborting after completion of the two branches, and the garbage collecting transition.

In Figure 4e the Petri net for a saga is shown. Outside of the transaction block there is no reverse flow, thus only places F_1 and F_2 are considered, while the other places are only needed inside the transaction block. Then a token in F_2 signals that the transaction has reached a consistent state again. This does not necessarily mean that the transaction was successful, it might as well have failed and executed its compensation. The nets for *skip*, *throw*, sequential and parallel composition of sagas are straightforward and thus omitted here.

Theorem 3. *Given a compensable process PP and the corresponding net N_{PP} with external places F_1, F_2 for the forward flow, R_1, R_2 for the reverse flow and I_1, I_2 for interrupts, we can state the following properties:*

1. *Every maximal execution of the net N_{PP} with initial marking F_1 is either of the form $f_{PP} : F_1 \rightarrow F_2$ (successful computation), or of the form $a_{PP} : F_1 \rightarrow R_2 + I_2$ (aborted computation);*
2. *Every maximal execution of the net N_{PP} with initial marking R_1 is of the form $r_{PP} : R_1 \rightarrow R_2$ (backward computation);*
3. *Every maximal execution of the net N_{PP} with initial marking $F_1 + I_1$ is of the form $i_{PP} : F_1 + I_1 \rightarrow R_2$ (interrupted computation).*

Proof. The proof is by structural induction (see appendix). □

Note that we call *interrupted* any computation that consumes the token I_1: it may as well happen that the net autonomously aborts (due to some *throww*), and the token I_1 is consumed by the garbage collection transition.

In order to show the correspondence of the denotational and operational semantics we need to introduce some kind of observation for the latter.

Definition 4. *Let PP be a compensable process and N_{PP} its corresponding net. For any $f \in T(N_{PP})$ we define the set $label(f)$ of action sequences as follows:*

$$label(A) = \{A\} \text{ for any basic activity } A$$
$$label(K) = \{K\} \text{ for any throw transition } K$$
$$label(f_1; f_2) = label(f_1)label(f_2)$$
$$label(f_1 + f_2) = label(f_1)|||label(f_2)$$
$$label(f) = \epsilon \text{ otherwise}$$

where juxtaposition and interleaving are defined elementwise.

It is immediate to check that the function *label* is well-defined, in the sense that it is invariant w.r.t. the equivalence axioms on $T(N)$. Moreover we define a function $filter(p)$ that removes every K from a label p. Using this definition we can now formulate the correspondence theorem.

Theorem 4 (Correspondence). *Let PP be a compensable process and N_{PP} the corresponding net. The correspondence of denotational and (maximal computations of the) operational semantics is given as follows:*

1. $(p\langle\checkmark\rangle, q\langle\checkmark\rangle) \in_5 PP$ iff there is a computation $f : F_1 \to F_2 \in \mathcal{T}(N_{PP})$ with $p \in label(f)$ and a computation $r : R_1 \to R_2 \in \mathcal{T}(N_{PP})$ with $q \in label(r)$.
2. $(p\langle!\rangle, q\langle\checkmark\rangle) \in_5 PP$ iff there is a computation $a : F_1 \to I_2 + R_2 \in \mathcal{T}(N_{PP})$ with label $pKq' \in label(a)$ for some q' such that $q = filter(q')$.
3. $(p\langle?\rangle, q\langle\checkmark\rangle) \in_5 PP$ iff there is a marking U and two computations $f : F_1 \to U, i : U + I_1 \to R_2 \in \mathcal{T}(N_{PP})$ such that $p \in label(f)$ and $q = filter(q')$ for some $q' \in label(i)$.

Proof. The proof is similar to the one of Theorem 3 (see appendix). □

Proposition 2. *Let PP be a compensable process and N_{PP} the corresponding net with external places $F_1, F_2, R_1, R_2, I_1, I_2$. Then: (i) N_{PP} with initial marking $F_1 + I_1$ is 1-safe, and (ii) N_{PP} with initial marking $R_1 + I_1$ is 1-safe.[3] Moreover, let P be a standard process and let N_P be its corresponding net with external places F_1, F_2, E and initial marking F_1. Then, N_P is 1-safe.*

Proof. The proof is by structural induction on PP and P. □

The main results can be extended to standard processes as follows.

Corollary 2. *Let P be a standard process and let N_P its corresponding net with external places F_1, F_2, E and initial marking F_1. Then: (i) any maximal execution of N_P leads either to F_2 or to E; (ii) $p\langle\checkmark\rangle \in_5 P$ iff there is a computation $f : F_1 \to F_2 \in \mathcal{T}(N_P)$ with $p \in label(f)$ ending in marking F_2 and (iii) $p\langle!\rangle \in_5 P$ iff there is a computation $a : F_1 \to E \in \mathcal{T}(N_P)$ with $p \in label(a)$.*

Logical Properties. We show now that Sagas $\{\![PP]\!\}$ satisfy some basic logical properties following their intuitive behavior. First, following [4], we define the concept of order of the activities in a Saga. To this end we need activities with a unique name. Also, we consider *throww* as a forward activity and use subscripts to distinguish between multiple occurrences of the same activity, like in $\{\![throww_1 | A_1 \div B; A_2 \div C; throww_2]\!\}$.

We let $\mathbf{A}(S)$ be the set of activities of a Saga (including *throww*).

Definition 5 (Order of a Saga). *The order of a Saga S is the least transitive relation $\prec_S$ such that:*

1. *if $A \div A'$ occurs in S then $A \prec_S A'$;*
2. *if $PP; QQ$ occurs in S then $A \prec_S B$ for each forward activity A occurring in PP and any forward activity B in QQ;*
3. *if $A \div A'$ and $B \div B'$ occur in S and $A \prec_S B$ then $B' \prec_S A'$.*

We let $pred_S(A) \triangleq \{B \in \mathbf{A}(S) \mid B \prec_S A\}$ be the set of the predecessors of the activity A w.r.t. the order $\prec_S$. We say a sequence $A_1 A_2 ... A_n$ *respects* the order $\prec_S$ if $A_i \prec_S A_j$ for any $1 \le i < j \le n$.

---

[3] This implies that N_{PP} is 1-safe when the initial marking is F_1 or R_1.

Theorem 5 (Completion). *Let* $S = \{[PP]\}$ *be a saga. If* PP *contains no* throww *activities, then it will succeed. In this case there exists a unique maximal computation* $f \in \mathcal{T}(N_{PP})$ *with initial marking* F_1 *and it leads to* F_2. *Furthermore,* $label(f)$ *is the set of possible interleavings of all forward activities in* $\mathbf{A}(S)$ *that respect* $\prec_S$.

Proof. By induction on the structure of the process inside the Saga S.   □

Theorem 6 (Successful compensation). *Let* $S = \{[PP]\}$ *be a saga. If* PP *contains at least a* throww *activity, then it will abort and it will be compensated. In this case all the maximal computations in the net* N_{PP} *with initial marking* F_1 *end in* $R_2 + I_2$. *Then, for any such computation* $a : F_1 \to R_2 + I_2$ *we have that each possible label in* $filter(label(a))$ *satisfies the conditions below:*

1. *activities in the label respect the order* $\prec_S$;
2. *any activity* A *such that* $A \prec_S throww_i$ *for all* $throww_i$ *in* $\mathbf{A}(S)$, *occurs in the label;*
3. *no forward activity* A *such that there exists* $throww_i$ *in* $\mathbf{A}(S)$ *with* $throww_i \prec_S A$, *occurs in the label;*
4. *if activity* A' *is the compensation of activity* A, *then* A *occurs in the label iff* A' *occurs in the label;*
5. *there exists at least one* $throww_i$ *such that all activities in* $pred_S(throww_i)$ *appear in the label and they precede each compensation activity* A' *appearing in the label.*

Moreover, for any action sequence q *satisfying conditions 1–5) above, there exists a maximal computation* $a : F_1 \to R_2 + I_2$ *such that* $q \in filter(label(a))$.

Proof. See appendix.   □

Conditions 1 to 4 correspond to some of the conditions in [4], while condition 5 does not hold in [4]: it characterizes the fact that our semantics allows only realistic traces, where compensations are not started before a fault is actually executed. Since faults are removed from labels, we consider that a fault can be executed when all the observable activities preceding it have been executed, thus enabling it. Consider the Saga $S = \{[C \div C' \mid (A; throww_1) \mid (B; throww_2)]\}$. Here we have $pred_S(throww_1) = \{A\}$ and $pred_S(throww_2) = \{B\}$. Then the trace $CC'AB$ can not be observed, since C' is preceded by neither A nor B. This trace is valid according to the semantics #4.

5   Discussion and Related Work

In this section we overview other process algebraic approaches that model long-running transactions and compensations. The concept of a Saga was first mentioned in the context of database theory in [12]. The process description language StAC [6] is among the first to transfer this concept to process algebras. It provides an implementation and an operational semantics given in a richer intermediate language called $StAC_i$. However the syntax included too many primitives, and its semantics was quite intricate, having to deal with spurious aspects.

Both Sagas [4] and cCSP [7] were developed as refinements of StAC, focusing just on a core set of operations combined in well-disciplined ways. In [11] it has been shown that Sagas can be implemented on top of an event-based middleware via a semantic preserving encoding. A more recent approach to sagas is presented in [15] (extended with nesting in [13]), but the compensation policy is expressed at a less abstract level, including spurious mechanisms closer to the implementation. Moreover it is not distributed, as the order of the execution of compensations depends on the total order of activities in the forward flow. We also mention [8], that shows an encoding of cCSP into the Conversation Calculus [22] (CC), even though CC does not include explicit compensation handlers.

This paper focused on different policies for executing compensations. However there are also different strategies for installing compensations, as shown in [14]. Roughly, we distinguish static, parallel and dynamic compensation definitions. Here we considered the static case, where compensations are known from the beginning. In the parallel case, new items of compensation can be added at runtime, in parallel w.r.t. the old ones. The dynamic case allows fully general updating of compensations. In contrast to our approach which is more focused on the control flow between components, [14] focuses on communicating components. This kind of proposals is based on extensions of the π-calculus, such as webπ [16], dcπ [21] and ATc [1], of the Join calculus, such as cJoin [5], or of CCS, such as CommTrans [10].

6   Conclusion

We have formally defined a novel policy of distributed compensation with interruption and compared it with existing policies. The new one improves them by allowing autonomous activation of compensation (as opposed to policies #1 and #3) and by discarding unrealistic traces (as opposed to policies #2 and #4). Moreover, the operational semantics defined in terms of Petri nets paves the way to the straightforward derivation of richer semantic domains than traces, where causal dependencies between events are recorded. This can have e.g. practical consequences in the tracking of the faults that triggered a compensation.

In the future we plan to extend the core language of parallel Sagas with choice and some form of iteration. For the denotational semantics this can be easily accomplished. For the Petri net semantics, choice can be handled with some care, while iteration might require a more sophisticated variant of Petri nets, where the finite sequence of actions performed while unfolding the iteration must be recorded in the net to derive a proper compensation: a finite 1-safe net will not be able to store such information, unless it is equipped with some mechanism for dynamic generation of new net items. Together with the above extension, we plan to develop a small suite of prototypical tools based on Maude, to support the theoretical studies and the experimentation. This includes the generation of Petri net layouts from sagas processes and the use of an LTL model checker.

References

1. Bocchi, L., Tuosto, E.: A Java Inspired Semantics for Transactions in SOC. In: Wirsing, M., Hofmann, M., Rauschmayer, A. (eds.) TGC 2010, LNCS, vol. 6084, pp. 120–134. Springer, Heidelberg (2010)
2. Boronat, A., Meseguer, J.: MOMENT2: EMF model transformations in Maude. In: JISBD 2009, pp. 178–179 (2009)
3. Bruni, R., Butler, M., Ferreira, C., Hoare, T., Melgratti, H., Montanari, U.: Comparing Two Approaches to Compensable Flow Composition. In: Abadi, M., de Alfaro, L. (eds.) CONCUR 2005. LNCS, vol. 3653, pp. 383–397. Springer, Heidelberg (2005)
4. Bruni, R., Melgratti, H., Montanari, U.: Theoretical foundations for compensations in flow composition languages. In: POPL 2005, pp. 209–220. ACM (2005)
5. Bruni, R., Melgratti, H.C., Montanari, U.: Nested Commits for Mobile Calculi: Extending Join. In: IFIP-TCS 2004, pp. 563–576 (2004)
6. Butler, M.J., Ferreira, C.: A Process Compensation Language. In: Grieskamp, W., Santen, T., Stoddart, B. (eds.) IFM 2000. LNCS, vol. 1945, pp. 61–76. Springer, Heidelberg (2000)
7. Butler, M., Hoare, T., Ferreira, C.: A Trace Semantics for Long-Running Transactions. In: Abdallah, A.E., Jones, C.B., Sanders, J.W. (eds.) CSP25. LNCS, vol. 3525, pp. 133–150. Springer, Heidelberg (2005)
8. Caires, L., Ferreira, C., Vieira, H.T.: A Process Calculus Analysis of Compensations. In: Kaklamanis, C., Nielson, F. (eds.) TGC 2008. LNCS, vol. 5474, pp. 87–103. Springer, Heidelberg (2009)
9. Clavel, M., Durán, F., Eker, S., Lincoln, P., Martí-Oliet, N., Bevilacqua, V., Talcott, C.: All About Maude - A High-Performance Logical Framework. LNCS, vol. 4350. Springer, Heidelberg (2007)
10. de Vries, E., Koutavas, V., Hennessy, M.: Communicating Transactions. In: Gastin, P., Laroussinie, F. (eds.) CONCUR 2010. LNCS, vol. 6269, pp. 569–583. Springer, Heidelberg (2010)
11. Ferrari, G.L., Guanciale, R., Strollo, D., Tuosto, E.: Event-Based Service Coordination. In: Degano, P., De Nicola, R., Bevilacqua, V. (eds.) Montanari Festschrift. LNCS, vol. 5065, pp. 312–329. Springer, Heidelberg (2008)
12. Garcia-Molina, H., Salem, K.: Sagas. In: SIGMOD Conference, pp. 249–259 (1987)
13. Lanese, I.: Static vs dynamic sagas. In: ICE 2010. EPTCS, vol. 38, pp. 51–65 (2010)
14. Lanese, I., Vaz, C., Ferreira, C.: On the Expressive Power of Primitives for Compensation Handling. In: Gordon, A.D. (ed.) ESOP 2010. LNCS, vol. 6012, pp. 366–386. Springer, Heidelberg (2010)
15. Lanese, I., Zavattaro, G.: Programming Sagas in SOCK. In: SEFM 2009, pp. 189–198. IEEE (2009)
16. Laneve, C., Zavattaro, G.: Foundations of Web Transactions. In: Sassone, V. (ed.) FOSSACS 2005. LNCS, vol. 3441, pp. 282–298. Springer, Heidelberg (2005)
17. Meseguer, J., Montanari, U.: Petri nets are monoids. Inf. Comput. 88(2), 105–155 (1990)
18. Oasis: Web Services Business Process Execution Language Version 2.0 (2007), http://docs.oasis-open.org/wsbpel/2.0/OS/wsbpel-v2.0-OS.html
19. Reisig, W.: Petri Nets. Springer, Heidelberg (1985)
20. Spagnolo, G.: Analisi e confronto di politiche di compensazione distribuita nell'ambito di transazioni a lunga durata. Master's thesis, Università di Pisa (2010)

21. Vaz, C., Ferreira, C., Ravara, A.: Dynamic Recovering of Long Running Trans-
    actions. In: Kaklamanis, C., Nielson, F. (eds.) TGC 2008. LNCS, vol. 5474, pp.
    201–215. Springer, Heidelberg (2009)
22. Vieira, H.T., Caires, L., Seco, J.C.: The Conversation Calculus: A Model of Service-
    Oriented Computation. In: Gairing, M. (ed.) ESOP 2008. LNCS, vol. 4960, pp.
    269–283. Springer, Heidelberg (2008)

A   Proofs of Main Results

Proof (Theorem 3). The proof is by structural induction. The theorem holds trivially for compensation pairs as well as for the primitives *skipp* and *throww*. For
the   sequential   composition   the   proof   follows   from   the   induction
hypothesis.

   Consider parallel composition. Assume a token is in F_1. Using $fork$ we create
a marking $F_1' + F_1'' + MEX$. We have then a case analysis according to the
behavior of the subnets. If both succeed then the $join$ transition creates a token
in F_2 and removes all the other tokens as desired. Note that a transition like
$f_{PP_1} + f_{PP_2}$, due to the distributive law given in [17], corresponds to all the
possible interleavings of the two transitions. Assume now that PP_1 aborts first
(the case where PP_2 aborts first is symmetric). By inductive hypothesis this
creates a token in I_1'', thus for PP_2 we go to the case of interrupted computation.
Thus we will end up with $R_2' + R_2''$. The thesis follows using the transition $Rjoin$.

   The case of backward computation is easy: the transitions $Rfork$ and $Rjoin$
are triggered, satisfying the hypothesis.

   Finally, consider the case of interrupted computation: if I_1 is still there after
MEX has been consumed, then X is taken, producing a token in R_1. This goes
back to the previous case. Assume instead that transition Iin is taken, producing
as marking $F_1' + F_1'' + I_1' + I_1''$. We can apply inductive hypothesis to reach a
marking $R_2' + R_2''$, satisfying the hypothesis executing transition $Rjoin$.   □

Proof (Theorem 4). The proof is by induction on the structure of process (which
corresponds to an induction on the structure of the corresponding net), with a
case analysis similar to the one of Theorem 3.

   The theorem holds trivially for compensation pairs, *skipp* and *throww*.

   Let us consider sequential composition. We have two kinds of traces: the ones
where PP succeeds, and the ones where it does not. Let us consider the first
case. By inductive hypothesis we have a computation $f_{PP} : F_1 \rightarrow F_3$ with
$label(f_{PP}) = p$. Then we have again a case analysis according to the behavior of
QQ. We consider just the case of success, the other being similar. In this case, by
inductive hypothesis we have a computation $f_{QQ} : F_3 \rightarrow F_2$ with $label(f_{QQ}) = p'$
and then computations $r_{QQ} : R_1 \rightarrow R_3$ with $label(r_{QQ}) = q'$ and $r_{PP} : R_3 \rightarrow$
R_2 with $label(r_{PP}) = q$. Thus the two computations $f_{PP}; f_{QQ}$ and $r_{QQ}; r_{PP}$
satisfy the thesis. The case where PP does not succeed is trivial by inductive
hypothesis.

Let us consider now parallel composition. We have again two possibilities: either both PP_1 and PP_2 succeed, or not. Let us consider the first case. Operationally, first transition $fork$ is executed. Then the inductive hypothesis is applied to the two subnets for the forward flow. Finally transition $join$ is executed. The analysis of the backward flow is similar. It is easy to see that labels are the desired ones. Let us consider the other case. First transition $fork$ is executed. We have then a case analysis according to the behavior of the two subnets. Assume PP_1 aborts. By inductive hypothesis there is a computation $a_{PP_1} : F_1' \to I_2' + R_2'$. Assume that PP_2 is interrupted. Again by inductive hypothesis, there are computations $i'_{PP_2} : F_1'' \to N$ and $i''_{PP_2} : N + I_1'' \to R_2''$. The only constraint on the possible interleavings is that i''_{PP_2} may only start after I_2' has been produced. This is the behavior captured by function itp. In the case of double abort the two subnets start compensating on their own, thus there is no synchronization constraint to be satisfied (both the sets defining function itp become not empty). The two notifications are garbage collected. The case of external interrupt is similar. The only difference is that if the interrupt is processed after the two processes have finished their computations successfully, then transition X' is used. In the denotational semantics there is no clause corresponding to this, but this produces the same traces of two yielding computations (and we always have a yielding computation for each successful one). $\qquad\square$

Proof (Theorem 6). First we prove by structural induction (using Theorem 3) that any maximal computation starting from F_1 ends in $R_2 + I_2$ with no token ever appearing in F_2 (needed for proving property 3, below). Next, we take any $a : F_1 \to R_2 + I_2$ and $q \in filter(label(a))$ and show that 1–5) hold for q.

Properties 1) and 4) are proved by structural induction on PP.

For property 2), since $A \prec_S throww_i$ for all $throww_i$, there must exist PP' and QQ' such that $PP = C[PP'; QQ']$ for some context $C[\cdot]$, where PP' contains A and QQ' contains all $throww_i$. Then, we conclude by structural induction on the shape of the context $C[\cdot]$, by applying Theorem 5 to PP'.

For property 3), let $throww_i$ be such that $throww_i \prec_S A$, therefore there exist PP' and QQ' such that $PP = C[PP'; QQ']$ for some context $C[\cdot]$, where PP' contains $throww_i$ and QQ' contains A. Therefore by applying the first argument of this proof to PP' we know that activities in QQ' are never enabled.

For property 5) we proceed by contradiction. Suppose that there exists a compensation activity A' such that for any activity $throww_i$ an activity $B_i \in pred_S(throww_i)$ can be found such that A' precedes B_i in q. Without loss of generality, let A' be the leftmost such activity appearing in q. Hence $q = A_1 \cdots A_n A' q'$ for some forward activities $A_1, \ldots, A_n$ and sequence q' that contains all B_i's. But then the firing of $A_1 \cdots A_n$ leads to a marking where no $throww_i$ is enabled and therefore A' cannot be enabled, which is absurd.

For the last part, let q be any action sequence that satisfies conditions 1–5) and let $A_1 \cdots A_n$ be the subsequence of q formed by forward activities. By condition 4), their backward activities $A_1', \ldots, A_n'$ are the only other activities that appear in q and we let $A_{i_1}', \ldots, A_{i_n}'$ be the corresponding subsequence of q. By conditions 1–3) the sequence of transitions $A_1 \cdots A_n$ can be fired (possibly firing additional $fork$ and $join$) starting from F_1. By condition 5) there is a $throww_i$ such that all activities in $pred_S(throww_i)$ are in $A_1, \ldots, A_n$ and therefore the transition K associated with $throww_i$ is enabled after the firing of $A_1 \cdots A_m$ for some $m \leq n$, which is a prefix of q. Note that the propagation of interrupts by transitions lin, $lprop'$ and $lprop''$ can be delayed until A_n is fired. Therefore the sequence of transitions $A_1 \cdots A_n A'_{i_1} \cdots A'_{i_n}$ is fireable, which induces a computation $a : F_1 \to R_2 + I_2$. It remains to show that $q \in filter(label(a))$, which can be done by induction on the number of action switches needed to transform $A_1 \cdots A_n A'_{i_1} \cdots A'_{i_n}$ to q exploiting the functorial axiom. Note, in fact, that one has to swap only some forward actions A_i (for $i > m$) and some backward actions A'_j (possibly with the interrupt propagation transitions that enables A'_j), such that $A_i \not\prec_S A'_j$ (by condition 1). $\qquad\square$

Towards a First-Order Deontic Action Logic[*]

Pablo F. Castro[1] and Tom S.E. Maibaum[2]

[1] Universidad Nacional de Rio Cuarto, Departamento de Computación, Argentina
pcastro@dc.exa.unrc.edu.ar
[2] McMaster University, Department of Computing & Software, Hamilton, Canada
tom@maibaum.org

Abstract. In this article we describe a first-order extension of the deontic logic introduced in [1]. The main useful and interesting characteristic of this extended logic is that it not only provides the standard quantifiers of first-order logic, but it also has similar algebraic operators for actions as for the propositional version of [1]. Since the pioneering works of Hintikka and Kanger, little advance has been made in developing first-order deontic logics. Furthermore, to the best of our knowledge, the introduction of quantifiers in deontic action logics (i.e., deontic action logics where predicates are applied only to actions) has not been investigated in detail in the literature. This paper represents a significant step in addressing these problems. We also demonstrate the application of this novel logic to fault-tolerance by means of a simple example.

1 Introduction

Deontic logics are focused on the study of the logical properties of operators such as permission, obligation and prohibition. Since these notions are closely related to the concept of violation or error, some computer science researchers have proposed the use of this kind of logic to specify fault-tolerant systems (see [2]). We have proposed a deontic logic to reason about fault-tolerance in [1] and we have investigated the application of this logic to practical examples in [3]. The logic introduced in these papers is a propositional logic, extended with temporal operators. However, for more complex applications, first-order quantifiers might be needed, in particular when we need to reason about infinite domains or complex data structures.

Deontic logics can be divided into *ought-to-be* logics and *ought-to-do* logics. The former apply deontic operators to predicates, and these are perhaps the ones normally investigated by deontic logicians. On the other hand, *ought-to-do* deontic logics use deontic predicates to state properties of actions. These logics are very similar to dynamic logics [4], where we have formulae of the form: $\varphi \rightarrow [\alpha]\psi$, φ and ψ being predicates, and α being an action. This formula can be read as saying that *if φ is true, then after the execution of α, ψ becomes true*. These formulae can be used to specify systems in the same way as in Hoare logic, i.e., thinking of φ as a precondition and ψ as a postcondition of α. Some authors have proposed the reduction of deontic action logic to dynamic logic [5]; in this case, permitted actions are those for which there are some ways of executing them such that no violation is introduced. We have proposed a different deontic

[*] This work has been supported by NSERC and MRI through an ORF-RE grant.

T. Mossakowski and H.-J. Kreowski (Eds.): WADT 2010, LNCS 7137, pp. 61–75, 2012.
© Springer-Verlag Berlin Heidelberg 2012

action logic in [1,3], where deontic operators are not related (*a priori*) to modalities. Also, we consider boolean operators/combinators for actions. The union of actions can be thought of as a non-deterministic choice, the intersection as a parallel execution and the complement as the execution of some alternative action. In addition, we consider two versions of permission: $P(\alpha)$, which is called strong permission and asserts that any execution of α is allowed, and $P_w(\alpha)$, which is called weak permission and can be read as saying that only some executions of α are allowed. Using these operators we can define other deontic operators, e.g., obligation or prohibition. We have given a sound and complete axiomatic system for this logic in [1].

Some interesting questions arise when the first-order operators are introduced. For example, the proof of completeness in the propositional case relies on the fact that the underlying boolean algebra of terms (denoting actions) is atomic, and therefore the atoms in this algebra can be used to build a canonical model. It is not straightforward (at first sight) to ensure preservation of this property when the quantifiers are added; adding parameters to actions produces a boolean algebra of terms which is not atomic. To solve this problem, we have added to the logic a generalized boolean operator: $\bigsqcup_x \alpha(x)$, which can be understood as the non-deterministic choice between all the possible executions of α with different parameters. Some restrictions are needed to again obtain an atomic boolean algebra; in particular, we need to rule out the parallel execution of the same action with different parameters. (But we do not rule out the parallel execution of different actions.) Adding this restriction, we obtain an atomic boolean algebra of terms. We give an axiomatic system for this logic in section 3 and we outline the proofs of its consistency and completeness.

An inspiring discussion about the intuitive properties that quantified deontic operators should exhibit can be found in the seminal works of Hintikka [6] and Stinger [7]. We will rehearse this discussion investigating the properties that quantifiers and deontic operators enjoy in our logic.

We expect to use this logic to specify and verify fault-tolerant software or systems. Possible applications are systems that use data structures that cannot be dealt with using the propositional version of the logic, in particular, in those cases where complex or infinite domains are involved. In section 5 we exhibit an example of a fault-tolerant memory system specified with this logic; we intend to demonstrate by means of this small example the usefulness of this logic in practice.

The paper is structured as follows. In the next section we describe the syntax and semantics of the logic and we propose an axiomatic system, and then we prove the soundness and completeness of this deductive system. In section 4 we disccuss the main properties relating quantifiers with deontic operators. Finally, we show an example of application and describe some conclusions and future work.

2   Syntax and Semantics

We start by describing the syntax of the language. A language has a set of rigid and flexible functions and variables; in addition, it has a finite set of action symbols, each with an associated arity. For the sake of simplicity, we do not consider sorts and relation symbols in our logic, although it is straightforward to express these notions in the logic described below. Rigid symbols are intended to be used to denote elements of primitive datatypes, instead flexible terms are related to the notion of attribute. From now on, we assume that we have an enumerable set of variables denoted by X.

Definition 1. *A language or vocabulary is a tuple $\langle \Delta_0, F_0, R_0 \rangle$ where, Δ_0 is a finite set of action symbols, each of them with an associated arity. F_0 is an enumerable set of flexible function symbols, each of them with an associated arity. R_0 is an enumerable set of rigid function symbols, each of them with an associated arity.*

As usual the functions with arity 0 are called *constants*. Given a language L we can define the set of terms over L; this set is denoted by $T_L(X)$ (or $T(X)$ when L is clear by context) and it is defined as follows:

Definition 2. *Given a language $L = \langle \Delta_0, F_0, R_0 \rangle$, the set of terms over this language is defined as follows:*

1. *If $x \in X$, then $x \in T_L(X)$.*
2. *If $f \in F_0$ or $f \in R_0$ with arity n and $t_1, \ldots, t_n \in T_L(X)$, then $f(t_1, \ldots, t_n) \in T_L(X)$.*
3. *No other element belongs to $T_L(X)$.*

In a similar way we can define the set of action terms. Since the definition of action terms and formulae are mutually dependent, we first define the concept of formula, assuming a set $\Delta(X)$ of action terms over X, and then we define the set of action terms. The set Φ of formulae is defined as follows.

Definition 3. *Given a language $L = \langle \Delta_0, F_0, R_0 \rangle$, the set of formulae over this language is defined as follows:*

1. *If $t_1, t_2 \in T_L(X)$, then $t_1 = t_2 \in \Phi$.*
2. *If $\alpha_1, \alpha_2 \in \Delta(X)$, then $\alpha_1 =_{act} \alpha_2 \in \Phi$.*
3. *If $\varphi, \psi \in \Phi$, then $\neg \varphi$ and $\varphi \to \psi \in \Phi$.*
4. *If $\varphi \in \Phi$ and $\alpha \in \Delta(X)$, then $[\alpha]\varphi \in \Phi$, $\mathsf{P}(\alpha) \in \Phi$ and $\mathsf{P_w}(\alpha) \in \Phi$.*
5. *If $\varphi \in \Phi$ and $x \in X$, then $(\forall x : \varphi) \in \Phi$.*

The notions of bound variable, free variable and sentence (closed formula) are defined as usual in first-order logic (see [8], for example). Given a formula φ, we denote by $FV(\varphi)$ the set of free variables of φ, similarly we can define $FV(t)$, for terms. When convenient, we write $\varphi(x_1, \ldots, x_n)$ when $FV(\varphi) = \{x_1, \ldots, x_n\}$. If a term $t \in T(X)$ is made up of rigid symbols or variables, we say that it is *rigid*, otherwise we say that t is *flexible*. As explained above, flexible terms are used to represent the notion of programming variables and related notions, and rigid terms are used to represent primitive datatypes. Note that we have two equalities: one for standard terms ($=$) and one for actions ($=_{act}$).

On the other hand, the set of action terms is defined as follows.

Definition 4. *Given a language $L = \langle \Delta_0, F_0, R_0 \rangle$, the set of action terms (denoted by $\Delta(X)$) over this language is defined as follows:*

1. *$\emptyset, \mathbf{U} \in \Delta(X)$.*
2. *If $a \in \Delta_0$ with arity n and $t_1, \ldots, t_n \in T_L(X)$, then $a(t_1, \ldots, t_n) \in \Delta(X)$.*

3. If α and $\beta \in \Delta(X)$, then $\alpha \sqcup \beta$, $\alpha \sqcap \beta$ and $\overline{\alpha} \in \Delta(X)$.

4. If α is an action term and $x \in X$, then $(\bigsqcup_x \alpha) \in \Delta(X)$.

The constant actions $\emptyset$ and $\mathbf{U}$ denote an impossible action and the universal choice of any action, respectively. In the following we use the notation $FV(\alpha)$; the notion of free variable can be extended straightforwardly to action terms as follows:

1. $FV(a(t_1,\ldots,t_n)) = FV(t_1) \cup \cdots \cup FV(t_n)$.
2. $FV(\alpha \sqcap \beta) = FV(\alpha) \cup FV(\beta)$.
3. $FV(\alpha \sqcup \beta) = FV(\alpha) \cup FV(\beta)$.
4. $FV(\overline{\alpha}) = FV(\alpha)$.
5. $FV(\bigsqcup_x \alpha) = FV(\alpha) \setminus \{x\}$.

That is, we treat the $\bigsqcup$ operator as a quantifier. Note that we have not introduced the dual of this operator (i.e., $\bigsqcap$), but it can be obtained from $\bigsqcup$ using the complement. The intuition behind each action term is as explained in [1]. $\alpha \sqcup \beta$ can be thought of as being a non-deterministic choice between actions α and β. $\alpha \sqcap \beta$ is the parallel execution of actions α and β, and $\overline{\alpha}$ can be thought of as the execution of an alternative action to α with its actual parameters. In contrast to the propositional version of the logic, we have actions with parameters; this allows us to capture the usual notion of procedure or command of programming languages. The novel part of the action algebra is the quantifier over variables appearing in an action term. The intuition behind this construction is to allow non-deterministic choices between the parameters of an action. Roughly speaking, the action term: $\bigsqcup_{x_i} \alpha(x_1,\ldots,x_n)$ can be thought of as the execution of an action $\alpha(x_1,\ldots,x_n)$ where we choose non-deterministically the value of the i-th parameter. Sometimes, we use some syntactic sugar and, for example, instead of writing: $\bigsqcup_{x_i} a(x_1,\ldots,x_n)$ we write: $a(?,\ldots,x_n)$ where the symbol $?$ indicates that the corresponding parameter is non-deterministically selected. Obviously, we can quantify over many parameters of an action term. In this case, instead of writing: $\bigsqcup_{x_1} \cdots \bigsqcup_{x_n} \alpha(x_1,\ldots,x_n)$ we write: $\bigsqcup_{x_1,\ldots,x_n} \alpha(x_1,\ldots,x_n)$ and using the syntactic sugar we can write this in short form: $\alpha(?,\ldots,?)$.

Here we understand that, in α, every occurrence of x_i is replaced with the symbol $?$. Another useful action term is: $\bigsqcup_{y_1,\ldots,y_n} \alpha(y_1,\ldots,y_n) \sqcap \alpha[y_1 \setminus t_1,\ldots,y_n \setminus t_n]$ (recall that the expression $\alpha(x_1,\ldots,x_n)$ means that the free variables in α are $x_1,\ldots,x_n$) Roughly speaking, it says that the action α is executed with some parameters different from $t_1,\ldots,t_n$. We use the following notation in this case: $\alpha(? \neq t_1,\ldots,? \neq t_n)$.

Given a language $L = \langle \Delta_0, F_0, R_0 \rangle$, and a set $\{a_1,\ldots,a_n\} \subseteq \Delta_0$, we are interested in action terms of the form:

$$a_1(t_1^1,\ldots,t_{k_1}^1) \sqcap \cdots \sqcap a_n(t_1^n,\ldots,t_{k_n}^n) \sqcap \overline{b_1(?,\ldots,?)} \sqcap \cdots \sqcap \overline{b_m(?,\ldots,?)}$$

where $b_1,\ldots,b_m \in \Delta_0 \setminus \{a_1,\ldots,a_n\}$. Note that, in these kinds of terms, we divide the set of primitive actions Δ_0 into two sets: $\{a_1,\ldots,a_n\}$ and $\{b_1,\ldots,b_m\}$. The elements of the latter set appear negated (i.e., under a complement) indicating that these actions are not executed, while the elements in the former set are used to point out that these actions are executed with some determined parameters. We shall see later on that these are the atoms of the boolean algebra in the canonical model. The set of these action terms is denoted by $At(\Delta_0)$. For the sake of simplicity we denote these kinds of terms with letters: $\delta_0, \delta_1, \delta_2, \ldots$. From now on, we call them *atomic action terms*. There are some interesting properties of this kind of action terms. Intuitively, each atomic action term denotes a unique event in the semantic structure. Before going into the details, we need to introduce the concept of semantic structures.

In the following, we use the notation $\vec{d}$ to denote a tuple of elements $(a_1, \ldots, a_n) \in A^n$; using this notation, when convenient we use $f(\vec{x})$ to denote the application of a given function f to $(x_1, \ldots, x_n)$.

Definition 5. *Given a language* $L = \langle \Delta_0, F_0, R_0 \rangle$*, an L-Structure is a tuple* $\langle \mathcal{W}, \mathcal{R}, \mathcal{P}, \mathcal{I}, \mathcal{E}, \mathcal{D} \rangle$ *where:*

- $\mathcal{D}$ *is a domain of elements. From now on, for any element* $d \in \mathcal{D}$*, we denote by* $\mathbf{d} : \mathcal{W} \to \mathcal{D}$ *the constant function:* $\mathbf{d}(w) = d$*, for every* $w \in \mathcal{W}$*; and we use* $\mathbf{D}$ *to denote the set of these constant functions.*
- $\mathcal{W}$ *is a set of worlds or states.*
- $\mathcal{E} = \{\mathcal{E}_w \mid w \in \mathcal{W}\}$ *is a a collection of non-empty set of names of events.*
- $\mathcal{R} \subseteq \mathcal{W} \times \mathcal{W} \times \bigcup_{w \in \mathcal{W}} \mathcal{E}$ *is an* $\mathcal{E}$*-labeled relation between states. We require that, if* $(w, w', e), (w, w'', e) \in \mathcal{R}$*, then* $w' = w''$ *and* $(w, w', e) \Rightarrow e \in \mathcal{E}_w$*.*
- $\mathcal{I}$ *is an interpretation function such that:*
  - *For each n-ary flexible function symbol* f*,* $\mathcal{I}(f) : (\mathcal{W} \to \mathcal{D})^n \to (\mathcal{W} \to \mathcal{D})$*.*
  - *For each n-ary rigid function symbol* g*,* $\mathcal{I}(g) : (\mathcal{W} \to \mathcal{D})^n \to (\mathcal{W} \to \mathcal{D})$*, such that it satisfies:* $\mathcal{I}(g)(\mathbf{d_1}, \ldots, \mathbf{d_n}) = \mathbf{d}$*, for every* $d_1, \ldots, d_n \in \mathcal{D}$ *and some* $d \in \mathcal{D}$*.*
  - *For each n-ary primitive action symbol* a*,* $\mathcal{I}(a) : \mathcal{D}^n \times \mathcal{W} \to \wp(\bigcup \mathcal{E})$*, satisfying* $\mathcal{I}(a)(\vec{d}, w) \subseteq \mathcal{E}_w$ *for each* $w \in \mathcal{W}$ *and* $\vec{d} \in \mathcal{D}^n$*.*

  such that the following conditions are fulfilled:

  I1 *For every* $e \in \mathcal{E}$ *and* $w \in \mathcal{W}$*, there is no n-ary* $a \in \Delta_0$ *and* $\vec{d}, \vec{d'} \in \mathcal{D}^n$ *such that* $e \in \mathcal{I}(a)(\vec{d}, w) \cap \mathcal{I}(a)(\vec{d'}, w)$ *and* $\vec{d} \neq \vec{d'}$*.*

  I2 *For every* $a \in \Delta_0$*,* $w \in \mathcal{W}$ *and any* $\vec{d} \in \mathcal{D}^n$ *we have that:*

  $$\left| \mathcal{I}(a)(\vec{d}, w) \cap \overline{\bigcup \{\mathcal{I}(b)(\vec{d'}, w) \mid b \in \Delta_0 \setminus \{a\} \text{ and } \vec{d'} \in \mathcal{D}^m\}} \right| \leq 1$$

  I3 *If* $e \in \mathcal{I}(a)(\vec{d}, w) \cap \mathcal{I}(b)(\vec{d'}, w)$ *where* $a, b \in \Delta_0$ *and* $a \neq b$*, then we have:* $\{e\} = \bigcap \{\mathcal{I}(a)(\vec{d}, w) \mid a \in \Delta_0, \vec{d} \in \mathcal{D}^n \text{ and } e \in \mathcal{I}(a)(\vec{d}, w)\}$

  I4 $\bigcup \{\mathcal{I}(b)(\vec{d'}, w) \mid b \in \Delta_0 \text{ and } \vec{d'} \in \mathcal{D}^m \text{ and } w \in \mathcal{W}\} = \mathcal{E}_w$
- $\mathcal{P} \subseteq \mathcal{W} \times \bigcup \mathcal{E}$ *is a relation that states which events are permitted in each state, it satisfies* $\mathcal{P}(w, e) \Rightarrow e \in \mathcal{E}_w$*.*

Notice that flexible constant symbols are interpreted as functions of the type $\mathcal{W} \to \mathcal{D}$, these elements are called intensional objects in modal logic [9]. The intuition is that we use references (in the programming sense) instead of using values of the domain. However, note that variables and rigid constant terms are interpreted as constant functions (i.e., elements of $\mathcal{D}$). We extend this idea to other terms: every function symbol takes as parameter functions of the type $\mathcal{W} \to \mathbf{D}$ and returns a function of this type. Roughly speaking, the condition on the interpretation of rigid function symbols says that rigid functions return constants when evaluated over elements of the domain. Given a structure, we need to assign values to variables; this is done by an assignment $v : X \to \mathbf{D}$. Given such an assignment we can define a function $\mathcal{I}^v : T(X) \to (\mathcal{W} \to \mathcal{D})$ as follows: $\mathcal{I}^v(x) \stackrel{\text{def}}{=} v(x)$, $\mathcal{I}^v(f(t_1, \ldots, t_n)) \stackrel{\text{def}}{=} \mathcal{I}(f)(\mathcal{I}^v(t_1), \ldots, \mathcal{I}^v(t_n))$. Now, given any state $w \in \mathcal{W}$ we define a function $\mathcal{I}^{v,w} : T(X) \to \mathcal{D}$ as follows: $\mathcal{I}^{v,w}(t) \stackrel{\text{def}}{=} \mathcal{I}^v(t)(w)$, which evaluates any terms in a given state. In the following we denote by $v[x \mapsto j]$ a function which coincides with v at every point except in the variable x, for which $v[x \to j](x) = j$.

Now, we can extend this function to deal with action terms as follows:

- $\mathcal{I}^{w,v}(a(x_1, \ldots, x_n)) \stackrel{\text{def}}{=} \mathcal{I}(a)(\mathcal{I}^{w,v}(x_1), \ldots, \mathcal{I}^{w,v}(x_n), w)$,
- $\mathcal{I}^{w,v}(\alpha \sqcup \beta) \stackrel{\text{def}}{=} \mathcal{I}^{w,v}(\alpha) \cup \mathcal{I}^{w,v}(\beta)$,
- $\mathcal{I}^{w,v}(\alpha \sqcap \beta) \stackrel{\text{def}}{=} \mathcal{I}^{w,v}(\alpha) \cap \mathcal{I}^{w,v}(\beta)$,
- $\mathcal{I}^{w,v}(\overline{\alpha}) \stackrel{\text{def}}{=} \mathcal{E}_w - \mathcal{I}^{w,v}(\alpha)$,
- $\mathcal{I}^{w,v}(\bigsqcup_x \alpha) \stackrel{\text{def}}{=} \bigcup_{j \in \mathbf{D}} \mathcal{I}^{w,v[x \mapsto j]}(\alpha)$.

Roughly speaking, a structure provides a domain of discourse together with an interpretation for function symbols. This structure is similar to the ones used in first-order modal logics [9]. Note that we have some topological requirements on the interpretation of actions (which are interpreted as sets of events). Item **I1** says that it is not possible to execute the same action with different parameters at the same time. Although this requirement seems too strong at first sight, it is a consequence of the concurrency model that we assume: only events from different components can be executed concurrently in a system. Within a component, events must be executed sequentially. Here we follow the ideas of [10] where components are captured as logical theories, and events identify, or witness, a set of primitive actions of the language of a theory/component. Note that an event may witness more than one internal action of the component. So, in this sense, our computational model includes concurrency between different actions within a component. Moreover, the seemingly strict restriction that disallows concurrent execution of the same action with different parameters within a component can be overcome by using a number of standard modeling techniques. For instance, if we are modeling the action of paying taxes, and we consider two different persons *Mary* and *John*, we have that *pay(Mary)* $\sqcap$ *pay(John)* cannot be executed in a component. Alternately, we consider two different components *Mary* and *John* and we have one instance of action *pay* for each component, i.e., we have two actions *Mary.pay* and *John.pay*. The structuring of components can be achieved by means of morphisms as done in [10].

I2 says that the isolated execution of an action generates either a unique event or no event at all. We assume that any non-determinism comes from the combination of external and internal actions. **I3** says that the parallel execution of all the actions which generate an event produces this unique event. In other words, for each event there is a maximal set of actions which *observe* this event. Finally, **I4** expresses the requirement that the set of all the events are generated by the actions in the vocabulary. Further commments and intuitions about requirements **I2-I4** can be found in [1].

Our first property says that the interpretation of atomic action terms is either the empty set or a singleton set.

Theorem 1. *Given a language L and a L-Structure M, for any atomic action term δ, valuation v and world w, we have that $|I^{v,w}(\delta)| \leq 1$.*

Proof. *The proof is straightforward using conditions I1, I2 and I3, and taking into account that atoms in the algebra of actions are monomials made up of primitive actions or complements of primitive actions.*

The relation of satisfaction $\vDash$ between models, valuations and formulae is defined inductively as follows:

- $w, \mathcal{M}, v \vDash t_1 = t_2$ iff $I^{v,w}(t_1) = I^{v,w}(t_2)$.
- $w, \mathcal{M}, v \vDash a_1 =_{act} a_2$ iff $I^{v,w}(a_1) = I^{v,w}(a_2)$.
- $w, \mathcal{M}, v \vDash \neg\varphi$ iff not $w, \mathcal{M}, v \vDash \varphi$.
- $w, \mathcal{M}, v \vDash \varphi \rightarrow \psi$ iff $w, \mathcal{M}, v \vDash \neg\varphi$ or $w, \mathcal{M}, v \vDash \psi$ or both.
- $w, \mathcal{M} \vDash [\alpha]\varphi$ iff for all $e \in I^v(\alpha)$, if $w \xrightarrow{e} w'$, then $w', \mathcal{M}, v \vDash \varphi$.
- $w, \mathcal{M}, v \vDash \forall x : \varphi$ iff for every $d \in \mathcal{D}$ we have $w, \mathcal{M}, v[x \mapsto d] \vDash \varphi$.
- $w, \mathcal{M}, v \vDash P(\alpha) \Leftrightarrow \forall e \in I^{v,w}(\alpha) : \mathcal{P}(w, e)$.
- $w, \mathcal{M}, v \vDash P_w(\alpha) \Leftrightarrow \exists e \in I^{v,w}(\alpha) : \mathcal{P}(w, e)$.

The obligation operator can be defined using the other deontic constructs as follows: $O(\alpha) = P(\alpha) \wedge \neg P_w(\overline{\alpha})$. Roughly speaking, an action is obliged when it is allowed to be executed in any context and it is forbidden to execute any other action.

3   An Axiomatic System

In this section we exhibit an axiomatic system for the logic described in the previous section. We write $\varphi[x \setminus t]$ for the formula obtained by replacing all the free occurrences of x by the term t. We say that t is free for x in φ when x does not occur in the scope of a quantifier Qy and y is free in t. We consider the following set of axioms:

- The set of propositional tautologies, the equational theory of boolean algebra, and axioms for equality for $=_{act}$ and $=$, see below.

A1. $\forall x : \alpha \sqsubseteq \bigsqcup_x \alpha$, for all actions α.
A2. $(\forall x : \alpha \sqsubseteq \beta) \rightarrow \bigsqcup_x \alpha \sqsubseteq \beta$, for all actions α and β, where x is not free in β.
A3. $(\forall x : P(\alpha)) \rightarrow P(\bigsqcup_x \alpha)$
A4. $(\exists x : P_w(\alpha)) \rightarrow P_w(\bigsqcup_x \alpha)$
A5. $(\forall x : [\alpha]\varphi) \rightarrow [\bigsqcup_x \alpha]\varphi$
A6. $[\emptyset]\varphi$
A7. $\langle\alpha\rangle\varphi \wedge [\alpha]\psi \rightarrow \langle\alpha\rangle(\varphi \wedge \psi)$
A8. $[\alpha \sqcup \alpha']\varphi \leftrightarrow [\alpha]\varphi \wedge [\alpha']\varphi$
A9. $P(\emptyset)$
A10. $P(\alpha \sqcup \beta) \leftrightarrow P(\alpha) \wedge P(\beta)$
A11. $\neg P_w(\emptyset)$
A12. $P_w(\alpha \sqcup \beta) \leftrightarrow P_w(\beta)$
A13. $P(\alpha) \wedge \alpha \neq \emptyset \rightarrow P_w(\alpha)$
A14. $P_w(\delta) \rightarrow P(\delta)$, where $\delta \in At(\Delta_0)$
A15. $a_1(?, \ldots, ?) \sqcup \cdots \sqcup a_n(?, \ldots, ?) =_{act} U$
A16. $\langle\delta\rangle\varphi \rightarrow [\delta]\varphi$, where $\delta \in At(\Delta)$

A17. $\forall \vec{x}, \vec{y} : \vec{x} \neq \vec{y} \rightarrow a(\vec{x}) \sqcap a(\vec{y}) =_{act} \emptyset$, for all $a \in \Delta_0$.
Subs. $\varphi[\alpha] \wedge (\alpha =_{act} \alpha') \rightarrow \varphi[\alpha \setminus \alpha']$
FOLSub. $(\forall x : \varphi) \rightarrow \varphi[x \setminus t]$, where t is free for x in φ.
Barcan. $(\forall x : [\alpha]\varphi) \rightarrow [\alpha](\forall x : \varphi)$

And the following deduction rules:

MP. If $\vdash \varphi$ and $\vdash \varphi \rightarrow \psi$, then $\vdash \psi$
GEN. If $\vdash \varphi$, then $\vdash [\alpha]\varphi$
FOL-GEN. If $\vdash \varphi$, then $\vdash \forall x : \varphi$

Axioms **Subs**, **FOLSub** and rule **FOL-GEN** are usual for first-order logics. Rule **Barcan** is standard in first-order modal logics; it expresses that the domain of discourse is not affected by modalities. Axioms **A3-A17** expresses basic properties of modalities and permissions; axioms **A6-A16** are also common for the propositional part of the logic, see [1]; for further comments and intuitions about these axioms. The novel axioms are **A1** and **A2**; these axioms say that the actions of the form $\bigsqcup \alpha$ are least upper bounds. On the other hand, axioms **A3-A5** express the relationship between quantifiers and the universal non-deterministic choice. We come back to these properties in section 4. For equalities involving only rigid terms we consider the theory LNI described in [9], whereas for equalities involving some flexible terms (and for $=_{act}$) we consider the theory CI of contingent identity, which contains reflexivity and a restricted version of the schema of replacement of equals for equals.

It is interesting to note the role of atomic action terms. Axiom **A14** expresses that, if an atomic action term is weakly allowed, then it is strongly allowed. In the semantic model this is true since atomic action terms are interpreted as a unique event. Another way of reading this axiom is as saying that each event is either allowed or forbidden. This is a benefit with respect to the work of Segerberg [11] where a boolean algebra of actions is used, but the property that, if an event is not allowed, then it is forbidden is not expressible, since the algebra lacks atoms.

3.1 Soundness and Completeness

In this section we prove the soundness and the completeness of the axiomatic system presented above. The relationship $\vdash \subseteq \wp(\Phi) \times \Phi$ is defined as usual in modal logic, taking into account that only the deduction rule **MP** is allowed to be applied to assumptions.

Theorem 2. *The axiomatic system is sound, i.e.,* $\vdash \varphi$ *implies* $\vDash \varphi$.
*Proof. We first prove the validity of the axioms. Axioms **A6-A13** are straightforward following the proofs given for the propositional version of the logic [1]. Axioms **A14** and **A16** are direct taking into account that the interpretation of δ terms are singleton or empty sets (theorem 1). Axiom **A15** is valid because of condition **I4**, and **A17** because of condition **I1**. Finally, **A3** and **A4** are straightforward using the semantical definition of strong and weak permission.*

*On the other hand, **FOLSub** is a standard axiom of first-order logics. The validity of **Subs** can be proven in the same way that it is proved for the propositional case (see [1]). The algebra of actions in any model is defined over sets using union, intersection and set complement, and therefore it is a boolean algebra. For axioms **A1** and **A2**, note* *that the operator* $\bigsqcup$ *is interpreted as the union of a collection of sets, which obviously satisfies* **A1** *and* **A2**. *The other axioms are standard for quantified normal modal logics.*

Now, we tackle the more difficult problem of completeness. First, we define a canonical model using well-known techniques coming from modal first-order logic: we extend the language with an enumerable number of new variables and we use the set of terms as the domain of discourse[1] (the basic idea is presented in [9]). As usual, we use maximal consistent sets of formulae in this language (which satisfy standard requirements) as states, and then we define relations between these states in such a way that the relationship between states is consistent with the properties of the necessity modality and the action operators; the atoms of the boolean algebra of action terms (modulo some equational theory) are the labels of these relations. In the following we sketch this proof.

Let us define a boolean algebra whose elements are equivalence classes of action terms obtained using the equivalence relation $=_{act}$ (i.e., we build a Lindenbaum algebra) modulo some equational theory denoted by Ξ. For any action term α consider $[\alpha]_\Xi = \{\beta \mid \Xi \vdash \alpha =_{act} \beta\}$ (i.e., its equivalence class under the relation $=_{act}$ and the equational theory Ξ). We denote this set of elements by Δ/Ξ, when Ξ is empty (i.e., we only consider the basic axioms for $=$) we write $\Delta/ =_{act}$. Now, we define the following operators over this set of elements: $[\alpha]_\Xi \sqcup_\sqcap [\beta]_\Xi \overset{\text{def}}{=} [\alpha \sqcup \beta]_\Xi$, $[\alpha]_\Xi \sqcap_\sqcap [\beta]_\Xi \overset{\text{def}}{=} [\alpha \sqcap \beta]_\Xi$, $-[\alpha]_\Xi \overset{\text{def}}{=} [\overline{\alpha}]_\Xi$. It is straightforward to prove that $\langle \Delta/\Xi, \sqcup_\sqcap, \sqcap_\sqcap, [\mathbf{U}]_\Xi, [\emptyset]_\Xi \rangle$ is a boolean algebra. It is important to note that $[\bigsqcup_x \alpha]_\Xi$ is the least upper bound of the set $\{[\alpha[x \setminus t]]_\Xi \mid t \in T_L(X)\}$. Let us prove this result.

Theorem 3. $[\bigsqcup_x \alpha]_\Xi$ *is the least upper bound of the set* $\{[\alpha[x \setminus t]]_\Xi \mid t \in T_L(X)\}$.

Proof. *We have* $\vdash \forall x : \alpha \sqsubseteq \bigsqcup_x \alpha$, *and therefore by* **FOLSub** *we obtain* $\vdash (\alpha \sqsubseteq \bigsqcup_x \alpha)[x \setminus t]$ *for any t. Since x is not free in* $\bigsqcup_x \alpha$, *this is equivalent to:* $\vdash \alpha[x \setminus t] \sqsubseteq \bigsqcup_x \alpha$. *On the other hand, suppose that* $\vdash \alpha[x \setminus t] \sqsubseteq \beta$ *for any term t. (Let us suppose that x is not free in* β, *otherwise we can replace x by another variable to obtain a term which is equivalent to* β.) *Therefore we have* $\vdash \alpha \sqsubseteq \beta$, *and using* **FOL-GEN** *we obtain* $\vdash \forall x : \alpha \sqsubseteq \beta$, *and therefore by axiom* **A2** *we get:* $\vdash \bigsqcup_x \alpha \sqsubseteq \beta$.

Another useful result is the following. This result shows that the Lindenbaum algebra defined above is atomic.

Theorem 4. *Given any language* $L = \langle \Delta_0, F, R \rangle$. *The Lindenbaum boolean algebra described above is atomic. Furthermore, the set of all atoms is given by:* $\{[\delta]_\Xi \mid \delta \in At(\Delta_0)/\Xi\}$.

Proof. *Each action term is equivalent to a term in DNF, where each clause is a monomial or a non-deterministic choice over the parameters of a primitive action term. These kinds of terms are bounded above by* δ-*terms. On the other hand, note that if we join a primitive action term to a* δ-*term, this new action term becomes equal to the impossible action, since the primitive action term appears negated or with a different parameter in the* δ-*term.*

---

[1] In [9] only variables are used as the domain of the canonical model, we extend this idea to encompass the rigid and flexible terms.

Using these properties, we tackle the completeness of the axiomatic system. We use these atoms for giving the canonical model. Given a language L, consider the collection of maximal consistent sets of formulae, i.e.:

$$\text{MCS} = \{ \Gamma \mid \Gamma \text{ is a maximal consistent set of formulae } \}.$$

We use the standard technique of adding an infinite enumerable set of variables to L: $y_0, y_1, y_2, \ldots$ We call this language L^*. It is not hard to see that the elements of MCS are consistent sets in the language L^*, and also L^* preserves the theorems of L (this is a standard result of first-order logics). As usual, for each formula φ of L we can consider a formula: $\neg \forall x : \varphi \rightarrow \neg \varphi[x \backslash c]$, where c is one of the new constant symbols. We denote this set of formulae by Θ. Now, for each $\Gamma \in \text{MCS}$ we can consider the set $\Gamma \cup \Theta$ and extend it until we obtain a maximal consistent set in L^*; we call this set Γ^*. Now, we define a new collection of sets of formulae as follows: $\text{MCS}^* = \{ \Gamma \cup \Theta \mid \Gamma \in \text{MCS} \}$.

The idea is that we provide a canonical model for L^*; the completeness of the calculus follows. But since the theorems of L are preserved for L^* (see below) we obtain the completeness of L too. In the following we consider a fixed enumeration of terms such that rigid terms appear before flexible terms.

Definition 6. $\mathcal{M}_C$ *is made up of:*

- $\mathcal{W}_C \overset{\text{def}}{=} \{ \Gamma^* \mid \Gamma^* \in MCS \}$
- $\mathcal{E}_C \overset{\text{def}}{=} \{ \mathcal{E}_w \mid w \in \mathcal{W}_C \}$, *where for each $w \in \mathcal{W}$ we define:*
  $\mathcal{E}_w = \{ [\delta]_\Xi \mid [\delta]_\Xi \in At_{L^*}(\Delta)/\Xi \text{ and } \Xi \text{ is the set of equations in } w \}$
- $\mathcal{D}_C \overset{\text{def}}{=} T(X \cup \{ y_1, y_2, \ldots \})$.
- $\mathcal{R}_C \overset{\text{def}}{=} \bigcup \{ \mathcal{R}_{\alpha,w,w'} \mid w, w' \in \mathcal{W}_C \wedge \alpha \in \Delta \wedge (\forall \varphi \in \Phi : [\alpha]\varphi \in w \Rightarrow \varphi \in w') \}$, *where*
  $\mathcal{R}_{\alpha,w,w'} \overset{\text{def}}{=} \{ w \overset{[\delta]}{\rightarrow} w' \mid \forall [\delta] \in \mathcal{E}_w \text{ and } (w \vdash \delta \sqsubseteq \alpha) \}$.
- $\mathcal{P}_C \overset{\text{def}}{=} \bigcup \{ \mathcal{P}_{w,\alpha} \mid w \in \mathcal{W}_C \wedge P(\alpha) \in w \}$, *where:* $\mathcal{P}_{w,\alpha} \overset{\text{def}}{=} \{ (w, [\delta]_{BA}) \mid \forall [\delta] \in \mathcal{E}_w \text{ and } (w \vdash \delta \sqsubseteq \alpha) \}$.
- $\mathcal{I}_C(a_i)(x_1, \ldots, x_n, w) \overset{\text{def}}{=} \{ [\delta] \in \mathcal{E}_w \mid (w \vdash \delta \sqsubseteq a_i(x_1, \ldots, x_n)) \}$.
- *If f is function symbol (rigid or flexible), and $t_1^* : \mathcal{W} \rightarrow \mathcal{D}, \ldots, t_n^* : \mathcal{W} \rightarrow \mathcal{D}$ then*
  $\mathcal{I}_C(f)(t_1^*, \ldots, t_n^*)(w) \overset{\text{def}}{=} t$, *with t being the first term in the enumeration such that $t = f(t_1^*(w), \ldots, t_n^*(w)) \in w$.*

Some notes about this canonical model may be useful. The set $\mathcal{E}_C$ is made up of sets $\mathcal{E}_w$ and each of these sets contains all the atoms of the action term algebra modulo the equational theory in w, and these elements are used as labels of the transition system. $\mathcal{W}_C$ is the set of worlds, where each world is a maximal consistent set of formulae, as explained above. $\mathcal{D}_C$ is made up suitable terms in the domain. The relationship $\mathcal{R}_C$ is defined using the sets $\mathcal{E}_w$ and taking into account the formulae of the form $[\alpha]\varphi$ in each world. The predicate $\mathcal{P}_C$ is defined using the deontic formulae belonging to each world. On the other hand, the interpretation $\mathcal{I}_C$ for each action gives us the corresponding set of atoms of the boolean algebra of actions. The conditions **I1-I4** are satisfied as a consequence of the properties of the Lindenbaum Algebra; this is shown for the propositional case in [1], and a similar proof can be obtained for this canonical model. Note that the enumeration given ensures us that rigid terms will denote the same value in every state, no equation between rigid terms can be added in any state as a consequence of the system LNI. The following lemma says that valuations over the canonical model preserve the order of the algebra of actions.

Lemma 1. *For every valuation v, we have: $\forall \alpha \in \Delta, \forall [\delta] \in I_C^{v,w}(\alpha) :\ w \vdash \delta \sqsubseteq \alpha[x_1 \setminus t_1, \ldots, x_n \setminus t_n]$, where $x_1, \ldots, x_n$ are the free variables appearing in α and we have $v(x_1)(w) = t_1, \ldots, v(x_n)(w) = t_n$.*

Proof. *The proof is by structural induction over the action term α. For the primitive action terms the proof is straightforward by definition of I. For the operators $\sqcup, \sqcap$ and $-$ the proof is as shown for the propositional logic in [1]. Let us prove the theorem for the least upper bound operators.*

Suppose that $\alpha = \bigsqcup_{x_i} \alpha'$. Suppose that $[\delta] \in I_C^{v,w}(\alpha)$. By definition of I_C we have that $[\delta] \in I^{v[x \mapsto i],w}(\alpha')$ for some term $t = i(w)$. But then, by induction, we obtain that

$$w \vdash [\delta] \sqsubseteq \alpha[x_i \setminus t, x_1 \setminus t_1, \ldots, x_n \setminus t_n].$$

*On the other hand, using axiom **A1**, **FOL-Sub** and **FOL-Gen** we obtain: $\vdash \alpha[x_i \setminus t, x_1 \setminus t_1, \ldots, x_n \setminus t_n] \sqsubseteq (\bigsqcup_x \alpha)[x_1 \setminus t_1, \ldots, x_n \setminus t_n]$ and then using the equation above together with the transitivity of $\sqsubseteq$ we obtain: $\vdash \delta \sqsubseteq (\bigsqcup_x \alpha)[x_1 \setminus t_1, \ldots, x_n \setminus t_n]$*

In the following, we use the valuation v_C, which given a variable x and a state s, it returns the first term t such that $x = t \in w$.

Now, we prove the following truth lemma:

Lemma 2. *For every φ, $\varphi \in w$ iff $w, \mathcal{M}_C, v_C \vDash \varphi$.*

Proof. *The proof is by induction on φ. The base cases are as follows.*

Suppose $t_1 = t_2 \in w$; by definition of the canonical model we know that $I_C^{v_C,w}(t_1)$ returns the first term t such that $t = t_1 \in w$, but we get that $t = t_2 \in w$ and therefore $I_C^{v_C,w}(t_1) = t$, so $w, \mathcal{M}_C, v_C \vDash t_1 = t_2$. The inductive cases are as follows. The standard operators are dealt with as usual. We provide proofs for the new ones.

*For strong permission, suppose that we have $\mathsf{P}(\alpha) \in w$; an induction on the structure of α shows that α is equal to $\delta_1' \sqcup \cdots \sqcup \delta_n'$ where each δ' is a atomic action term, or is obtained from a δ term using the big choice ($\bigsqcup$). Now, since $\mathsf{P}(\alpha) \in w$, we have that $\mathsf{P}(\delta_1' \sqcup \cdots \sqcup \delta_n') \in w$; using axiom **A10** we get $\mathsf{P}(\delta_i') \in w$, for each δ_i'. If δ_i' is equal to some δ_i(an atomic action term) then we get, by definition of $\mathcal{M}_C$: $\mathcal{M}_C, w, v_C \vDash \mathsf{P}(\delta_i')$. If $\delta_i' = \bigsqcup_x \delta_i$, for some δ_i, we get, by axiom **A10** and **FOLGen**, that $(\forall x : \mathsf{P}(\delta_i)) \in w$, and by definition of $\mathcal{M}_C$ we obtain: $\mathcal{M}_C, w, v_C \vDash \forall x : \mathsf{P}(\delta_i)$ and this by axiom **A3** implies $\mathcal{M}_C, w, v_C \vDash \mathsf{P}(\bigsqcup_x \delta_i)$. That is, we have $\mathcal{M}_C, w, v_C \vDash \mathsf{P}(\delta_1' \sqcup \cdots \sqcup \delta_n')$, and therefore $\mathcal{M}_C, w, v_C \vDash \mathsf{P}(\alpha)$. The other direction is similar.*

*The proof for weak permission is analogous to the proof given for strong permission, but using axiom **A4** instead of axiom **A3**.*

*Now, suppose $[\alpha]\varphi \in w$; therefore, for any $\vdash \delta \sqsubseteq \alpha$, we have, by axiom **A8**, $[\delta]\varphi \in w$. By definition we know that for every w' such that $w \overset{[\delta]}{\rightarrow} w'$ we have $\varphi \in w'$; by induction this implies $w', \mathcal{M}_C, v_C \vDash \varphi$, and therefore $w, \mathcal{M}_C, v_C \vDash [\delta]\varphi$. Using the boolean property introduced above that an action can be rewritten as the union of its atoms, we get*

$w, \mathcal{M}_C, v_C \vDash [\alpha]\varphi$. Now, suppose $[\alpha]\varphi \notin w$; by properties of maximal consistent sets, we have $\langle\alpha\rangle\neg\varphi \in w$ and, therefore, by the properties of the canonical model (see [1]), there is a state w' such that $w \overset{[\delta]}{\to} w'$ such that $w', \mathcal{M}_C, v_C \vDash \neg\varphi$, and then, we obtain $w, \mathcal{M}_C, v_C \vDash \langle\delta\rangle\neg\varphi$; here, using axiom **A8**, we have $w, \mathcal{M}_C, v_C \vDash \langle\alpha\rangle\neg\varphi$.

If $\forall x : \varphi \in w$, let $v_{C_{x \mapsto t}}$ be a variation of v_C. By axiom **FOLSub** we have $\varphi[x \setminus t] \in w$, and therefore $w, \mathcal{M}_C, v_C \vDash \varphi[x \setminus t]$; this implies that: $w, \mathcal{M}_C, v_C[x \mapsto t] \vDash \varphi$. Since this is true for every variation of $v[x \mapsto t]$, we have that: $w, \mathcal{M}_C, v_C \vDash \forall x : \varphi$. The other direction is similar using **FOL-GEN**.

We have obtained a canonical model for the language L^*, but since L^* is conservative with respect to the theorems of L (which can be verified checking the axioms and the deduction rules), we also obtain the completeness of the axiomatic system in L.

4 Related Logics

In this section we briefly compare the logic developed in this paper with related logics. To the authors' knowledge, the generalized non-deterministic choice and the generalized parallel execution have not been investigated in the literature of dynamic or boolean logics [4,12]. It is worth mentioning that dynamic logic [4] provides a non-deterministic assignment $x :=?$ which can be thought of as an instance of the more general operator introduced here (we have followed Harel's notation throughout this paper). Deontic action logics with boolean operators have been widely investigated in the literature [11,5,13,14,15]. However, it seems that first-order operators have only been used in [14,15], where a sound and complete axiomatic system is not provided.

On the other hand, some notes about the interplay between deontic operators and quantifiers may be useful for understanding the characteristics of the logic obtained. Hintikka and Kanger agreed that some properties relating quantifiers and deontic operators are not intuitively true. The logics proposed by these authors are more expressive than the one introduced here. For example, the predicate $Ax : O(Px)$, is a quantification over actions; the intuitive meaning of this expression is: *every action of type P is obliged to be executed*. In the same way, we can write: $O(Ax : Px)$ which must be read as: *it is obliged that every act of type A is performed*. The formula $Ax : O(Px) \to O(Ax : Px)$ is discarded with intuitive examples of the style: *In some settings, everyone ought to pay fines, but it is not true, in every deontically perfect world, that every one should pay fines*. Our logic is different: we can only quantify over parameters. It is straightforward to prove the following properties: (1) $\exists x : P_w(\alpha(x)) \leftrightarrow P_w(\bigsqcup_x \alpha(x))$, (2) $\forall x : P(\alpha(x)) \leftrightarrow P(\bigsqcup_x \alpha(x))$. These valid formulae expose the existential (formula (1)) and universal character (formulae (2)) of weak and strong permission, respectively. Note that similar properties for obligation are not true in our setting. That is, neither of the following formulae is a theorem in our logic: (3) $\exists x : O(\alpha(x)) \leftrightarrow O(\bigsqcup_x \alpha(x))$, (4) $\forall x : O(\alpha(x)) \leftrightarrow O(\bigsqcup_x \alpha(x))$.

Let us illustrate this by means of a simple example. Let push be the action of pushing an element onto a stack; the formula $\forall x : O(\text{push}(x))$ means that we are obliged to push every element onto the stack; this equivalent to the formula: $\forall x : P(\text{push}(x)) \wedge \forall x : \neg P_w(\overline{\text{push}(x)})$, which means that we are allowed to push any x onto the stack, and it is forbidden to not push some element onto the stack. In contrast, $O(\bigsqcup_x \text{push}(x))$ is equivalent to: $P(\bigsqcup_x \text{push}(x)) \land \neg P_w(\bigsqcap_x \text{push}(x))$, which can be read as saying that we are allowed to push any x onto the stack and that some element should be pushed onto the stack. A similar analysis can be done for the existential formula. It is important to remark that our model of parallelism implies that statements of the type $\forall x : O(\text{push}(x))$ can only be fulfilled when either pushing an element is impossible or the domain is a singleton. Let us remark that the logic is intended to be used for specifying computer systems, where components are captured as logical theories (as done in [10]), and therefore categorical constructions can be used to manage components (see [16]). A more expressive logic can be obtained if requirement **I1** is dropped, but in this case it is not clear whether strong completeness could be preserved; we leave this as further work.

5 An Example

In this section we describe a simple example to illustrate the use of this logic in practice. We consider the scenario of a fault-tolerant memory. Roughly speaking, we have a database where errors may occur while writing, and therefore inconsistent data may be stored. To avoid this situation, we have two memories which are coordinated with respect to writing: if they disagree in some position, then a rollback is executed to return the memories to a safe state. In the example shown below, we consider types or sorts to make the specification more appealing, although we have not considered types or sorts in our logic; as usual we can express them using predicates. We have the following types:

- Val denotes the set of values that can be written in the memories.
- Pos denotes the set of positions of the memories.

We consider components m_1 and m_2 that are instances of a specification **Mem**. We consider the following actions:

- $m_k.\text{Set}(i : \text{Pos}, v : \text{Value})$ denotes the action of writing the memory m_k in position i with value v.
- $m_i.\text{RBack}$ denotes the action of taking memory m_i to a safe state.
- $m_i.\text{Fail}$ denotes the failure of memory m_i.

We also have flexible functions $m_k.\text{Get} : \text{Pos} \to \text{Val}$, these functions return the value in a given position of a memory. We use the usual notation in computer science, indicating the parameters of the actions by means of names of variables. We also consider predicates Init and V, which are used to indicate the beginning of time and a state of error, respectively. In the following we exhibit some axioms and their intuitive reading:

A1 $\forall i \in \text{Pos}, v \in \text{Val} : [m_k.\text{Set}(i, v) \sqcap \overline{m_k.\text{Fail}}]m_k.\text{Get}(i) = v$, where $k = 1, 2$; this axiom says that, when there is no error, the action Get behaves correctly.

A2 $[m_k.\text{Fail}]V$, where $k = 1, 2$; this axiom expresses that after a failure we have a violation.

A3 $V \to O(m_1.\text{RBack} \sqcap m_2.\text{Rback})$, this axiom says that, when we have a violation, then we ought to execute the roll back in both memories. If executed, this ensures that we go into a safe state.

A4 $\forall i \in Pos, v_1, v_2 \in Val : v_1 \neq v_2 \rightarrow F(m_1.Set(i, v_1) \sqcap m_2.Set(i, v_2))$, this formula says that it is forbidden to write different values at the same position in the two memories.

A5 $\forall i \in Pos, v \in Val : F(m_2.Set(i, v) \sqcap \overline{m_1.Set(i, v)})$, this axiom expresses the requirement that it is forbidden to write in m_2 and not in m_1.

A6 $\forall i \in Pos, v \in Val : F(\overline{m_1.Set(i, v)} \sqcap m_2.Set(i, v))$, this axiom is the symmetric case of axiom **A5**.

A7 $\forall i : Pos, v \in Val : m_k.Get(i) = v \rightarrow [m_k.Set(i, v) \sqcup m_k.Fail]m_k.Get(i) = v$, this axiom says that when the memories are not written and there is no failure, then the state of the memories is preserved.

A8 $Init \rightarrow \neg V$, this axiom says that at the beginning there is no violation.

A9 $F(\alpha) \rightarrow [\alpha]V$, this axiom says that, if we execute a forbidden action, then we go into an error state. Note that it is an axiomatic schema.

A10 $\forall i \in Pos : [m_1.Rback \sqcap m_2.Rback]m_1.Get(i) = m_2.Get(i)$, this axiom says that, after a rollback both memories reach the same state.

The predicate $F(\alpha)$ is the prohibition predicate, and it is defined as $F(\alpha) \overset{def}{=} \neg P_w(\alpha)$. Axioms **A4**-**A6** are interesting as they contain deontic prescriptions. Roughly speaking, these axioms state what the ideal behaviour of the system is, although some errors or failures may cause the system to behave in a different way.

The deontic component of the logic allows us to perform different analyses over the specification. For example, we can prove properties over the scenarios where the system behaves as expected (that is, when the deontic prescriptions are followed). In this case a desirable property is that, when there is no violation, both memories contain the same data. On the other hand, we can also prove properties over erroneous or faulty behaviors (for example, when data is written in only one memory); in this case one interesting property to prove is that both memories reach a safe state (i.e., a state where they contain the same information). We have shown some examples of proofs and specifications in [1,3]. In the case of the first-order logic shown here, in order to prove interesting properties, it is necessary to have some implementation of the axiomatic system in a semi-automatic prover; this will facilitate the task of proving properties. We leave this as further work.

6   Conclusions and Further Remarks

In this paper we have presented a first-order extension of a deontic action logic whose propositional version we have presented in [1]. Deontic action logics and modal action logics (or dynamic logics) have been extensively investigated by several authors. However, first-order extensions of deontic logics have not enjoyed any substantial advance since the seminal works of Hintikka and Kanger. This is in part because the intuitive properties of deontic predicates and quantifiers are hard to capture in a formal setting, see [17] for detailed discussions about these difficulties. Here we restrict ourselves to first-order quantification, and then we show how a sound and complete calculus is obtained. It is important to clarify the fact that we intend to use this logic for the specification and verification of computing systems, in particular for the specification and verification of fault-tolerant systems. We have shown a small example of application where a fault-tolerant memory system is specified. The modal operators allow us to express pre and post-conditions, as done in dynamic logics, and the deontic operators allow us to distinguish the expected or ideal behaviour of our system from the unexpected or erroneous behaviour. On the other hand, we intend to use techniques similar to those introduced in [10] to be able to modularize specifications. This will allow us to facilitate the analysis of specifications, and to obtain a more appealing software engineering framework. Note that the restriction that we cannot execute in parallel the same action (with different parameters) can be circumvented if we consider two instances of a specification; each one can be thought of as being an instance of the component running on its own processor. Summarizing, modularization techniques are needed to apply this logic framework to more complex and interesting examples.

References

1. Castro, P.F., Maibaum, T.: Deontic action logic, atomic boolean algebra and fault-tolerance. Journal of Applied Logic 7(4), 441–466 (2009)
2. Wieringa, R.J., Meyer, J.J.: Applications of deontic logic in computer science: A concise overview. Deontic Logic in Computer Science, Normative System Specification (1993)
3. Castro, P.F.: Deontic Action Logics for the Specification and Analysis of Fault-Tolerance. PhD thesis, McMaster University, Department of Computing and Software (2009)
4. Harel, D., Kozen, D., Tiuryn, J.: Dynamic Logic. MIT Press (2000)
5. Meyer, J.: A different approach to deontic logic: Deontic logic viewed as variant of dynamic logic. Notre Dame Journal of Formal Logic 29 (1988)
6. Hintikka, J.: Quantifiers in deontic logic. In: Societas Scientiarum Fennica, Commentationes Humanarum Litterarum (1957)
7. Kanger, S.: New foundations for ethical theory. In: Deontic Logic: Introductory and Systematic Readings, Dordrecht (1971)
8. Emerson, E.: A Mathematical Introduction to Logic. Academic Press (1972)
9. Hughes, G.E., Cresswell, M.J.: A New Introduction to Modal Logic. Routledge (1996)
10. Fiadeiro, J.L., Maibaum, T.: Temporal theories as modularization units for concurrent system specification. Formal Aspects of Computing 4, 239–272 (1992)
11. Segerberg, K.: A deontic logic of action. Studia Logica 41, 269–282 (1982)
12. Gargov, G., Passy, S.: A note on boolean logic. In: Petkov, P.P. (ed.) Proceedings of the Heyting Summerschool. Plenum Press (1990)
13. Broersen, J.: Modal Action Logics for Reasoning about Reactive Systems. PhD thesis, Vrije University (2003)
14. Khosla, S., Maibaum, T.: The prescription and description of state-based systems. In: Banieqnal, B., Pnueli, H.A. (eds.) Temporal Logic in Computation. Springer, Heidelberg (1985)
15. Kent, S., Quirk, B., Maibaum, T.: Specifying deontic behaviour in modal action logic. Technical report, Forest Research Project (1991)
16. Castro, P.F., Aguirre, N.M., López Pombo, C.G., Maibaum, T.S.E.: Towards managing dynamic reconfiguration of software systems in a categorical setting. In: Cavalcanti, A., Deharbe, D., Gaudel, M.-C., Woodcock, J. (eds.) ICTAC 2010. LNCS, vol. 6255, pp. 306–321. Springer, Heidelberg (2010)
17. Makinson, D.: Quantificational reefs in deontic waters. In: Hilpinen, R. (ed.) New Studies in Deontic Logic, Dordrecht (1981)

Casl-Mdl, Modelling Dynamic Systems with a Formal Foundation and a UML-Like Notation

Christine Choppy[1] and Gianna Reggio[2]

[1] LIPN, UMR CNRS 7030 - Université Paris 13, France
Christine.Choppy@lipn.univ-paris13.fr
[2] DISI, Università di Genova, Italy
gianna.reggio@disi.unige.it

Abstract. In this paper we present a part of Casl-Mdl, a visual modelling notation based on Casl-Ltl (an extension for dynamic system of the algebraic specification language Casl). The visual constructs of Casl-Mdl have been borrowed from the UML, thus existing editors may be used. A Casl-Mdl model is a set of diagrams but it corresponds to a Casl-Ltl specification, thus Casl-Mdl is a suitable means to easily read and write large and complex Casl-Ltl specifications. We use as a running example a case study that describes the functioning of a consortium of associations.

1 Introduction

The aim of our work is to reshape the formal specification language Casl-Ltl [11] (a Casl [7] extension for dynamic systems) as a visual modelling notation, and this requires to provide a visual syntax to the Casl-Ltl specifications. This work is motivated by the fact that in our opinion the currently available modelling notations have some problematic aspects, for example the lack of a formal semantics if not of a well-defined syntax.

We decided to attempt this experiment for the following reasons:
- Casl-Ltl is very suitable to specify/model different kinds of dynamic systems, and at different levels of abstraction. Indeed, it has been used to specify the use-case based requirements [2], the main features (the domain, the requirements and the machine) for the basic problem frames [3], and recently it has been used to specify the services in the field of SOA (Service Oriented Architecture) [6].
- A modelling (specification) method for Casl-Ltl was developed [4], where the modelling is guided by the use of simple ingredients/concepts as data types, dynamic system, elementary interaction, and cooperations between systems.
- Casl-Ltl is extremely expressive (it includes a powerful first-order temporal logic), allows different styles of specification/modelling (e.g., property oriented and constructive [4]), while still based on a limited number of constructs,
- it is not object-oriented, and this may be an advantage, whenever the models are not used for developing object-oriented software (e.g., when using SOA),

T. Mossakowski and H.-J. Kreowski (Eds.): WADT 2010, LNCS 7137, pp. 76–97, 2012.

- and obviously it has a well-defined semantics, and there exists software tools to help the formal verification of its specifications.

However, the textual, quite verbose syntax, may prevent to use CASL-LTL for large specifications, needless to say that this threatens the acceptance by the non-academic modellers. A visual syntax for CASL-LTL will help to keep the dimension of the models quite reasonable, and obviously will ease the understanding and the production of models, even by people without a deep know-how in logics and algebraic specifications.

In [4] we already made a first attempt to give a visual presentation to some CASL-LTL specifications, using ad hoc graphical symbols and icons. The attempt was successful for what concerns the compactness of the specifications and the ease to produce them, but further applications and extensive experimentations were prevented by the lack of supporting software tools, e.g., an editor. Obviously these tools could be developed but to produce high quality tools requires really a large effort. Moreover, the graphics of this first attempt of a visual syntax needed to be improved for a better legibility.

In [10] we find the same concern to ease the use of formal specifications by providing a graphical/visual notation for it, and they use both class diagrams and constraints diagrams to represent Z specifications. In [9], constraint diagrams and VisualOCL are compared as a means to visualize OCL expressions. Clearly, our approach here is not to propose a formal semantics to the UML [12], thus we do not refer to the numerous works of that field [1].

Our present work with CASL-MDL relies on borrowing visual constructs of the UML to build a visual syntax for CASL-LTL. This choice has some advantages:
- the graphical constructs are widely known, and were introduced in the UML by pre-existing notations, and now have been tested by a huge number of users for a long time; thus they are familiar and may be easily understood
- some peculiar characteristics of the UML, as its flexibility and the easiness to define variants of itself, and the very loose static semantics, make it possible to define the concrete visual syntax of CASL-LTL as a variant of the UML (that is as a UML profile)
- software tools for editing the UML models are widely available, and many good ones are free.

However, an issue is that, when looking at a CASL-MDL model, some people may be confused between UML and CASL-MDL, and perhaps may use CASL-MDL with the UML intuition. To overcome this we use the profile mechanism to stress the semantic differences; for example a CASL predicate is depicted as a UML operation without return value stereotyped by <<pred>>. A definitive answer about the fact that the borrowed syntax may lead to confusions can only be given by means of rigorous experiments, as proposed by the empirical software engineering, where people knowing both UML and CASL-MDL would be asked to interpret and to produce some models in a controlled way.

Let us remind of a famous occurrence of "syntax borrowing" that resulted in a big success, i.e., the definition of the Java programming language, where the syntax of the C language was kept on purpose whenever possible even if their semantics are totally different. In this case it seems that no confusion arose and also that the familiar aspects of the new language helped its acceptance.

Not any specifications of CASL-LTL will have a visual counterpart, but only a subset, however large enough to include all the specifications produced following the method of [4].

In CASL-MDL we have a type diagram, introducing the datatypes and the dynamic types, which are types of dynamic system (either simple or structured), used in the model. Constraints allow one to express properties on the introduced types (i.e., on the corresponding values or dynamic systems) using first-order and temporal logics. Interaction diagrams express visually properties on the interactions among the components of a structured dynamic systems using the same constructs as the UML sequence diagrams. The behaviour of the dynamic systems of a given type is modelled by interaction machines, using the same constructs as the UML state machines.

The definition of CASL-MDL is an ongoing work and in this paper we will present only the type diagrams and the interactions diagrams, while in [5] we describe also the interaction machines, the constraints, and the constructive definition of data types. Up to know we have not introduced in CASL-MDL diagrams for modelling the workflow, as the UML activity diagrams or the BPMN process diagrams, which would be very useful for using CASL-MDL for business modelling and the modelling of business processes; we are currently working on that.

In Sect. 2 we introduce the CASL-MDL models, in Sect. 3 and in Sect. 4 the type diagrams and the interaction diagrams respectively, and finally in the Sect. 5 the conclusions and the future works.

In the paper we use as a running example the modelling of ASSOC, a case study that describes the functioning of a consortium of associations where associations have boards with a chair and several members, and board meetings take place, to communicate informations or to take decisions via voting. ASSOC has been used as a paradigmatic case study to present a method for the business modelling based on the UML, and thus we think that it may be a good workbench to test the modelling power of CASL-MDL. Fragments of the model of ASSOC will be used to illustrate the various CASL-MDL constructs, an organic presentation of this model can be found in [5].

2   CASL-MDL Models

A CASL-MDL model represents the modelled item in terms of values and of dynamic systems, and we use the term *"entity"* to denote something that may be a value or a dynamic system; similarly an *entity type* defines a type of entities. In Fig. 1 we present the structure of a CASL-MDL model, by means of its "conceptual" metamodel expressed using the UML[1]. The corresponding concrete syntax will be expressed by means of a UML profile, allowing to use the UML tools for editing and for model transformations (e.g., into the corresponding textual CASL-LTL specifications). Thus CASL-MDL has both a conceptual and a

---

[1] In the UML the black diamond denotes composition and the big arrow specialization.

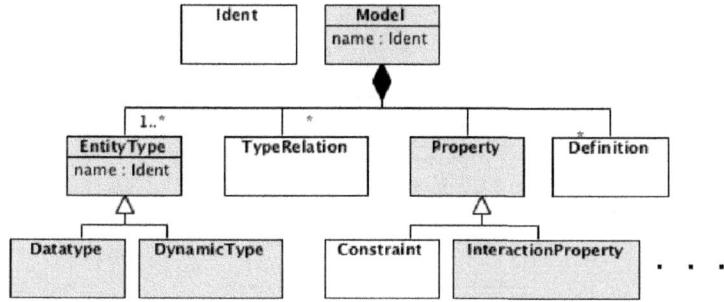

Fig. 1. Structure of the CASL-MDL models ("conceptual" metamodel)

concrete metamodel, the first will be used by the human to grasp the notation, and the latter by the computers to produce and elaborate the CASL-MDL models.

A CASL-MDL model consists of *entity type* declarations (EntityType), of *relationships* between entity types such as extension and subtyping, of *properties* about some of those entities and of *definitions* describing completely some of those entities. In this paper for lack of space we consider only the highlighted parts.

A CASL-MDL model corresponds to a CASL-LTL specification with at least a sort for each declared entity type, whereas the properties are a set of axioms and the definitions in subspecifications built by the CASL-LTL "free" construct.

Translation

TModel : Model → CASL-LTL-Specification

TModel(mod) =

spec mod.name = TETypes(mod.entityType)[2] **then axioms** TProps(mod.property)

The translation of the entity types (at least one must be present in a CASL-MDL model) yields a CASL-LTL specification declaring all the sorts corresponding to the types, plus some auxiliary sorts, and obviously all the declared operations and predicates.

A property in CASL-MDL corresponds to some CASL-LTL formulas on some of the entities introduced in the model, which will be used to extend the specification resulting from the type declarations. A CASL-MDL model having only properties will in the end correspond to a loose CASL-LTL specification.

A property may be a constraint consisting of a CASL-LTL formula written textually, similarly to the UML constraints expressed using the OCL, but in CASL-MDL constraints are suitable to express also properties on the behaviour of the dynamic systems, whereas OCL roughly corresponds to first-order logic. In CASL-MDL it is also possible to visually present some properties having a specific form, for examples some formulas on the interactions among the parts

---

[2] In the UML the name of the target class with low case initial letter is used to navigate along an association, thus mod.entityType denotes the set of the elements of class EntityType associated with mod.

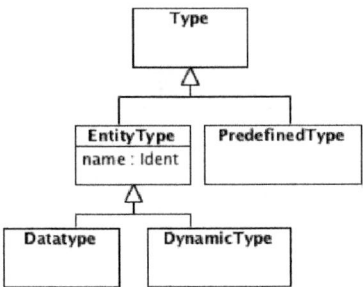

Fig. 2. Structure of Type and the Entity types (metamodel)

of a structured system may be expressed visually by diagrams denoted as UML sequence diagrams, and other formulas may be represented by diagrams similar to the UML activity diagrams. In this paper we consider only the properties of kind constraint and interaction properties.

Visually a CASL-MDL model is a set of diagrams including at least a Type-Diagram presenting the entity types together with the associated constraints, and part of the definitions, whereas the other diagrams correspond to the remaining kind of definitions and to the properties having a visual counterpart. In this paper a CASL-MDL model consists of a type diagram made by entity type declarations and constraints and of a set of interaction properties.

The TypeDiagram may become quite large and thus hard to read and to produce, so in CASL-MDL it is possible to split a TypeDiagram in several ones to describe parts of the types and of the constraints. Furthermore some features, as operations and predicates, of a type may be present in one diagram and others in another one. This possibility is like the one offered by the UML with several class diagrams in a model (a class may appear in several of them, and some of its features - operations and attributes - are in one diagram and some in another).

3   Entity Types and Type Diagrams

A type may be either predefined or an entity type (declaration) which, as shown in Fig. 2, defines a datatype or a dynamic type. In Sect. 3.1 we describe the datatypes, and in Sect. 3.2 the dynamic types.

The predefined datatypes of CASL-MDL are those introduced by the CASL libraries and includes the datatypes, e.g., NAT, INT, LIST and SET.

Translation
  TETypes : EntityType* → CASL-LTL-Specification
TETypes($et_1 \ldots et_n$) =
  LIBRARY **then**
  Basic(et_1.name) **and** ... **and** Basic(et_n.name) **then**
  Detail(et_1); ... Detail(et_n);
where LIBRARY is a CASL specification corresponding to all the predefined datatypes (parameterized or not) defined by the CASL libraries [7].

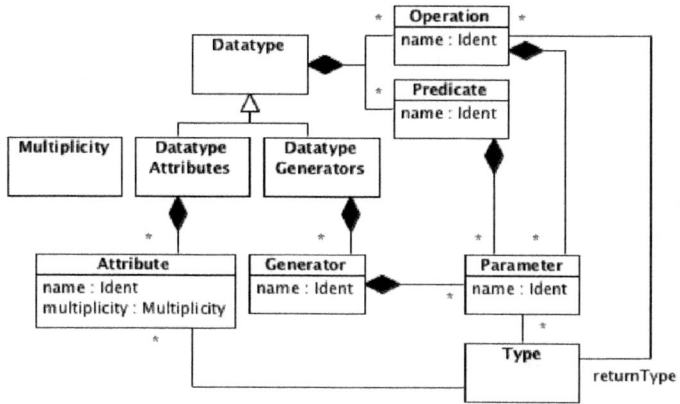

Fig. 3. Datatype Structure (metamodel)

The translation of a set of entity types consists of a CASL-LTL specification corresponding to the predefined types, enriched with the basic specifications of all the types of the model (defined by the function Basic) and after with the details of each type defined by the Detail function. The Basic function introduces the sort corresponding to the identifier passed as argument. Splitting the translation of a CASL-MDL type allows one to have that a type in the type diagram may use all the other types present in the same diagram to define its features.

3.1 Datatypes

CASL-MDL allows to declare new datatypes using the construct Datatype, and their metamodel is presented in Fig. 3[3].

The datatypes may have predicates and operations, which must have at least an argument typed as the datatype itself, and the operations have a return type.

The structure of a datatype of CASL-MDL may be defined in two different ways, using either *generators* or *attributes*.

In the first case the datatype values are denoted using generators (as in CASL).[4] The arguments of the generators may be typed using the predefined types (corresponding to those of the CASL library) and the user defined datatypes and dynamic types present in the same TypeDiagram.

The other possibility is to define the datatype values in terms of attributes, similarly to UML. An attribute attr: T of a datatype D corresponds to a CASL operation _ .attr: D → T. In this case there is a standard generator named as the type itself having as many arguments as the attributes, but it is introduced when defining the datype by an appropriate definition.

---

[3] Note that for the UML diagrams we follow the convention that a multiplicity equal to 1 is omitted, thus an attribute has exactly one type.

[4] We prefer to use the term generator instead of constructor used in the OO world to make clear that in our notation we have datatypes with values and not classes with objects.

<<datatypeA>> DataA
attr1 : T1
...
attrn : Tn
<<pred>> pr(T1', ..., Tk') ...(...)
opr(T1", ..., Tm") : T" ....(...)

(a) Schematic datatype with attributes

<<datatypeG>> DataG
<<gen>> gen(T1, ..., Th) ...(...)
<<pred>> pr(T1', ..., Tk') ...(...)
opr(T1", ..., Tm") : T" .....(...)

(b) Schematic datatype with generators

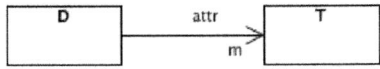

(c) Alternative visual presentation of an attribute

Fig. 4. Visual notation for datatypes

Fig. 4 presents the visual notation for the two forms of datatypes by means of two schematic examples, one with attributes and one with generators ($<<$ pred $>>$ marks the predicates and $<<$ gen $>>$ the generators).

The attributes may have a multiplicity, and its meaning is that the type of the attribute is a set of the associated type and that its values satisfy an implicit constraint [5] about the size of their set values (e.g., multiplicity 0..1 means that the attribute may be typed by the empty set or by a singleton, * that may be typed by any set also empty, and 1..* by any nonempty set). Multiplicity 1 is omitted and corresponds to type the attribute with the relative type. This construct of the CASL-MDL motivates the implicit definition of the finite sets for each type in the translation of the entity types given in the following.

Obviously anonymous casting operations converting values into singleton sets and vice versa are available.

An attribute attr [m]: T of a dataype D may be also visually presented by means of an oriented association as in Figure 4(c).

The modellers are free to use plain attributes or their visual counterpart, but notice that using the arrows shows the structuring relationships among the various types.

Notice that it is possible that only the name of the datatype is provided (no generator or attribute, no predicate or operation), and visually it is simply represented by a box including the name of the datatype.

Translation

Basic : Datatype $\rightarrow$ CASL-LTL-Specification

Basic(dat) = FINITESET[**sort** dat.name]

The basic part of the translation of a datatype is the CASL specification of the finite sets of elements of sort dat.name (sort dat.name is declared in the specification). The need for an implicit declaration of a finite set type for each datatype (as well as for the dynamic types) is motivated by the possibility to associate a multiplicity to the attributes, which corresponds to implicitly declare their type as a set.

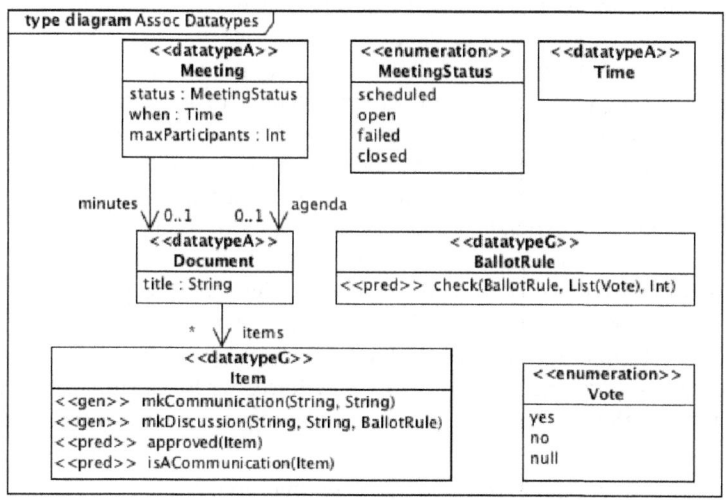

Fig. 5. ASSOC: Type Diagram containing some dataypes

Detail : DatatypeAttributes → CASL-LTL-Specification
Detail($datA$) = TAttributes($datA$.attribute, $datA$.name) ;
    TPredicates($datA$.predicate) ; TOperations($datA$.operation) ;
Below we give part Detail of the translation of the schematic example of
datatype with attributes of Fig. 4(a).

op   _._$attr1 : DataA → T1$; **%%** an operation corresponding to an attribute
    ...
pred $pr : T1' × \ldots × Tk'$; **%%** a predicate ...
op $opr : T1'' × \ldots × Tm'' → T''$; **%%** an operation ...

Notice that at this point the standard generator for the sort $DataA$ has not
been introduced, the type has only some selector like operations corresponding
to the attributes (this allows to refine the datatype with more attributes).

Detail : DatatypeGenerators → CASL-LTL-Specification
Detail($datG$) = TGenerators($datG$.generator, $datG$.name) ;
    TPredicates($datG$.predicate) ; TOperations($datG$.operation) ;

Below we give part Detail of the translation of the schematic example of
datatype with generators of Fig. 4(b).

type $DataG ::= gen(T1; \ldots Th) | \ldots$;
pred $pr : T1' × \ldots × Tk'$; **%%** a predicate ...
op $opr : T1'' × \ldots × Tm'' → T''$; **%%** an operation ...

ASSOC Model: Datatypes
Fig. 5 presents a Type Diagram of the CASL-MDL model of ASSOC containing only datatypes. It includes some enumerated types, precisely MeetingStatus

and Vote (they are a special case of datatype having only generators without arguments considered as literal [5]).

Time is a datatype where no detail is given (it just corresponds to the introduction of the type name). Similarly, no generator is available for BallotRule however a predicate, check, given the votes and the number of voters says if the voting result was positive or not (Int and List are the predefined CASL datatypes for integers and lists). There are some generators for the Item datatype, together with some predicates. Then there are two examples of datatypes with attributes. A Document has a title and some items (possibly zero), and this is expressed by the textual attribute title typed by the predefined String and by items represented by an arrow. A Meeting always has a status, a date and the maximum number of participants (textual attributes in the picture), and optionally it may have an agenda and/or minutes (visual attributes with multiplicity 0..1).

Here there is the CASL-LTL specification fragment corresponding to part Detail of those types translation.

free type   *Vote* ::= *yes* | *no* | *null*; %% enumerated type
free type   *MeetingStatus* ::= *scheduled* | *open* | *failed* | *closed*;
          %% at this stage no generator available for the sort *BallotRule*
pred   *check* : *BallotRule* × *List*[*Vote*] × *Int*;
type   *Item* ::= *mkCommunication*(*String*; *String*)
          | *mkDiscussion*(*String*; *String*; *BallotRule*);
          %% An item is a communication or a discussion with a ballot rule
pred   *isACommunication* : *Item*;
pred   *approved* : *Item*;
op   _.*status* : *Meeting* → *MeetingStatus*; %% corresponds to an attribute . . .
op   _.*agenda* : *Meeting* → *Set*(*Document*); . . .
axiom   ∀ *m* : *Meeting* • *size*(*m.agenda*) ≤ *1* ∧ *size*(*m.minutes*) ≤ *1*

Notice that in this part of the translation there is nothing concerning the datatype Time, since the corresponding sort has been already introduced in the basic part of the translation of the types (FINITESET[**sort** *Time*]).

3.2   Dynamic Types

In CASL-LTL and thus in CASL-MDL the dynamic systems represent any kind of dynamic entities, i.e., entities with a dynamic behaviour without making further distinctions (such as reactive, proactive, autonomous, passive behaviour, inner decomposition in subsystems), and are formally considered as labelled transition systems, that we briefly summarize below.

A *labelled transition system* (*lts* for short) is a triple (*State*, *Label*, →), where *State* denotes the set of states and *Label* the set of transition labels, and →⊆ *State* × *Label* × *State* is the *transition relation*. A triple (*s*, *l*, *s*′) ∈→ is said to be a *transition* and is usually written $s \xrightarrow{l} s'$.

Given an *lts* we can associate with each s_0 ∈ *State* a tree (*transition tree*) with root s_0, such that, when it has a node *n* decorated with *s* and $s \xrightarrow{l} s'$, then it has a node *n*′ decorated with *s*′ and an arc decorated with *l* from *n* to

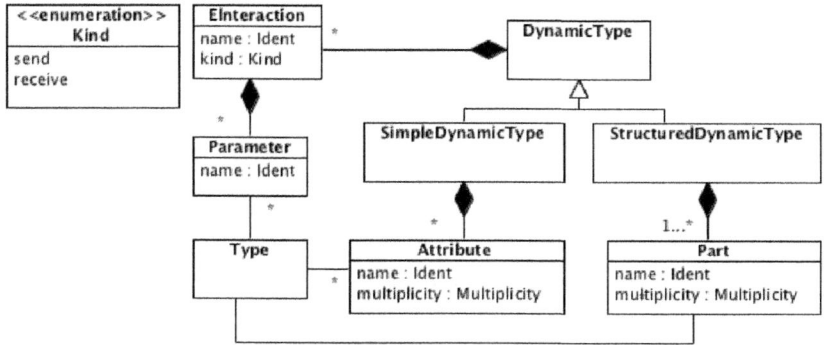

Fig. 6. Dynamic Type Structure (metamodel)

n'. A dynamic system is thus modelled by a transition tree determined by an *lts* (*State, Label,* $\rightarrow$) and an initial state $s_0 \in State$.

CASL-LTL has a special construct **dsort** *state* **label** *label* to declare the two sorts *state* and *label*, and the associated predicate

$\_\_\text{--}\_\_\text{-->}\_\_ : state \times label \times state$

for the transition relation.

Thus a value of a dynamic sort corresponds to a dynamic system, precisely to the labelled transition tree having such value as root, and thus a CASL-LTL specification with a dynamic sort may be truly considered as a dynamic type.

The labels of the transitions of a dynamic system are named in this paper *interactions* and are descriptions of the information flowing in or out the system during the transitions, thus they truly correspond to interactions of the system with the external world[5].

In Fig. 6 we present the structure of the CASL-MDL declaration of dynamic types (i.e., types of dynamic systems) by means of its metamodel,[6] and later we will detail the two cases of simple and structured dynamic types.

Simple Dynamic Types. The simple dynamic systems do not have dynamic subsystems, and in the context of this work, the interactions of the simple systems are either of kind sending or receiving (with a naming convention !_*xx* and ?_*yy*, for sending and receiving interactions resp.) and are characterized by a name and a possibly empty list of typed parameters. These simple interactions correspond to basic acts of either sending out or of receiving something, where the something is defined by the arguments. Obviously, a send act will be matched by a receive act of another simple system and vice versa, and again quite obviously the matching pairs of interactions !_*xx*($v_1, \ldots, v_n$) and ?_*xx*($v_1, \ldots, v_n$).

---

[5] Obviously, a transition may also correspond to some internal activity not requiring any exchange with the external world, in that case the transition is labelled by a special *TAU* value.

[6] DynamicType is a specialization of Type (see also Fig. 2) which has a link to Part.

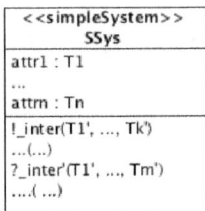

Fig. 7. A schematic Simple Dynamic Type

The states of simple systems are characterized by a set of typed attributes (precisely the states of the associated labelled transition system), similarly to the case of datatypes with attributes (and, as for each attribute, there is the corresponding operation). A dynamic type DT has also an extra implicit attribute __.id: ident_DT containing the identity of the specific considered instance; the identity values are not further detailed. Obviously the identity is preserved by the transitions and no structured dynamic system will have two subsystems with the same identity. Notice how the treatment of the identity in CASL-MDL is completely different from the one of the UML, where the elements of the type associated with a class are just their identities, because CASL-MDL is not object-oriented.

Fig. 6 shows that a simple dynamic type (i.e., a type of simple systems) is determined by a set of elementary interactions (EInteraction) and by a set of attributes; notice that it has also a name since SimpleDynamicType specializes EntityType, see Fig. 2.

In Fig. 7 we present the visual notation for the simple dynamic types by the help of a schematic example.

Translation

Basic : SimpleDynamicType → CASL-LTL-Specification

Basic($simpDT$) =

FINITESET[**sort** $simpDT$.name]   **and** IDENT **with** $ident$ ↦ ident_$simpDT$.name

The basic translation of a simple dynamic type includes also the declaration of a datatype for the identity of the dynamic systems having such type.

Detail : SimpleDynamicType → CASL-LTL-Specification

Detail($simpDT$) =

dsort $simpDT$.name **label** label_$simpDT$.name

op __.id : $simpDT$.name → ident_$simpDT$.name

TAttributes($simpDT$.attribute, $simpDT$.name);

TEInteractions($simpDT$.eInteraction, label_$simpDT$.name);

ASSOC Model: Simple Dynamic Types

Fig. 8 presents a type diagram including two declarations of simple dynamic types. Notice that the type Member has other elementary interactions, e.g.,

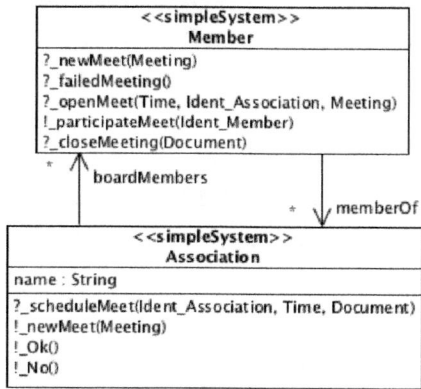

Fig. 8. ASSOC Example: a type diagram including simple dynamic types

!_vote(Item,Vote,Ident_Member) concerning taking part in a meeting not reported here, they are visible in the complete type diagram [5]).

The simple dynamic type Association models the various associations, characterized by a name and by their members (given by the attributes name and members, the latter represented visually as an arrow). We have used a dynamic system and not a datatype since we are interested in the dynamic behaviour of an association. The elementary interaction ?_scheduleMeeting corresponds to receive a request to schedule a new meeting of the association board, and the last two parameters correspond to the meeting date and agenda, whereas the first, typed by Ident_Association is the identity of the association itself. !_Ok and !_Ko correspond respectively to answer positively and negatively to that request.

Part Detail of the translation of the simple type Association is as follows.

dsort *Association* **label** label_*Association*

op __.*id* : *Association* → ident_*Association*

op __.*name* : *Association* → *String*

op ?_*scheduleMeeting* : ident_*Association* × *Time* × *Document* →
label_*Association*

op !_*Ok*, !_*Ko*, *TAU* :→ label_*Association*

TAU is a special implicit element used to label the transitions that do not require any exchange of information with the external world, thus without any interaction. Notice that the sorts *Association* and ident_*Association* have been already introduced by the basic part of the type translation.

Structured Dynamic Types. We recall that a structured system (cf. Fig. 6) is characterized by its parts, or subsystems (that are in turn other simple or structured dynamic systems), and has its own elementary interactions and name.

In Fig. 9 we present the visual syntax by the above schematic structured dynamic type; its parts are depicted by the dashed boxes (in this case all of them have multiplicity one); DType1, DType2, ..., DTypeN are dynamic types (i.e., types corresponding to dynamic systems, simple or structured, defined in

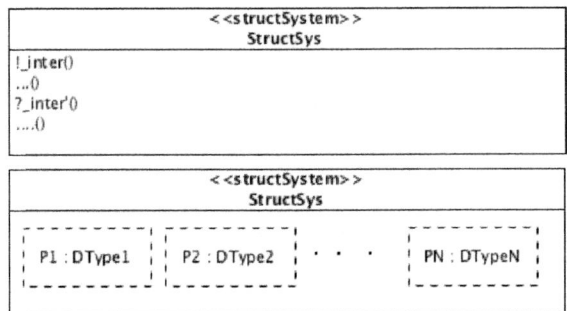

Fig. 9. A schematic Structured Dynamic Type

the same model) and P1, P2, . . . , PN are the optional names of the parts. At this
level we only say that there will be at least those parts, but nothing is said about
the way they interact with each other and on the behaviour of the whole system.
We use two different boxes for the elementary interactions and the structure in
terms of parts to keep the internal structuring encapsulated.

A structured dynamic type has a predefined predicate *isPart* checking if it
has a part having a given identity.

Translation
Basic : StructuredDynamicType → CASL-LTL-Specification
Basic(*structDT*) =
   FINITESET[**sort** *structDT*.name] **and**
   IDENT **with** *ident* ↦ ident_*structDT*.name **and** LOCALINTERACTIONS

LOCALINTERACTIONS specifies the local interactions sets of the structured dynamic systems defined by *structDT*, where a local interaction is a pair consisting
of the identity and of an elementary interaction of one of the parts of *structDT*;
the local interactions are added to the labels of the associated labelled transition
system to record the activities of the parts.

Detail : StructuredDynamicType → CASL-LTL-Specification
Detail(*structDT*) =
   dsort *structDT*.name **label** label_*structDT*.name
   op __.id : *structDT*.name → ident_*structDT*.name
   pred *isPart* : *structDT*.name × *ident_all*
   TParts(*structDT*.part, *structDT*.name);
   TEInteractionsStruct(*structDT*.eInteraction, label_*structDT*.name,
        localInteractions_*structDT*.name);

ident_all is an extra auxiliary sort having as subsorts the identity sorts of all the
dynamic systems in the model.

ASSOC Model: Structured Dynamic System
The whole world of ASSOC is modelled as a structured dynamic system ASSOC
having as parts the associations, the members and the chairs, any number of

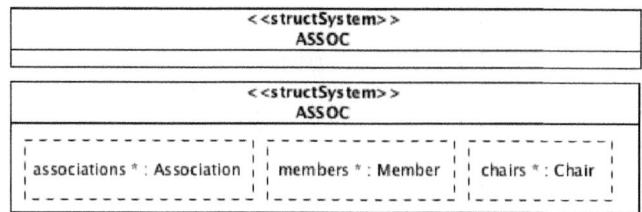

Fig. 10. ASSOC Example: a type diagram including a structured dynamic type

them (see the multiplicity * on the three parts). ASSOC is a closed system, that is it does not interact with its external world and so it has no elementary interactions, and all the transitions of the associated labelled system will be labelled by the special null interaction *TAU*.

The Casl-Ltl specification fragment corresponding to the detail part of the translations of the structured dynamic type ASSOC is given below.

dsort $ASSOC$ **label** label_$ASSOC$
op _._$id : ASSOC \rightarrow$ ident_$ASSOC$
op $associations : ASSOC \rightarrow Set[Association]$
op $members : ASSOC \rightarrow Set[Member]$
op $chairs : ASSOC \rightarrow Set[Chair]$
pred $isPart : ASSOC \times ident\_all$
op $TAU :$ localInteractions_$ASSOC \rightarrow$ label_$ASSOC$
where LocalInteractions= FiniteSet[LocalInteraction] and
LocalInteraction =
free type $= LocalInteraction ::=$
$< \_\_ \_\_ > ($ident_$Association;$ label_$Association) \mid$
$< \_\_ \_\_ > ($ident_$Member;$ label_$Member) \mid$
$< \_\_ \_\_ > ($ident_$Chair;$ label_$Chair)$

4   Interaction Properties

The metamodel of Casl-Mdl interaction properties is given in Fig. 11.

An interaction property describes the way parts of a structured dynamic system (that are in turn dynamic systems) interact. Thus, first of all it should be *anchored* to a specific structured dynamic system represented by an expression typed by a structured dynamic type, which may have free variables, corresponding to express a property on more than one dynamic system. Furthermore an interaction property includes a *context* defining the other free variables (universally and existentially quantified) that may appear in it.

In Casl-Mdl, contrary to UML sequence diagrams, an interaction property explicitly states if it expresses a property of all possible lives of the anchor, or if there exists at least one life of the anchor satisfying that property. It also states whether the property about the interactions must hold in all possible instants of

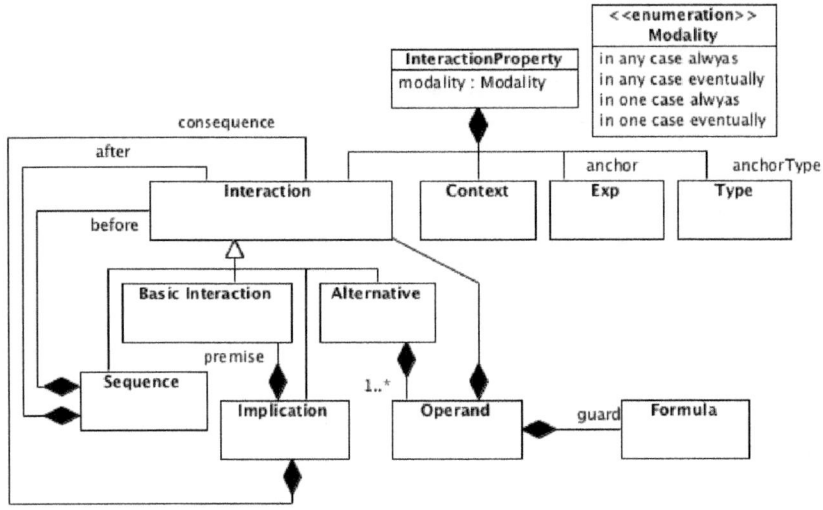

Fig. 11. Interaction Properties structure (metamodel)

those lives, or if eventually there will be an instant in which it will hold. Thus an interaction property has a *modality*, that may assume four values, see Fig. 11.

The Interaction part expresses the required pattern on the interactions among the parts of the anchor and it may be a basic interaction, or a structured interaction built by some combinators (in this paper we consider only alternative, sequential composition and implication).

As shown in Fig. 12, an interaction property is visually presented by reusing the UML sequence diagrams (any v1:T1,...,vn:Tn, one v' 1:T' 1,...,v' m:T' m is the context).

The BasicInteraction, defined in Fig. 13, is the simplest form of Interaction and just corresponds to assert that a series of *elementary interaction occurrences* happen in some order among some generic roles for dynamic systems parts of the anchor (*lifelines*), where an interaction occurrence is the simultaneous performing of a pair of matching input and output elementary interactions by two lifelines.

A lifeline is characterized by a name (just an identifier) and a (dynamic) type and defines a role for a participant to the interaction. An elementary interaction occurrence connects two lifelines in specific points (represented by the lifeline

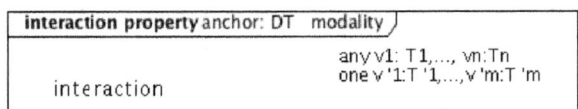

Fig. 12. Visual presentation of a generic CASL-MDL interaction property

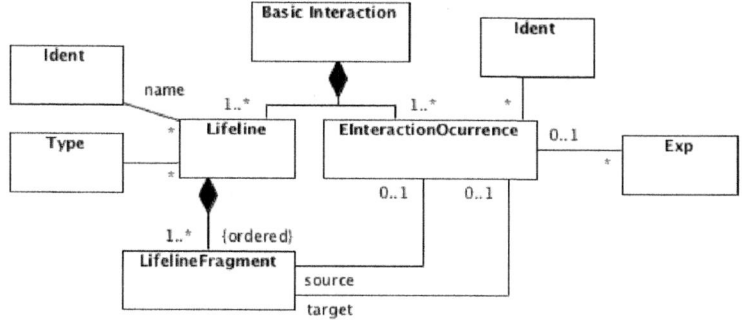

Fig. 13. Structure of Basic Interactions (metamodel)

fragments); the ordering of the interaction points of the various lifelines must determine a partial order on the interaction occurrences. An interaction occurrence is characterized by the name of an elementary interaction s.t. the source type owns it with kind "send" and the target type owns the matching one with the kind "receive", and a set of arguments represented by expressions whose types are in accord with the parameters of the two elementary interactions.

Visually a lifeline is depicted as a box containing its name and type, and by a dashed line summarizing all its fragments, whereas an interaction occurrence is depicted as a horizontal arrow with filled head from the source lifeline to the target one. An elementary interaction occurrence arrow is labelled by inter(exp1...,expn) where !_inter is the send interaction of T1, ?_inter the receive interaction of T2, and exp1..., expn are expressions whose types are in order those of the arguments of !_inter, that are the same of those of ?_inter. Fig. 14 shows a generic case of two lifelines and of an elementary interaction occurrence.

As in the UML the relative distance between two elementary interaction occurrences has no meaning, similarly the only guaranteed ordering is among the the occurrences attached to a single lifeline (due to the ordering of its fragments), whereas in the other cases the visual ordering between two occurrences has no meaning. In Fig. 15 we show two different basic interactions that are, however, perfectly equivalent determining both the partial order listed at the bottom; notice that there are many other ones visually different but still equivalent.

An interaction property corresponds to a Casl-Ltl formula.

Translation

TIntProp: InteractionProperty → Casl-Ltl-Formula

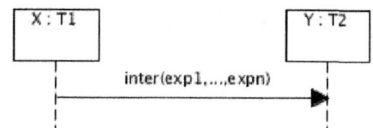

Fig. 14. Generic example of elementary interaction occurrence

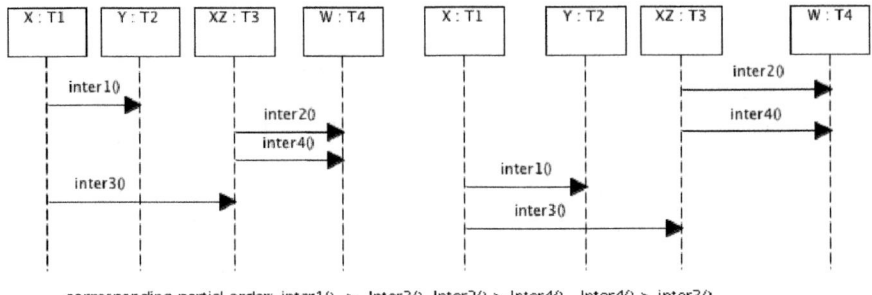

Fig. 15. Two perfectly equivalent basic interactions

TIntProp(iPr) =

$\forall$ *freeVars* TContext(iPr.context) • ($\wedge_{x \in iPr.\text{lifeline}}$ $isPart(x.id, iPr.\text{anchor})$) $\Rightarrow$
    TModal(iPr.modality, iPr.anchor, TInteract(iPr.interaction, *true*))

where *freeVars* are all the free variables appearing in the anchor expression and those corresponding to the lifelines.

    TModal: Modality × Exp × CASL-LTL-PathFormula → CASL-LTL-Formula

TModal(in any case always, $dexp, PF$) = $in\_any\_case(dexp, always\ PF)$
    similarly for the other three cases

    TInteract: Interaction × CASL-LTL-PathFormula → CASL-LTL-PathFormula

    The translation of an interaction is defined by cases, depending on its particular type, and takes as argument a path-formula that will play the role of a continuation; this technical trick allows to correctly translate sequential compositions of interactions.

TInteract($basicInt, cont$) =

$\vee_{eIOc_{i_1} \ldots eIOc_{i_n}}$ admissible ordering of $eIOc_1, \ldots, eIOc_n$
    TIntOcc($eIOc_{i_1}$) $\wedge$ *eventually* (TIntOcc($eIOc_{i_2}$) $\wedge$ (*eventually* ...

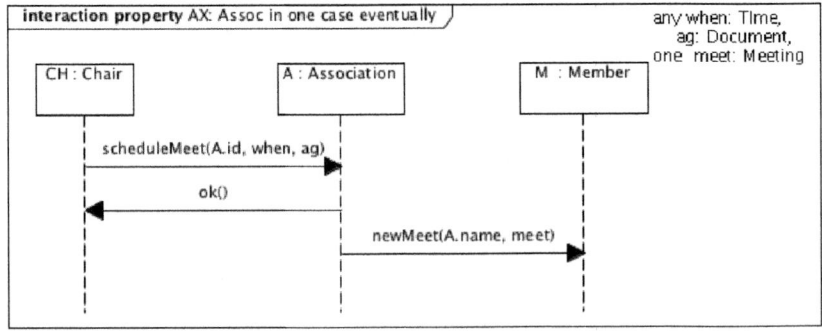

Fig. 16. ASSOC: scheduling a new meeting (successful case)

$$(\mathsf{TIntOcc}(eIOc_{i_n}) \wedge \textit{eventually cont})\dots))$$

where $basicInt.\text{eInteractionOccurence} = eIOc_1,\dots,eIOc_n$

$\mathsf{TIntOcc}$: $\mathsf{InteractionOccurrence} \to \textsc{Casl-Ltl-PathFormula}$

$\mathsf{TIntOcc}(eIOc) =$

$$(x.id:!\_inter(exp1,\dots,expn) \wedge y.id:?\_inter(exp1,\dots,expn))^7$$

where $eIOc$ has the form in Fig. 14.

Fig. 16 shows an interaction property with a basic interaction modelling a successful scheduling a new meeting. This diagram presents a sample of a possible way to execute the successful scheduling of a meeting, precisely the chair asks the association to schedule a new meeting passing the date and the agenda, the association answers ok, and then informs the board members of the new meeting.

Fig. 11 presents also the structured interactions. We can see that it is possible to express:

– the sequential composition of two interactions, with the intuitive meaning to require that the interaction pattern described by the before argument is followed by the interaction pattern described by the after argument;

– the choice among several guarded alternatives, subsuming conditional and nondeterministic choices; one of the interaction patterns corresponding to the alternatives with the true guard must be performed, if no guards is true it corresponds to require nothing on the interactions;

– the fact that the happening of some elementary interactions matching a given pattern (represented by a basic interaction) must be followed mandatory by some elementary interactions matching another pattern.

The visual representation of these structured interactions is illustrated in Fig. 17 and Fig. 18.

To model that the answer of the association may be also negative (elementary interaction ko) we need the structured interactions built with the sequential and alternative combinators, and this corresponds to give just some samples of successful and of failed executions, whereas to represent that after a request of scheduling a new meeting there will be surely an answer by the association we need the implication combinator. Fig. 17 and Fig. 18 presents the interaction properties, with a structured interaction part, corresponding to those cases. In Fig. 17 we have the sequential combination of a basic interaction consisting just of the elementary interaction occurrence scheduleMeet(A.id,when,ag) followed by the alternative among two basic interactions, where the guards are both true corresponding to the pure nondeterministic choice. Again this diagram presents sample of the execution of the scheduling procedure, making explicit that there are two possibilities, a successful one and a failing one; but this diagram does not require that any request to an association will be followed by an answer. Fig. 18 instead shows that an occurrence of the elementary interaction scheduleMeet(A.id,when,ag) will be eventually either followed by an occurrence of ko() or of ok(). Notice that the modality of this interaction property is different, it says that whenever the scheduling request occurs it will be followed by an answer.

---

[7] Recall that $\_.id$ is the standard attribute returning the identity of a dynamic system, and that id: $interact$ is a local interaction atom.

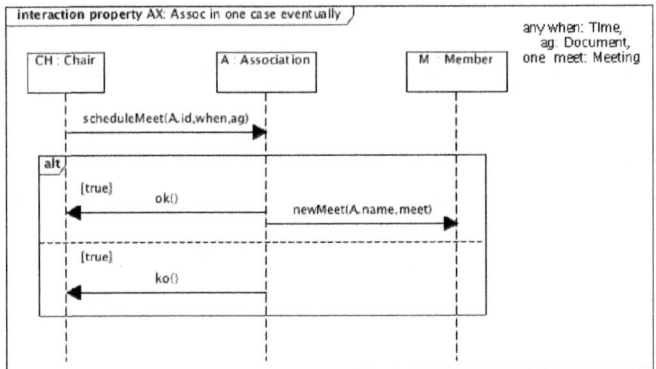

Fig. 17. ASSOC: scheduling a new meeting (sequence and alternative combinator)

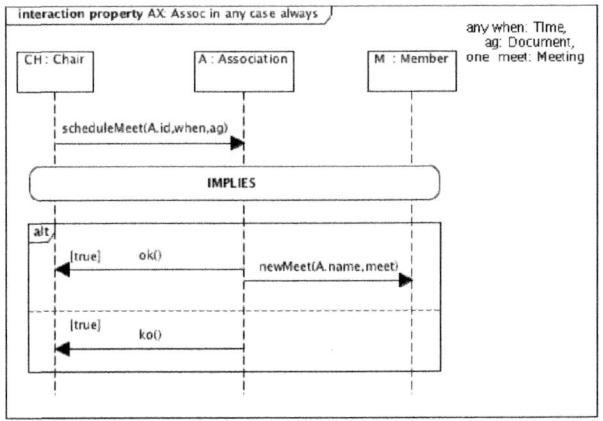

Fig. 18. ASSOC: scheduling a new meeting (implies combinator)

Translation

TInteract: Interaction × CASL-LTL-PathFormula → CASL-LTL-Formula

TInteract($altInt, cont$) =

$\wedge_{J \subseteq \{1,...,n\}}$ (($\wedge_{j \in J}$ op_j.guard $\wedge \wedge_{i \in \{1,...,n\}-J} \neg$ op_i.guard) $\Rightarrow$
     $\vee_{j \in J}$ TInteract(op_j.interaction, $cont$))

where $altInt$.operand = $op_1, \ldots, op_n$

TInteract($seqInt, cont$) = TInteract($seqInt$.before, TInteract($seqInt$.after, $cont$))

TInteract($implInt, cont$) =

$\wedge_{eIOc_{i_1}...eIOc_{i_n}}$ admissible ordering of $eIOc_1,...,eIOc_n$
     (TIntOcc($eIOc_{i_1}$) $\Rightarrow$ next always (TIntOcc($eIOc_{i_2}$) $\Rightarrow$ next always ($\ldots$
          (TIntOcc($eIOc_{i_n}$) $\Rightarrow$ next eventually TInteract($implInt$.consequence,
$cont$))$\ldots$)))

where $implInt$.premise.eInteractionOccurence = $eIOc_1, \ldots, eIOc_n$

Here there is the CASL-LTL formula corresponding to the interaction property of Fig. 17 after some simplifications due to the fact that the guards are both equal to true:

$\forall$ *AX: Assoc, when: Time, ag: Document, CH: Chair, A: Association, M: Member*
$\exists$ *meet: Meeting* •
$(isPart(CH.id, AX) \land isPart(A.id, AX) \land isPart(M.id, AX)) \Rightarrow$
   in_one_case$(AX, eventually$
   $(CH.id :!\_scheduleMeet(A.id, when, ag) \land A.id :?\_scheduleMeet(A.id, when, ag) \land$
   $(eventually$
       $(A.id :!\_ok() \land CH.id :?\_ok() \land eventually$
           $(A.id :!\_newMeet(A.name, meet) \land M.id :?\_newMeet(A.name, meet)))$
       $\lor (A.id :!\_ko() \land CH.id :?\_ko())))$

The CASL-LTL formula corresponding to the interaction property of Fig. 18 can be found in [5].

5   Conclusions and Future Work

In this paper we present a part of CASL-MDL, a visual modelling notation based on CASL-LTL (the extension for dynamic system of the algebraic specification language CASL developed by the COFI initiative). The visual constructs of CASL-MDL have been borrowed to the UML, so as to use professional visual editors; in this paper for example we used Visual Paradigm for UML[8].

A CASL-MDL model is a set of diagrams but it corresponds to a CASL-LTL specification, thus CASL-MDL is a suitable means to easily read and write large and complex CASL-LTL specifications; furthermore the quite mature technologies for UML model transformation may be used to automatize the transformation of the CASL-MDL models into the corresponding CASL-LTL specifications.

CASL-MDL may be used by people familiar with CASL-LTL to produce in an easier way specifications written with it with the help of an editor. However, the corresponding specifications are readable and can be modified directly, for example if there is the need of fine tuning for automatic verification.

We present here a part of CASL-MDL, the type diagram and the interaction diagrams, [5] presents also constraints, definitions for datatypes (which make precise their structure and the meaning of their operations and predicates), definitions of structured dynamic types, which fix their structures and the way their parts interact among them, and interaction machines, which are diagrams visually similar to the UML state machines, modelling the behaviour of the simple dynamic types.

---

[8] http://www.visual-paradigm.com/product/vpuml/

We are currently working out the relationships among the types, and consider the introduction of workflow-like diagrams similar to the UML activity diagrams to visualize formulas on the behaviour of groups of dynamic systems.

UML is the most relevant visual modelling notation, thus it is important to asses the common aspects and the differences with CASL-MDL.

CASL-MDL and UML are visually alike, but they are quite different, first of all because CASL-MDL is not object-oriented and has a simple "native" formal semantics, and because the semantics of syntactically similar constructs is not exactly the same. Consider for example the CASL-MDL interaction diagrams visually similar to the UML sequence diagrams; the interaction diagrams allow also to express implications among the interactions (message exchanges in the UML), thus they are more powerful than the UML sequence diagrams, and closer to the live charts of Harel and Damm [8]. The appendix compares in a tabular form the features of CASL-MDL and of UML.

We think that a careful investigation of the differences and relationships between CASL-MDL and UML may have as a result a better understanding of some of the UML constructs and perhaps some suggestions for possible evolutions.

As regards the relationships between the UML and CASL-MDL let us note that CASL-MDL is not a semantics of the UML expressed in CASL-LTL, and that it is not true that a CASL-MDL model may be transformed into an equivalent UML model.

Acknowledgements. We warmly thank Maura Cerioli for a careful reading of a draft of this paper, and for her valuable comments. We would also like to thank Hubert Baumeister and the anonymous referees for their helpful comments.

References

1. Cengarle, M.V., Knapp, A., Mühlberger, H.: Interactions. In: Lano, K. (ed.) UML 2 Semantics and Applications, pp. 205–248. John Wiley & Sons (2009)
2. Choppy, C., Reggio, G.: Improving Use Case Based Requirements Using Formally Grounded Specifications. In: Wermelinger, M., Margaria-Steffen, T. (eds.) FASE 2004. LNCS, vol. 2984, pp. 244–260. Springer, Heidelberg (2004)
3. Choppy, C., Reggio, G.: A UML-Based Approach for Problem Frame Oriented Software Development. Journal of Information and Software Technology 47, 929–954 (2005)
4. Choppy, C., Reggio, G.: A formally grounded software specification method. Journal of Logic and Algebraic Programming 67(1-2), 52–86 (2006)
5. Choppy, C., Reggio, G.: CASL-MDL, modelling dynamic systems with a formal foundation and a UML-like notation (full report). Technical report, Université Paris 13, and Università di Genova (2010),
   http://www-lipn.univ-paris13.fr/~choppy/REPORTS/casl-mdl-report.pdf
6. Choppy, C., Reggio, G.: Service Modelling with Casl4Soa: A Well-Founded Approach - Part 1 (Service in isolation). In: Symposium on Applied Computing, pp. 2444–2451. ACM (2010)
7. Mosses, P.D. (ed.): CASL Reference Manual. LNCS, vol. 2960. Springer, Heidelberg (2004)

8. Damm, W., Harel, D.: LSCs: Breathing Life into Message Sequence Charts. Formal Methods in System Design 19(1), 45–80 (2001)
9. Fish, A., Howse, J., Taentzer, G., Winkleman, J.: Two visualisations of OCL: A comparison. Technical Report VMG.05.1, University of Brighton (2005)
10. Kim, S.-K., Carrington, D.: Visualization of formal specifications. In: Proceedings of the Sixth Asia Pacific Software Engineering Conference, APSEC 1999, pp. 38–45. IEEE Computer Society (1999)
11. Reggio, G., Astesiano, E., Choppy, C.: CASL-LTL: A CASL Extension for Dynamic Reactive Systems Version 1.0–Summary. Technical Report DISI-TR-03-36 (2003)
12. UML Revision Task Force. OMG UML Specification, http://www.uml.org

A   Comparison between Casl-Mdl and UML

	CASL-MDL	UML
datatypes		
user defined attribute style	+	+
user defined with constructors a la ML	+	
explicit predicates	+	
partial operations	+	exceptions? OCL ?
property oriented definition	invariants, pre-post conditions on operations plus any kind of first order formulas about operations, constructors and attributes	invariants pre-post conditions on operations
constructive definition	rule-based definitions of operations and constructors	methods associated with operations
dynamic entities		
	dynamic systems	active objects
communication mechanism	execution of groups of matching elementary interactions	operation call and signal sending
property oriented definition	branching time temporal logic formulas (e.g., invariants, safety and liveness)	invariants, pre-post conditions on operations
constructive definition	interaction machine (reactive, proactive, passive and internal behaviour)	state machine (reactive behaviour)
objects	as a special kind of passive dynamic systems	native objects
structured dynamic entities	structured dynamic systems	standard community of all objects, structured classes
specification of the interaction among components of structured entities	interaction properties (possibility of expressing liveness and safety properties)	sequence diagrams (samples of message exchanges)
workflow	under development	activity diagrams
....		

Lambda Expressions
in CASL Architectural Specifications

Mihai Codescu

DFKI GmbH Bremen
Mihai.Codescu@dfki.de

Abstract. CASL architectural specifications provide a way to specify the structure of the implementations of software systems. Their semantics has been introduced in two manners: the first is purely model-theoretic and the second attempts to discharge model semantics conditions statically based on a diagram of dependencies between components (extended static semantics). In the case of lambda expressions, which are used to define the way generic units are built, the two semantics do not agree. We present a number of situations of practical importance when the current situation is unsatisfactory and propose a series of changes to the extended static semantics to remedy this.

1 Introduction

An idealized view on the process of software development would be to start with a requirement specification (most likely structured) and then to proceed with an architectural design describing the expected structure of the implementation (which can be different from the one of the specification). Architectural specifications in CASL [3] have been introduced as means of providing structure for the implementation: each architectural specification contains a number of components together with a linking procedure which describes how to combine the components to obtain an implementation of the overall system. (In contrast, the models of a structured specification are monolithic and have no more structure than models of basic specifications).

In the figure on the right, SP is the initial specification, $U_1, \ldots U_n$ are the components of the architectural specifications with their specifications $SP_1, \ldots, SP_n$ and k is the linking procedure involving the units, while the refinement relation is denoted $\rightsquigarrow$. The specification of each component can then play the role of requirement

$$SP \rightsquigarrow k \begin{cases} U_1 : SP_1 \\ \quad \vdots \\ U_n : SP_n \end{cases}$$

specification and the entire process repeats until specifications that can be easily translated into a program are reached. The only interaction allowed between components is the one contained in the architectural specification they are part of, that acts as an interface for them; this allows for a separation of implementation tasks, which can be performed independently.

The semantics of architectural specifications relies on compatibility checks between units as prerequisite for combining them. The intuitive idea is that

T. Mossakowski and H.-J. Kreowski (Eds.): WADT 2010, LNCS 7137, pp. 98–117, 2012.

shared symbols must be interpreted in the same way for two models to be put together. The rules have been presented in two ways: the first is to define a basic static semantics and model semantics in a purely model-theoretical fashion and the compatibility checks are required in the model semantics whenever needed, while the second is an extended static semantics analysis which builds a graph of dependencies between units and discards the compatibility conditions statically. We briefly recall the two semantics and the relationships between them in Section 2. Units of an architectural specification can be *generic* [10], with the intended intuitive meaning that the implementation of the result specification depends on the implementations of the arguments (e.g. some auxiliary functions). Generic units are built using generic unit expressions, written in CASL using the λ-notation: $\lambda X_1 : SP_1, \ldots, X_n : SP_n . UT$, where UT is a unit term which contains $X_1, \ldots X_n$.

The motivation of this paper is rather technical: the extended static semantics rule for generic unit expression does not keep track of the dependencies between the units used in the unit term UT. This is unsatisfactory for a number of reasons that we give in detail in Section 3: first, the completeness theorem for extended static semantics (Theorem 5.4 in [4]) no longer holds when the language is extended with definitions of parametric units. Moreover, *unit imports* are known to be introducing complexity in semantics and verification of architectural specifications. One way to reduce complexity is to replace unit imports with an equivalent construction as below, provided that M is made visible locally in the anonymous architectural specification:

<table>
<tr><td></td><td></td><td>**units** M : SP1;</td></tr>
<tr><td>**units** M : SP1;</td><td></td><td>N : **arch spec** {</td></tr>
<tr><td>   N : SP2 **given** M;</td><td>is equivalent to</td><td>   **units** F : SP1 → SP2</td></tr>
<tr><td>   ...</td><td></td><td>   **result** F[M]};</td></tr>
<tr><td></td><td>...</td><td></td></tr>
</table>

If N would be a generic unit, then the result of the architectural specification in the right side would be a λ-expression and the two constructions would no longer be equivalent because they treat differently the dependency between M and N. In Section 4 we present our proposed changes for the extended static semantics of architectural specifications, followed by a discussion in Section 4.1 on how the completeness result can be extended to cover lambda expressions as well. Section 4.2 further extends the changes to *parametric* architectural specifications i.e. those having lambda expressions as result, while in Section 5 we present a larger example motivating the introduction of the new rules, involving refinement of units with imports. Section 6 concludes the paper.

2   CASL Architectural Specifications

As mentioned above, CASL architectural specifications describe how the implementation is structured into component units. Each unit is given a name and assigned a specification; the intended meaning is to provide a model of the specification. Units can be *generic*, taking a list of specifications as arguments and having a result specification; such units denote partial functions that take as arguments models of the parameter specifications and return a model of the result specification. The result is required to preserve the parameters (*persistency*), with the intuition that the program of the parameter must not be re-implemented, and the function is only defined on *compatible* models, meaning that the implementation of the parameters must be the same on common symbols. Units are combined in unit expressions with operations like renaming, hiding, amalgamation and applications of generic units. Again, terms are only defined for compatible models, in the sense that common symbols must be interpreted in the same way. Let us mention that architectural specifications are independent of the underlying formalism used for basic specifications, which is modelled as an institution [5].

An architectural specification consists of a list of unit definitions and declarations followed by a result unit expression. Fig. 1 presents a fragment of the grammar of the CASL architectural language that is relevant for the examples of this paper; the complete grammar can be found in [4]. Notice that we allow the specification of a unit to be itself architectural (named or anonymous) and that for units declarations there is an optional list of imported units (marked with $<\_>$). The list must be empty when USP is architectural. Moreover, in Fig. 1 A is a unit name, S is a specification name, SP is a structured specification and σ is a signature morphism. We denote $\iota_{\Sigma \subseteq \Sigma'}$ the injection of Σ in Σ' when Σ' is a union of signatures with Σ among them.

$$
\begin{aligned}
ASP ::=\ &\textbf{units}\ UDD_1 \ldots UDD_n \\
&\textbf{result}\ UE \\
UDD ::=\ &UDEFN \mid UDECL \\
UDECL ::=\ &A : USP\ < \textbf{given}\ UT_1, \cdots UT_n > \\
USP ::=\ &SP \mid SP_1 \times \cdots \times SP_n \to SP \mid \\
&\textbf{arch spec}\ S \mid \textbf{arch spec}\ \{ASP\} \\
UDEFN ::=\ &A = UE \\
UE ::=\ &UT \mid \lambda\ A_1 : SP_1, \ldots,\ A_n : SP_n \bullet UT \\
UT ::=\ &A \mid A\ [FIT_1] \ldots [FIT_n] \mid UT\ \textbf{and}\ UT \mid UT\ \textbf{with}\ \sigma : \Sigma \to \Sigma' \mid \\
&UT\ \textbf{reduction}\ \sigma : \Sigma \to \Sigma' \mid \textbf{local}\ UDEFN_1 \ldots UDEFN_n\ \textbf{within}\ UT \\
FIT ::=\ &UT \mid UT\ \textbf{fit}\ \sigma : \Sigma \to \Sigma'
\end{aligned}
$$

Fig. 1. Restricted language of architectural specifications

The CASL semantics produces for any specification a signature and a class of models over that signature. This is not different for architectural specifications: the basic static semantics yields an architectural signature, while the model semantics produces an architectural model. We give definitions of this notions and a brief overview of the two semantics below.

An architectural signature consists of a unit signature for the result together with a static unit context, describing the signatures of each unit. A unit signature can be either a plain signature or a list of signatures for the arguments and a signature for the result. Starting with the initial empty static unit context, the static semantics for declarations and definitions adds to it the signature of each new unit and the static semantics for unit terms and expressions does the type-checking in the current static context. For any architectural specification ASP, we denote $|ASP|$ the specification obtained by removing everything but the signature from the specifications used in declarations.

Model semantics is assumed to be run only after a successful run of the basic static semantics and it produces an architectural model over the resulting architectural signature. Model semantics of an individual unit is either simply a model of the specification, for non-generic units, or a partial function taking compatible models of the argument specifications to a model of the result specification. The result is required to protect the parameters when reduced back to a model of the corresponding signature. Generic units can be interpreted as total functions by introducing an additional value $\perp$ - this ensures consistency of generic unit specifications in $|ASP|$ whenever the unit specification is already consistent in an architectural specification ASP and is called *partial model semantics* in [4], Section IV:5. An architectural model over an architectural signature consists of a result unit over the result unit signature and a collection of units over the signatures given in the static context, named by their unit names. Model semantics produces a unit context, which is a class of unit environments - maps from unit names to units, and a unit evaluator, which is a map that gives a unit when given a unit environment in the unit context. The analysis starts with the unit context of all environments and each declaration and definition enlarges the unit context, adding a new constraint. Finally, the semantics of unit terms produces a unit evaluator for a given unit context.

$$P_{st}(F) = \tau : \Sigma \to \Sigma'$$
$$C_{st} \vdash T \triangleright \Sigma^A$$
$$\sigma : \Sigma \to \Sigma^A$$
$$\frac{(\sigma_R, \tau_R, \Sigma_R) \text{ is the pushout of } (\sigma, \tau)}{P_{st}, C_{st} \vdash F[T \text{ fit } \sigma] \triangleright \Sigma_R}$$

$$C \vdash T \triangleright UEv$$
$$\text{for each } E \in C, UEv(E)|_\sigma \in domE(F) \textbf{ (i)}$$
$$\text{for each } E \in C, \text{ there is a unique } M \in Mod(\Sigma_R) \text{ such that}$$
$$M|_{\tau_R} = UEv(E) \text{ and } M|_{\sigma_R} = E(F)(UEv(E)|_\sigma) \textbf{ (ii)}$$
$$\frac{UEv_R = \{E \mapsto M | E \in C, M|_{\tau_R} = UEv(E), M|_{\sigma_R} = E(F)(UEv(E)|_\sigma)\}}{C \vdash F[T \text{ fit } \sigma] \triangleright UEv_R}$$

Fig. 2. Basic static and model semantics rules for unit application

Fig. 2 presents the basic static semantics and model semantics rules for unit application (notice that we simplify to the case of units with just one argument). The static semantics rule produces the signature of the term T and returns as signature of $F[T]$ the pushout Σ_R of the span (σ, τ), where τ is the unit signature of F stored in the list of parameterized unit signatures P_{st}.

The model semantics rule first analyzes the argument T and gives a unit evaluator UEv. Then, provided that the conditions (i) the actual parameter actually fits the domain and (ii) the models $UEv(E)$ and $E(F)(UEv(E)|_\sigma)$ can be amalgamated to a Σ_R-model M hold, the result unit evaluator UEv_R gives the amalgamation M for each $E \in C$.

Typically one would expect that conditions (ii) would be discarded statically. For this purpose, an extended static semantics was introduced in [11], where the dependencies between units are tracked with the help of a diagram of signatures. The idea is that we can now verify that the interpretation of two symbols is the same by looking for a "common origin" in the diagram, i.e. a symbol which is mapped via some paths to both of them. We will present in this paper only the relevant rules of extended static semantics in Section 3. We are going to make use of the following notions. A *diagram* D is a functor from a small category to the category of signatures of the underlying institution. In the following, let D be a diagram. A family of models $\mathcal{M} = \{M_p\}_{p \in Nodes(D)}$ indexed by the nodes of D is *consistent with* D if for each node p of D, $M_p \in Mod(D(p))$ and for each edge $e : p \to q$, $M_p = M_q|_{D(e)}$. A *sink* α on a subset K of nodes consists of a signature Σ together with a family of morphisms $\{\alpha_p : D(p) \to \Sigma\}_{p \in K}$. We say that D *ensures amalgamability* along $\alpha = (\Sigma, \{\alpha_p : D(p) \to \Sigma\}_{p \in K})$ if for every model family $\mathcal{M}$ consistent with D there is a unique model $M \in Mod(\Sigma)$ such that for all $p \in K$, $M|_{\alpha_p} = M_p$.

The two semantics of architectural specifications are related by a soundness result [11]: if the extended semantics of an architectural specification is defined, then so is the basic semantics and the latter gives the same result. In [4], completeness is also proved for a simplified variant of the architectural language[1] and with a modified model semantics. We will discuss this in more detail in Section 4.1.

3   Semantics of Generic Unit Expressions

We present now the extended static semantic rule for generic unit expressions, with the help of a typical example of a dependency between the unit term of a lambda expression and the generic unit defined by it. Such dependencies are not tracked in the diagram built with the rules for extended static semantics defined in [4].

Example 1. Let us consider the CASL architectural specification from Fig. 3. The unit term $L1[A1]$ **and** $L2[A2]$ is ill-formed w.r.t. the rules of extended static semantics for architectural specifications because in the diagram in the Fig. 4 (built using the extended static semantics rules for generic unit expressions and unit applications, which are presented in Fig. 5 and Fig. 6 respectively) the sort s can not be traced to a common origin (which should be the node M).   □

The rule for analysis of generic unit expressions (Fig. 5) introduces a node p for the unit term of the lambda expression that keeps track of the sharing information of the terms involved. However, this node p is not further used in

---

[1] It is nevertheless argued that the generalization to the full features of CASL architectural language is of no genuine complexity, excepting the case of imports. Our approach covers the imports as well.

spec S = **sort** s
spec S1 = **sort** $s1$
spec S2 = **sort** $s2$
arch spec ASP =
units M : S; A1 : S1; A2 : S2;
       L1 = λ X1 : S1 • M **and** X1;
       L2 = λ X2 : S2 • M **and** X2;
result L1 [A1] **and** L2 [A2]

Fig. 3. Lost sharing

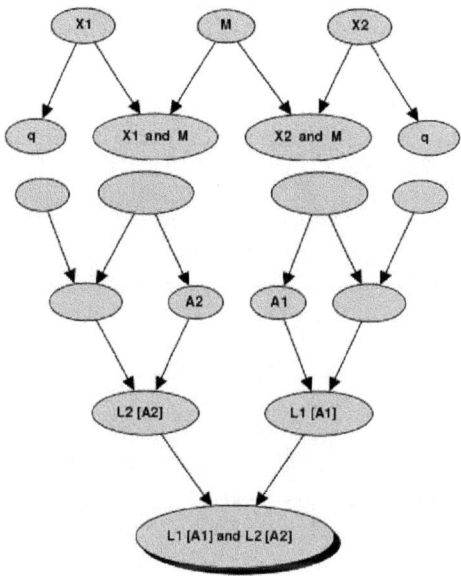

Fig. 4. Diagram of ASP

application of lambda expressions. In the extended static context, the entry corresponding to the lambda expression only contains a new node labeled with the empty signature, denoted z in Fig. 5, as node of imports, and this new node is isolated. Notice also that the purpose of inserting the node q and the edges from nodes p_i to q is to ensure compatibility of the formal parameters when making the analysis of the unit term.

     Using this version of the rules raises a series of problems. First, there is no methodological justification for making terms like the one in our example illegal by not keeping track of the unit M in the lambda expressions. Moreover, ASP has a denotation w.r.t. the basic semantics (it is easy to see that the specification type-checks) and $|ASP|$ has a denotation w.r.t. the model semantics (there is

$$\frac{\begin{array}{c} \Gamma_s \vdash \texttt{UNIT-BIND-1} \rhd (UN_1, \Sigma_1) \ldots \Gamma_s \vdash \texttt{UNIT-BIND-n} \rhd (UN_n, \Sigma_n) \\ \Sigma_a = \langle \Sigma_1, .., \Sigma_n \rangle \text{ and } \Sigma = \Sigma_1 \cup ... \cup \Sigma_n \\ UN_1, \ldots, UN_n \text{ are new names} \\ D' \text{ extends } dgm(C_S) \text{ by new node } q \text{ with } D'(q) = \Sigma, \\ \text{nodes } p_i \text{ and edges } e_i : p_i \to q \text{ with } D'(e_i) = \iota_{\Sigma_i \subseteq \Sigma} \text{ for } i \in 1, \ldots, n \\ C'_s = (\{\}, \{UN_1 \to p_1, \ldots, UN_n \to p_n\}, D') \\ \Gamma_s, C_s + C'_s \vdash \texttt{UNIT-TERM} \rhd (p, D'') \\ D'' \text{ ensures amalgamability along } (D''(p), \langle id_{D''(p)}, \iota_{\Sigma_i \subseteq D''(p)} \rangle_{i \in 1, \ldots, n}) \\ D''' \text{ extends } D'' \text{ by new node } z \text{ with } D'''(z) = \emptyset \end{array}}{\begin{array}{c} \Gamma_s, C_s \vdash \texttt{unit-expr UNIT-BIND-1}, .., \texttt{UNIT-BIND-n UNIT-TERM} \rhd \\ (z, \Sigma_a \to D''(p), D''') \end{array}}$$

Fig. 5. Extended static semantics rule for unit expressions (CASL Ref. Manual)

$$\frac{\begin{array}{c} C_s = (P_s, B_s, D) \\ P_s(UN) = (p^I, (\Sigma_1, ..., \Sigma_n \to \Sigma)) \\ \Sigma^F = D(p^I) \cup \Sigma_1 \cup ... \cup \Sigma_n \\ \Sigma_i, \Gamma_s, C_s \vdash \texttt{FIT-ARG-i} \rhd (\sigma_i : \Sigma_i \to \Sigma_i^A, p_i^A, D_i) \text{ for } i \in 1, \ldots, n \\ D_1, .., D_n \text{ disjointly extend } D \\ \text{let } D^A = D_1 \cup .. \cup D_n \\ \Sigma^A = D(p^I) \cup \Sigma_1^A ... \cup \Sigma_n^A \\ \sigma^A = (id_{D(p^I)} \cup \sigma_1 \cup .. \cup \sigma_n) : \Sigma^F \to \Sigma^A \\ \sigma^A(\Delta) : \Sigma \to (\Sigma^A \cup \Sigma^A(\Delta)), \text{ where } \Delta : \Sigma^F \to \Sigma \text{ is the signature extension} \\ \Sigma^R = \Sigma^A \cup \Sigma^A(\Delta) \\ D^A \text{ ensures amalgamability along } (\Sigma^A, \langle \iota_{D(p^I) \subseteq \Sigma^A}, \iota_{\Sigma_i^A \subseteq \Sigma^A} \rangle_{i \in 1, \ldots, n}) \\ D' \text{ extends } D^A \text{ by new node } q^B, \text{ edge } e^I : p^I \to q^B \text{ with } D'(e^I) = \iota_{D(p^I) \subseteq \Sigma)}, \\ \text{nodes } p_i^F \text{ and edges } e_i^F : p_i^F \to q^B \text{ with } D'(e_i^F) = \iota_{\Sigma_i \subseteq \Sigma} \\ \text{and } e_i : p_i^F \to p_i^A \text{ with } D'(e_i) = \sigma_i \text{ for } i \in 1, \ldots, n \\ D' \text{ ensures amalgamability along } (\Sigma^R, \langle \sigma^A(\Delta), \iota_{\Sigma_i^A \subseteq \Sigma^R} \rangle_{i \in 1, \ldots, n}) \\ D'' \text{ extends } D' \text{ by new node } q, \text{ edge } e' : q^B \to q \text{ with } D''(e') = \sigma^A(\Delta) \\ \text{and edges } e_i' : p_i^A \to q \text{ with } D''(e_i') = \iota_{\Sigma_i^A \subseteq \Sigma^R} \text{ for } i \in 1, \ldots, n \end{array}}{\Gamma_s, C_s \vdash \texttt{unit-appl UN FIT-ARG-1}, .., \texttt{FIT-ARG-n} \rhd (q, D'')}$$

Fig. 6. Extended static semantics rules for unit application (CASL Ref. Manual)

no problem in amalgamating M with a model of specifications $S1$ or $S2$, since there are no shared symbols, and when making the amalgamation of $L1[A1]$ with $L2[A2]$ the symbol s is interpreted in the same way by construction). Thus, since one would expect that the completeness result of [4] should still hold for the entire architectural language, ASP should have a denotation w.r.t. the extended static semantics.

Another reason to consider the current rules unsatisfactory is the relation between units with imports and generic units. A unit declaration with imports has been informally explained in the literature as a generic unit instantiated once, like in the following example.

Example 2. The following unit declarations, taken from the architectural specification of a steam boiler control system (Chapter 13 of [2]):

B : BASICS;
MR : VALUE → MESSAGES_RECEIVED **given** B;

can be expressed as a generic unit instantiated once (notice that the linear
visibility of units, required in [4], is assumed to be extended):

B : BASICS;
MR : **arch spec** {
    units F : BASICS × VALUE → MESSAGES_RECEIVED
    result λ X : VALUE • F [B] [X]};   □

The two declarations in Example 2 are not equivalent because the former traces
the dependency between MR and B while the latter does not. However it has
been noticed that to be able to write down refinements of units with imports
using the CASL refinement language designed in [8], this equivalence must become
formal. This can only be the case if the second construction also tracks the
dependency of B with MR.

Notice that in general the unit imported may be written as a more complex
unit term and then its specification is no longer available directly. Moreover, as
remarked in [7], it is not always possible to find a specification that captures
exactly the class of all models that may arise as the result of the imported unit
term. It is however possible to use the proof calculus for architectural specifications defined in [6] and Section IV.5.3 of [4] to generate a structured specification
that includes this model class among its models. Another advantage of making
the equivalence formal is that the completeness result for extended static semantics and the proof calculus for architectural specifications cover imports as
well, since they can now be regarded only as "syntactic sugar" for the equivalent
construction.

4   Adding Dependency Tracking

The proposed changes are based on the following observation: in the rule for unit
application(Fig. 6), new nodes are needed for the formal parameters and for the
result (labeled p_i^F and q^B respectively). However, for lambda expressions the
nodes p_i and p in Fig. 5 have already been introduced with the same purpose.
This symmetry can be exploited when making the applications of a lambda
expression and we will therefore need to keep track of the mentioned nodes.

Recall that an extended static unit context consists of a triple (P_s, B_s, D),
where $B_s \in UnitName \rightarrow Item$ and stores the corresponding nodes in the
diagram for non-generic units, $P_s \in UnitName \rightarrow Item \times ParUnitSig$ and
stores the parameterized unit signature of a generic unit together with the node
of the imports, such that both B_s and P_s are finite maps and have disjoint
domains and D is the signature diagram that stores the dependencies between
units.

Firstly, we need to modify the definition of extended static unit contexts such that P_s maps now unit names to pairs in $[Item] \times ParUnitSig$, to be able to store the nodes of the parameters and of the result for lambda expressions. Notice that a lambda expression must have at least one formal parameter, so the list of items contains either the node of the union of the imports in the case of generic units or at least two elements in the case of definitions of lambda expressions. Moreover, unit declarations of form UN : $\mathbf{arch\,spec}\,ASP$ where ASP is an architectural specification whose result unit is a lambda expression also should store the nodes for parameters and the result. The rule changes needed for this latter case are not straightforward and will be addressed separately in section 4.2. In Section 4.2 we will also make use of this list of nodes for a different purpose, namely tracking dependencies between different levels of visibility for units.

$$\frac{\begin{array}{c} \Gamma_s \vdash \texttt{UNIT-BIND-i} \rhd (UN_i, \Sigma_i) \text{ for } i \in 1, \ldots, n \\ \Sigma_a = \langle \Sigma_1, .., \Sigma_n \rangle \text{ and } \Sigma = \Sigma_1 \cup ... \cup \Sigma_n \\ UN_i \text{ are new names} \\ D' \text{ extends } dgm(C_S) \text{ by new node } q \text{ with } D'(q) = \Sigma, \\ \text{nodes } p_i \text{ with } D'(p_i) = \Sigma_i \\ \text{and edges } e_i : p_i \to q \text{ with } D'(e_i) = \iota_{\Sigma_i \subseteq \Sigma} \text{ for } i \in 1, \ldots, n \\ C'_s = (\{\}, \{UN_i \to p_i | i \in 1, \ldots, n\}, D') \\ \Gamma_s, C_s + C'_s \vdash \texttt{UNIT-TERM} \rhd (r, D'') \\ D'' \text{ ensures amalgamability along}(D''(r), \langle id_{D''(r)}, \iota_{\Sigma_i \subseteq D''(r)} \rangle) \\ \cancel{D''' \text{ extends } D'' \text{ by new node } z \text{ with } D'''(z) = \emptyset} \\ D''' \text{ removes from } D'' \text{ the node } q \text{ and its incoming edges}\end{array}}{\begin{array}{c} \Gamma_s, C_s \vdash \texttt{unit-expr UNIT-BIND-1}, .., \texttt{UNIT-BIND-n UNIT-TERM} \rhd \\ ([r, p_1, .. p_n], \Sigma_a \to D'''(r), D''') \end{array}}$$

Fig. 7. Modified extended static semantics rule for unit expressions

Fig. 7 presents the modified static semantics rule for generic unit expressions, which introduces new nodes p_i for the parameters and a node q to ensure their compatibility during the analysis of the unit term. Then, the result node of the unit term p together with the nodes for parameters are returned as result of the analysis of the lambda expression, together with the diagram resulting by removing the node q and the edges from the nodes p_i to q from the diagram obtained after the analysis of the unit term. The reason why the node q must be removed is that the nodes of the formal parameters will be connected to the actual parameters and their compatibility must be rather checked than ensured.

We also have to make a case distinction in the rule of unit application. In the case of generic units, we can use the existing rule for unit applications. The rule for application of lambda expressions is similar with the one used in the first case, but it puts forward the idea that the nodes for formal parameters and result that were stored in the analysis of the lambda expression should be used when making the application. However, this requires special care, as we will illustrate with the help of some examples.

Example 3. Repeated applications of the same lambda expression. Let us consider the definition $F = \lambda X : SP \cdot X$ **and** M where we assume that SP and the specification of M do not share symbols and $M1, M2 : SP$. If we use the stored nodes for parameters and result at *every* application of F, we obtain the diagram in Fig. 8, resulting after applying F to $M1$ and $M2$. Notice that the edges from X to $M1$ and $M2$ respectively introduce a sharing requirement between the actual parameters, which is not intended.   □

The solution to this problem is to copy at every application the nodes introduced in the diagram during the analysis of the term of the lambda expression. The copy can be obtained starting with the stored nodes p_i by marking their copies as new formal parameter nodes and going along their outgoing edges: for each new node accessible from p_i, we introduce a copy of it in the diagram together with copies of its incoming edges - this last step copies also the dependencies of the unit term of the lambda expression with the outer units (in

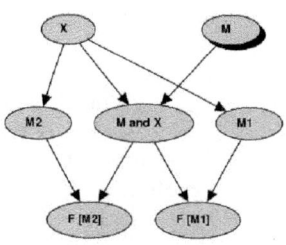

Fig. 8. Unwanted sharing

the example, the edge from the node of M to the node of M **and** X is copied). The copying stops when all nodes have been considered, and the copy of the result node is then marked as new result node. Let us denote the procedure described above *copyDiagram*, which takes as inputs the nodes for result and formal parameters of the lambda expression and the current diagram and returns the copied nodes for formal parameters and result and the new diagram. The procedure described works as expected because the diagram created during the analysis of the unit term of the lambda expression consists of exactly the nodes accesible from the formal parameter nodes and it has no cycles; moreover, no new dependencies involving these nodes are ever added in the diagram.

Example 4. Tracking dependencies of the actual parameters with the environment.

Let us consider the architectural specification in Fig. 9, where the actual parameter and the unit A used in the term of the lambda expression share the sort symbol s, which can be traced in the dependency diagram to a common origin, which is the node of P - see Fig. 10. This application should be therefore considered correct.   □

Refering to the rule in Fig. 6, the generic unit is given by the inclusion $\Delta :$ $\Sigma^F \to \Sigma$ of its formal parameters into the body and at application, the fitting arguments give a signature morphism $\sigma^A : \Sigma^F \to \Sigma^A$ from the formal parameters to the actual parameters. Then, $\Sigma^A \cup \Sigma^A(\Delta)$ results by making the union of the fitting arguments with the body translated along the signature extension $\sigma^A(\Delta) : \Sigma \to \Sigma^A \cup \Sigma^A(\Delta)$. Originally, an application has been considered not well-formed if the result signature is not a pushout of the body and argument signatures (this is hidden in the use of the notation $\sigma^A(\Delta)$, see [4]) and notice that this is indeed not the case in Example 4. We can drop this requirement in the case of lambda expressions and rely on the condition that the diagram should ensure amalgamability; indeed, in this case the application is correct if

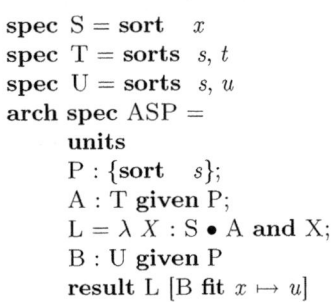

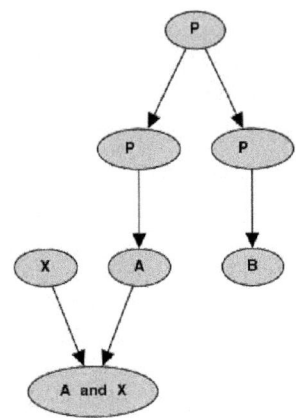

Fig. 9. Sharing between arguments and actual parameter

Fig. 10. Diagram of Example 4 before application

whenever a symbol is present both in the body and in the argument signatures, the symbol can be traced in the diagram to a common origin which need not be the node of the formal parameter, like in the case of sort s above.

Taking into account the observations in Examples 3 and 4, the rule of for application of lambda expressions is presented in Fig. 11.

$$C_s = (P_s, B_s, L_s, D_0)$$
$$L_s(UN) = ([p, p_1, \ldots, p_n], (\Sigma_1, \ldots, \Sigma_n \to \Sigma))$$
$$([r, f_1, \ldots, f_n], D) = copyDiagram([p, p_1, \ldots, p_n], D_0)$$
$$\Sigma^F = \Sigma_1 \cup \ldots \cup \Sigma_n$$
$$\Sigma_i, \Gamma_s, C_s \vdash \texttt{FIT-ARG-i} \rhd (\sigma_i : \Sigma_i \to \Sigma_i^A, p_i^A, D_i) \text{ for } i \in 1, \ldots, n$$
$$D_1, \ldots, D_n \text{ disjointly extend } D$$
$$\text{let } D^A = D_1 \cup \ldots \cup D_n$$
$$\Sigma^A = \Sigma_1^A \ldots \cup \Sigma_n^A$$
$$\sigma^A = (\sigma_1 \cup \ldots \cup \sigma_n) : \Sigma^F \to \Sigma^A$$
$$\sigma^A(\Delta) : \Sigma \to (\Sigma^A \cup \Sigma^A(\Delta)), \text{ where } \Delta : \Sigma^F \to \Sigma \text{ is the signature extension}$$
$$\text{and the pushout condition for } \Sigma^A \cup \Sigma^A(\Delta) \text{ is dropped}$$
$$\Sigma^R = \Sigma^A \cup \Sigma^A(\Delta)$$
$$D^A \text{ ensures amalgamability along } (\Sigma^A, \langle \iota_{\Sigma_i^A \subseteq \Sigma^A} \rangle_{i \in 1, \ldots, n})$$
$$D' \text{ extends } D^A \text{ with edges } e_i : f_i \to p_i^A \text{ with } D'(e_i) = \sigma_i, \text{ for } i \in 1, \ldots, n$$
$$D' \text{ ensures amalgamability along } (\Sigma^R, \langle \sigma^A(\Delta), \iota_{\Sigma_i^A \subseteq \Sigma^R} \rangle_{i \in 1, \ldots, n})$$
$$D'' \text{ extends } D' \text{ by new node } q, \text{ edge } e' : r \to q \text{ with } D''(e') = \sigma^A(\Delta)$$
$$\text{and edges } e_i' : p_i^A \to q \text{ with } D''(e_i') = \iota_{\Sigma_i^A \subseteq \Sigma^R}, \text{ for } i \in 1, \ldots, n$$

$$\Gamma_s, C_s \vdash \texttt{unit-appl UN FIT-ARG-1}, \ldots, \texttt{FIT-ARG-n} \rhd (q, D'')$$

Fig. 11. Extended static semantics rule for unit application of lambda expressions

Fig. 12 presents the diagram of the architectural specification *ASP* in Example 1 using the modified rules of Fig. 7 and 11[2]; notice that in this diagram the sort *s* can be traced to a common origin and thus the amalgamation is correct. Moreover, when making the application of the lambda expression, the diagram of the term *M* **and** *X* is copied such that no dependency between the actual parameters is incorrectly introduced by edges from the formal parameter node and copying the diagram does not duplicate the node *M*.

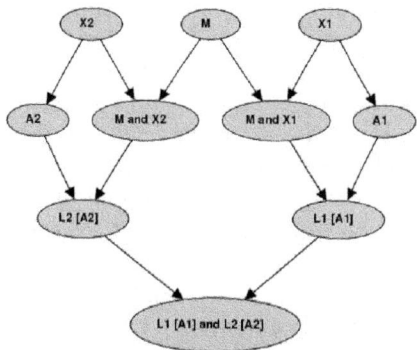

Fig. 12. Diagram of Example 1 with the new rules

4.1   Completeness of Extended Static Semantics

In this section we will extend the soundness and completeness result from [4] to the architectural specification language obtained by adding definitions of generic units to the original fragment language in Section IV.5 of [4], i.e. unit definitions assign to unit names unit expressions instead of unit terms. Comparing with the language in Fig. 1, the differences are that this language does not mix declarations and definitions of units, i.e. all declarations are done locally in the **local** ... **within** construction, unit declarations do not have imports and unit specifications are never architectural. Also we only restrict to lambda expressions with a single parameter. Notice that these differences do not modify the language in an essential way. The soundness and completeness result is formulated as follows.

Theorem 1. *For any architectural specification ASP in which no generic unit is applied more than once we have that ASP has a denotation w.r.t. the extended static semantics iff ASP has a denotation w.r.t. the static semantics and |ASP| has a denotation w.r.t. the partial model semantics.*

The requirement that no generic unit is applied more than once is a simplifying assumption for achieving a *generative* semantics, i.e. repeated applications of a generic unit to same arguments no longer yields the same result.

---

[2] Note that we omitted the nodes of the term of the lambda expression that are copied at each application and only kept the significant ones.

The theorem is proved using a quite technical lemma (Lemma 5.6 in [4]) which we don't present in full detail here. Intuitively, it says that the extended static semantics for a unit term is successful if and only if the static and model semantics are successful as well and if it is the case, the signatures match and the environment obtained in the model semantics can be represented as a family of models compatible with the diagram obtained in the extended static semantics. The proof of this lemma is done by induction on the structure of the unit term. In order to extend the proof to cover lambda expressions as well, we have two new cases to consider: applications of lambda expressions and local declarations of generic units. The new proof is quite long and tedious, but follows very closely the existing proof. Therefore, we only sketch here the proof idea. For applications of lambda expressions, we simply repeat the proof for unit applications but use this time the copies of the nodes for arguments and result that are stored in the context instead of introducing arbitrary distinguished ones. For local declarations of generic units, the proof is similar to the one of local declarations of non-generic units, only that now we have to spell out the rules for lambda expressions before applying the inductive step for the unit term in the lambda expression. The introduced dependency between the lambda expression and its unit term is essential when proving compatibility of the environment with the diagram.

4.2   Parametric Architectural Specifications

Further changes are needed when considering the complete language in Fig. 1. The result unit of an architectural specification ASP can be itself a lambda expression. In this case the architectural specification is called parametric. Notice that the grammar of the architectural language also covers the case when the specification of a unit is itself architectural (either named or anonymous). For such units, we must ensure that designated nodes for formal parameters and result exist in the diagram, since they are required in the rule of unit application of generic units.

Let us first consider the case of anonymous parametric architectural specifications. For the specification below, the static analysis of the architectural specification is currently done in the empty extended static context and thus the nodes for formal parameters and result, which are introduced when making the analysis of the result lambda expression, are no longer present in the diagram at the global level. Notice that the dependency between M and F must be tracked in the diagram in order to ensure correctness of the term $F[M1 \ \textbf{fit} \ t \mapsto u]$ and $F[M2 \ \textbf{fit} \ t \mapsto v]$.

spec S = **sort** s
spec T = **sort** t
spec U = **sort** u
spec V = **sort** v
arch spec ASP2 =
    units
    F : **arch spec** {

units M : S
result $\lambda\ X$: T • M **and** X
};
M1 : U; M2 : V;
result F [M1 **fit** $t \mapsto u$] **and** F [M2 **fit** $t \mapsto v$]

The way we overcome this problem is by making the analysis of the inner architectural specification in the existing global context instead of using an empty global context. After the analysis, we will keep in the global context the diagram resulting from the analysis of the locally-declared units. Thus, the nodes introduced locally become available for further references. Moreover, the units declared locally will not be kept in the global extended context, since we do not want to extend their scope. By making the analysis of the local specification in the global context, the visibility of units declared at the global level is extended to the local context as well (remember that we assumed this extension of visibility in Example 2) and the dependencies of the global units with the local environment are tracked by keeping the entire resulting diagram at the global level.

The second case to consider is the one of unit declarations of form U : **arch spec** ASP, when ASP is a named parametric architectural specification. In this case, ASP cannot refer to units other than those declared within itself and therefore its diagram does not carry any dependency information relevant for the global level. Therefore, instead of adding the diagram of ASP to the global diagram, we only need to introduce new nodes for formal parameters and edges to a new result node. This abstracts away the dependencies of the result node of ASP with the units declared locally (which we don't need) and only keeps the dependencies of the result node with the parameter nodes along the new edges, which will be then copied as diagram of the unit term of the lambda expression at each application of U.

The modifications of the extended static semantics rules are presented in figures 13 to 21 and can be summarized as follows. At the library level, the analysis of an architectural specification (Fig. 13) starts in the empty extended static unit context. The analysis of an architectural specification (Fig. 14), we need to extend the diagram for anonymous parametric architectural specifications (first rule) and named parametric architectural specifications (third rule). In the latter case, we also need to return the (new) nodes for formal parameters and result $(r, p_1, \cdots, p_n)$. The rule for basic architectural specifications (Fig. 15) analyzes the list of declarations and definitions in the context received as parameter rather than in the empty context like before. Thus the diagrams built locally will be added to the global diagram and the visibility of global units is extended. The rule for result unit (Fig. 17) makes a case distinction for each of the four alternatives in Fig. 1. When the specification of the unit is not architectural (first two rules), the imported units are analyzed, a new node p labelled with the signature union of all imports is introduced in the diagram and the dependency between the declared unit and the imports is tracked either via the edge from p to q in the first case, or by storing the node p as node of imports in the second case. When the specification of the unit is a parametric architectural specification (third rule), the nodes of formal parameters and results are saved and the unit will be applied using the rule for lambda expressions. Finally, when the specification of the unit is a non-parametric architectural specification (last rule), we set the pointer for the unit to the node of the result unit of the architectural specification to be able to trace its dependencies. Notice that in the last two cases there are no imports so the node p will always be labeled with the empty signature. The changes made for unit specifications (Figures 18 to 20) are just meant to propagate the results.

$$\frac{\begin{array}{c} \Gamma_S = (G_s, V_s, A_s, T_s) \\ ASN \text{ is a new name} \\ \Gamma_s, C_0 \vdash \texttt{ARCH-SPEC} \triangleright (nodes, A\Sigma, D') \end{array}}{\Gamma_S \vdash \texttt{arch-spec-defn } ASN \texttt{ ARCH-SPEC} \triangleright (G_s, V_s, A_s \cup \{ASN \mapsto A\Sigma\}, T_s)}$$

Fig. 13. Rule for architectural library items

$$\boxed{\Gamma_s, C_s \vdash \texttt{ARCH-SPEC} \triangleright (nodes, A\Sigma, D)}$$

$$\frac{\Gamma_s, C_s \vdash \texttt{BASIC-ARCH-SPEC} \triangleright (nodes, A\Sigma, D')}{\Gamma_s, C_s \vdash \texttt{BASIC-ARCH-SPEC qua ARCH-SPEC} \triangleright (nodes, A\Sigma, D')}$$

$$\frac{\begin{array}{c} ASN \in Dom(A_s) \\ A_s(ASN) = (S, \Sigma) \\ D' \text{ extends } dgm(C_s) \text{ with a new node } n \text{ such that } D'(n) = \Sigma \end{array}}{(G_s, V_s, A_s, T_s), C_s \vdash ASN \texttt{ qua ARCH-SPEC} \triangleright ([n], A_s(ASN), dgm(C_s))}$$

$$\frac{\begin{array}{c} ASN \in Dom(A_s) \\ A_s(ASN) = (S, \langle \Sigma_1, ..., \Sigma_n \rangle \to \Sigma) \\ D' \text{ extends } dgm(C_s) \text{ with new nodes } p_1, .., p_n, r \text{ and edges } p_i \to r \\ \text{such that } D'(p_i \to r) = \iota_{\Sigma_i \subseteq \Sigma} \end{array}}{(G_s, V_s, A_s, T_s), C_s \vdash ASN \texttt{ qua ARCH-SPEC} \triangleright ([r, p_1, ..., p_n], A_s(ASN), D')}$$

Fig. 14. Rules for architectural specifications

$$\boxed{\Gamma_s, C_s \vdash \texttt{BASIC-ARCH-SPEC} \triangleright (nodes, A\Sigma, D)}$$

$$\frac{\begin{array}{c} \Gamma_s, C_s^0 \vdash UDD^+ \triangleright C_s \\ \Gamma_s, C_s \vdash \texttt{RESULT-UNIT} \triangleright (nodes, U\Sigma, D) \end{array}}{\Gamma_s, C_s^0 \vdash \texttt{basic-arch-spec } UDD^+ \texttt{ RESULT-UNIT} \triangleright (nodes, (ctx(C_s), U\Sigma), D)}$$

Fig. 15. New extended static semantics rule for basic architectural specifications

$$\boxed{\Gamma_s, C_s \vdash \text{UNIT-DECL-DEFN}^+ \triangleright C_s'}$$

$$\Gamma_s, C_s^0 \vdash \text{UDD1} \triangleright (C_s)_1$$

$$\dots$$

$$\frac{\Gamma_s, (C_s)_{n-1} \vdash \text{UDDn} \triangleright (C_s)_n}{\Gamma_s, C_s^0 \vdash \text{UDD1}, \dots, \text{UDDn} \triangleright (C_s)_n}$$

Fig. 16. New extended static semantics rule for lists of declarations and definitions

$$\boxed{\Gamma_s, C_s \vdash \text{RESULT-UNIT} \triangleright (nodes, U\Sigma, D)}$$

$$\frac{\Gamma_s, C_s \vdash \text{UNIT-EXPR} \triangleright (p, U\Sigma, D)}{\Gamma_s, C_s \vdash \text{result-unit } \text{UNIT-EXPR} \triangleright ([p], U\Sigma, D)}$$

$$\frac{\Gamma_S, C_s \vdash \text{UNIT-EXPR} \triangleright (r : fs, U\Sigma, D)}{\Gamma_s, C_s \vdash \text{result-unit } \text{UNIT-EXPR} \triangleright (r : fs, U\Sigma, D)}$$

Fig. 17. New extended static semantics rule for result unit expressions

$$\boxed{\Gamma_S, C_s \vdash \text{ARCH-UNIT-SPEC} \triangleright (nodes, U\Sigma, D)}$$

$$\frac{\Gamma_s, C_s \vdash \text{ARCH-SPEC} \triangleright (nodes, (S, U\Sigma), D')}{\Gamma_s, C_s \vdash \text{ARCH-SPEC qua ARCH-UNIT-SPEC} \triangleright (nodes, U\Sigma, D')}$$

Fig. 18. New extended static semantics rule for architectural unit specifications

$$\boxed{\Gamma_s, C_s \vdash \text{unit-defn } UN \text{ UNIT-EXPR} \triangleright C_s'}$$

$$\Gamma_s, C_s \vdash \text{UNIT-EXPR} \triangleright ([p], \Sigma, D)$$
$$UN \text{ is a new name}$$
$$\overline{\Gamma_s, C_s \vdash \text{unit-defn } UN \text{ UNIT-EXPR} \triangleright (\{\}, \{UN \mapsto (p, \Sigma)\}, D)}$$

$$\Gamma_s, C_s \vdash \text{UNIT-EXPR} \triangleright (r : fs, U\Sigma, D)$$
$$UN \text{ is a new name}$$
$$\overline{\Gamma_s, C_s \vdash \text{unit-defn } UN \text{ UNIT-EXPR} \triangleright (\{UN \mapsto (r : fs, U\Sigma)\}, \{\}, D)}$$

Fig. 19. New rule for unit definitions

$$\boxed{\Sigma, \Gamma_s, C_s \vdash \text{UNIT-SPEC} \triangleright (nodes, U\Sigma, D)}$$

$$\frac{\Gamma_s, C_s \vdash \text{ARCH-UNIT-SPEC} \triangleright (nodes, U\Sigma, D')}{\Sigma, \Gamma_s, C_s \vdash \text{ARCH-UNIT-SPEC qua UNIT-SPEC} \triangleright (nodes, U\Sigma, D')}$$

Fig. 20. New extended static semantics rule for arch unit specs as unit specs

5   An Application: Refinement of Units with Imports

This section illustrates the use of the new semantics rules for architectural specifications with the help of a case study example - the specification of a warehouse
system by Baumeister and Bert [1]. The system keeps track of stocks of products and of orders and allows adding, canceling and invoicing orders, as well as
adding products to the stock.

Fig. 22 presents the specifications involved and the relations between them.
The specifications ORDER, PRODUCT and STOCK specify the objects of the
system. The main purpose for the INVOICE specification is to specify an operation for invoicing an order for a product in the stock. The QUEUES and
ORDER_QUEUES specifications specify different types of queues (pending, in-
voiced) for orders. The WHS specification is the top-level specification, with the

$$\boxed{\Gamma_s, C_s \vdash \text{UNIT-DECL} \rhd (C'_s, D)}$$

$$C_s \vdash \text{UNIT-IMPORTED} \rhd (p, D)$$
$$C = C_s + (\{\}, \{\}, D)$$
$$D(p), \Gamma_s, C \vdash \text{UNIT-SPEC} \rhd ([], \Sigma, D')$$
$$UN \text{ is a new name}$$
$$D'' \text{ extends } D' \text{ by a new node } q \text{ with } D''(q) = D'(p) \cup \Sigma$$
$$\text{and edge } e : p \to q \text{ with } D''(e) = \iota_{D'(p) \subseteq D''(q)}$$

$$\overline{\Gamma_s, C_s \vdash \text{unit-decl } UN \text{ UNIT-SPEC UNIT-IMPORTED} \rhd (\{\}, \{UN \mapsto q\}, D'')}$$

$$C_s \vdash \text{UNIT-IMPORTED} \rhd (p, D)$$
$$C = C_s + (\{\}, \{\}, D)$$
$$D(p), \Gamma_s, C \vdash \text{UNIT-SPEC} \rhd ([], \langle \Sigma_1, .., \Sigma_n \rangle \to \Sigma_0, D')$$
$$UN \text{ is a new name}$$

$$\overline{\Gamma_s, C_s \vdash \text{unit-decl } UN \text{ UNIT-SPEC UNIT-IMPORTED} \rhd}$$
$$(\{UN \mapsto (p, \langle \Sigma_1, .., \Sigma_n \rangle \to \Sigma_0 \cup \Sigma^I)\}, \{\}, D')$$

$$C_s \vdash \text{UNIT-IMPORTED} \rhd (p, D)$$
$$C = C_s + (\{\}, \{\}, D)$$
$$D(p), \Gamma_s \vdash \text{UNIT-SPEC} \rhd (r : fp, \langle \Sigma_1, .., \Sigma_n \rangle \to \Sigma, D')$$
$$UN \text{ is a new name}$$

$$\overline{\Gamma_s, C_s \vdash \text{unit-decl } UN \text{ UNIT-SPEC UNIT-IMPORTED} \rhd}$$
$$(\{UN \mapsto (r : fp, \langle \Sigma_1, .., \Sigma_n \rangle \to \Sigma)\}, \{\}, D')$$

$$C_s \vdash \text{UNIT-IMPORTED} \rhd (p, D)$$
$$C = C_s + (\{\}, \{\}, D)$$
$$D(p), \Gamma_s \vdash \text{UNIT-SPEC} \rhd ([n], \Sigma, D')$$
$$UN \text{ is a new name}$$

$$\overline{\Gamma_s, C_s \vdash \text{unit-decl } UN \text{ UNIT-SPEC UNIT-IMPORTED} \rhd}$$
$$(\{\}, \{UN \mapsto ([n], \Sigma)\}, D')$$

Fig. 21. New rules for unit declarations

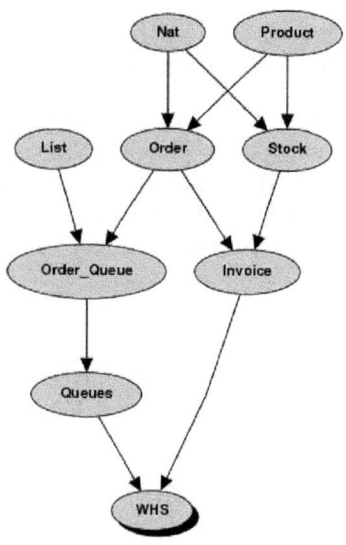

Fig. 22. Structure of the specification of the warehouse system

main operations of the system. The next step is to come up with a more concrete realization of ORDER, that allows to distinguish between different orders on the same quantity of a product by introducing labels. This results in specifications ORDER', INVOICE' and WHS'. The specification WHS' of the warehouse system is then further refined to an architectural specification describing the structure of the implementation of the system. Moreover, NAT and LIST are the usual specifications of natural numbers and lists.

The modular decomposition of the warehouse system is recorded in the architectural specification below:

arch spec WAREHOUSE =
    units NATALG : NAT; PRODUCTALG : PRODUCT;
    ORDERFUN : PRODUCT → ORDER' **given** NATALG;
    ORDERALG = ORDERFUN [PRODUCTALG];
    STOCKFUN : PRODUCT → STOCK **given** NATALG;
    STOCKALG = STOCKFUN [PRODUCTALG];
    INVOICEFUN : {ORDER' **and** STOCK} → INVOICE';
    QUEUESFUN : ORDER → QUEUES;
    WHSFUN : {QUEUES **and** INVOICE'} → WHS'
    result WHSFUN[QUEUESFUN [ORDERALG]
                    and INVOICEFUN [ORDERALG **and** STOCKALG]]

Using the refinement language introduced in [8], we can write this refinement chain in the following way:

refinement R =
        WHS **refined to**
        WHS' **refined to arch spec** WAREHOUSE

We can further proceed to refine each component separately. For example, let us assume we want to further refine ORDER' in such a way that the labels of orders are natural numbers and denote the corresponding specification ORDER".

The changes in the extended static semantics rules allow us to rephrase the declaration of ORDERFUN in an equivalent way using generic units[3]:

ORDERFUN :
        arch spec
        {**units** F : NAT × PRODUCT → ORDER'
        result lambda X : PRODUCT • F [NATALG] [X]
        };

Then we need to write a unit specification for the specification of ORDERFUN to be able to further refine it:

unit spec NATORDER' = NAT × PRODUCT → ORDER'

and another unit specification to store the signature after refinement as well:

unit spec NATORDER" = NAT × PRODUCT → ORDER"

The refinement is done along a morphism that maps the sort *Label* to *Nat*:

refinement R' =
        NATORDER' **refined via** *Label* $\mapsto$ *Nat* **to** NATORDER"

The CASL refinement language can be easily modified to allow the refinement of ORDERFUN without making use of the arbitrary name (in our case F) chosen for the generic unit [4]:

refinement R" = R **then** {ORDERFUN **to** R'}

6   Conclusions

We have presented and discussed a series of changes to extended static semantics of CASL architectural specifications, motivated by the unsatisfactory treatment of lambda expressions in the original semantics of CASL [4]. We have identified a number of practically important situations requiring lambda expressions to have dependency tracking with their unit term and we formulated the modified rules accordingly. We have also discussed briefly how the known completeness result

---

[3] Notice that this equivalence becomes visible at the level of *refinement signatures* as defined in [8].

[4] More exactly, the composition of refinement signatures must be slightly adapted to make this composition legal.

can now be successfully extended to the whole CASL architectural language; a full proof is very lengthy and follows the lines of the existing result; for this reason we have omitted it. Finally, we have presented an example of refinement of generic units with imports; without the changes introduced in this paper such a refinement could not have been expressed using the CASL refinement language. The implementation of the modified rules in the Heterogeneous Tool Set Hets [9] is currently in progress.

Acknowledgments. I would like to thank Till Mossakowski and Lutz Schröder for prompt and detailed comments. I am grateful to Andrzej Tarlecki for suggesting a series of significant technical improvements. This work has been supported by the German Research Council (DFG) under grant MO-971/2.

References

1. Baumeister, H., Bert, D.: Algebraic specification in CASL. In: Frappier, M., Habrias, H. (eds.) Software Specification Methods: An Overview Using a Case Study, ch. 12. FACIT (Formal Approaches to Computing and Information Technology), pp. 209–224. Springer, Heidelberg (2000)
2. Bidoit, M., Mosses, P.D. (eds.): CASL User Manual. LNCS, vol. 2900. Springer, Heidelberg (2004)
3. Bidoit, M., Sannella, D., Tarlecki, A.: Architectural specifications in CASL. Formal Aspects of Computing 13, 252–273 (2002)
4. Mosses, P.D. (ed.): CASL Reference Manual. LNCS, vol. 2960. Springer, Heidelberg (2004)
5. Goguen, J.A., Burstall, R.M.: Institutions: Abstract model theory for specification and programming. Journal of the Association for Computing Machinery 39, 95–146 (1992)
6. Hoffman, P.: Verifying Generative CASL Architectural Specifications. In: Wirsing, M., Pattinson, D., Hennicker, R. (eds.) WADT 2003. LNCS, vol. 2755, pp. 233–252. Springer, Heidelberg (2003)
7. Hoffman, P.: Architectural Specifications and Their Verification. PhD thesis, Warsaw University (2005)
8. Mossakowski, T., Sannella, D., Tarlecki, A.: A Simple Refinement Language for CASL. In: Fiadeiro, J.L., Mosses, P., Orejas, F. (eds.) WADT 2004. LNCS, vol. 3423, pp. 162–185. Springer, Heidelberg (2005)
9. Mossakowski, T., Maeder, C., Lüttich, K.: The Heterogeneous Tool Set, HETS. In: Grumberg, O., Huth, M. (eds.) TACAS 2007. LNCS, vol. 4424, pp. 519–522. Springer, Heidelberg (2007)
10. Sannella, D., Tarlecki, A.: Toward formal development of programs from algebraic specifications: implementations revisited. Acta Informatica 25, 233–281 (1988)
11. Schröder, L., Mossakowski, T., Tarlecki, A., Klin, B., Hoffman, P.: Semantics of Architectural Specifications in CASL. In: Hussmann, H. (ed.) FASE 2001. LNCS, vol. 2029, pp. 253–268. Springer, Heidelberg (2001)

A Proof Theoretic Interpretation of Model Theoretic Hiding

Mihai Codescu[1], Fulya Horozal[2], Michael Kohlhase[2],
Till Mossakowski[1], and Florian Rabe[2]

[1] DFKI Bremen
[2] Jacobs University Bremen

Abstract. Logical frameworks like LF are used for formal representations of logics in order to make them amenable to formal machine-assisted meta-reasoning. While the focus has originally been on logics with a proof theoretic semantics, we have recently shown how to define model theoretic logics in LF as well. We have used this to define new institutions in the Heterogeneous Tool Set in a purely declarative way.

It is desirable to extend this model theoretic representation of logics to the level of structured specifications. Here a particular challenge among structured specification building operations is hiding, which restricts a specification to some export interface. Specification languages like ASL and CASL support hiding, using an institution-independent model theoretic semantics abstracting from the details of the underlying logical system.

Logical frameworks like LF have also been equipped with structuring languages. However, their proof theoretic nature leads them to a theory-level semantics without support for hiding. In the present work, we show how to resolve this difficulty.

1 Introduction

This work is about reconciling the model theoretic approach of algebraic specifications and institutions [AKKB99, ST11, GB92] with the proof theoretic approach of logical frameworks [HHP93, Pau94].

In [Rab10, CHK+10], we show how to represent institutions in logical frameworks, notably LF [HHP93], and extend the Heterogeneous Tool Set [MML07] with a mechanism to add new logics that are specified declaratively in a logical framework.

In the present work, we extend this to the level of structured specifications, including hiding. In particular, we will translate the ASL-style structured specifications with institutional semantics [SW83, Wir86, ST88] (also used in CASL [Mos04]) into the module system MMT [RK10] that has been developed in the logical frameworks community.

Like ASL, MMT is a generic structuring language that is parametric in the underlying language. But where ASL assumes a model theoretic base language – given as an institution – MMT assumes a proof theoretic base language given in

T. Mossakowski and H.-J. Kreowski (Eds.): WADT 2010, LNCS 7137, pp. 118–138, 2012.

terms of typing judgments. If we instantiate MMT with LF (as done in [RS09]), we can represent both logics and theories as MMT-structured LF signatures. This is used in the LATIN project [KMR09] to obtain a large body of structured representations of logics and logic translations. An important practical benefit of MMT is that it is integrated with a scalable knowledge management infrastructure based on OMDoc [Koh06].

However, contrary to model theoretic structuring languages like ASL, structuring languages like MMT for logical frameworks have a proof theoretic semantics and do not support hiding, which makes them less expressive than ASL. Therefore, we proceed in two steps. Firstly, we extend LF+MMT with primitives that support hiding while preserving its proof theoretic flavor. Here we follow and extend the theory-level semantics for hiding given in [GR04]. Secondly, we assume an institution that has been represented in LF, and give a translation of ASL-structured specifications over it into the extended LF+MMT language.

The paper is organized as follows. In Sect. 2, we recall ASL- or CASL-style structured specifications with their institution-independent semantics; and in Sect. 3 we recall LF and MMT with its proof theoretic semantics. In Sect. 4, we extend MMT with hiding, and in Sect. 5, we define a translation of ASL style specifications into MMT and prove its correctness. Sect. 7 concludes the paper.

2   Structured Specifications

Institutions [GB92] have been introduced as a formalization of the notion of logical systems. They abstract from the details of signatures, sentences and models and assume that signatures can be related via signature morphisms (which carries over to sentences and models).

Definition 1. *An* **institution** *is a quadruple* $I = (Sig, Sen, Mod, \models)$ *where:*

- *Sig is a category of* signatures;
- *Sen* : *Sig* → $\mathcal{SET}$ *is a functor to the category* $\mathcal{SET}$ *of small sets and functions, giving for each signature* Σ *its set of* sentences $Sen(\Sigma)$ *and for any signature morphism* $\varphi : \Sigma \to \Sigma'$ *the* sentence translation *function* $Sen(\varphi) : Sen(\Sigma) \to Sen(\Sigma')$ *(denoted also* φ*)*;
- *Mod* : Sig^{op} → *Cat is a functor to the category of categories and functors Cat* [1] *giving for any signature* Σ *its category of models* $Mod(\Sigma)$ *and for any signature morphism* $\varphi : \Sigma \to \Sigma'$ *the* model reduct *functor* $Mod(\varphi) :$ $Mod(\Sigma') \to Mod(\Sigma)$ *(denoted* $\_|_\varphi$*)*;
- *a satisfaction relation* $\models_\Sigma \subseteq |Mod(\Sigma)| \times Sen(\Sigma)$ *for each signature* Σ

such that the following satisfaction condition holds:

$$M'|_\varphi \models_{\Sigma'} e \Leftrightarrow M' \models_\Sigma \varphi(e)$$

for each $M' \in |Mod(\Sigma')|$ *and* $e \in Sen(\Sigma)$*, expressing that truth is invariant under change of notation and context.*

---

[1] We disregard here the foundational issues, but notice however that *Cat* is actually a so-called quasi-category.

For an institution I, a *presentation* is a pair (Σ, E) where Σ is a signature and E is a set of Σ-sentences. For a set E of Σ-sentences, $Mod_\Sigma(E)$ denotes the class of all Σ-models satisfying E. For a class $\mathcal{M}$ of Σ-models, $Th_\Sigma(\mathcal{M})$ denotes the set of all sentences that hold in every model in $\mathcal{M}$, and for a set of Σ-sentences E, we write $Cl_\Sigma(E)$ for $Th_\Sigma(Mod_\Sigma(E))$. Presentations provide the simplest form of specifications, and we refer to them as *basic specifications*.

Working with basic specifications is only suitable for specifications of fairly small size. For practical situations as in the case of large systems, they would become impossible to understand and use efficiently. Moreover, a modular design allows for reuse of specifications. Therefore, algebraic specification languages provide support for structuring specifications.

The semantics of (structured) specifications can be given as a signature and either (i) a class of models of that signature (*model-level semantics*) or (ii) a set of sentences over that signature (*theory-level semantics*). In the presence of structuring, the two semantics may be different in a sense that will be made precise below. The first algebraic specification language, Clear [BG80], used a theory-level semantics; the first algebraic specification language using model-level semantics for structured specifications was ASL [SW83, Wir86], whose structuring mechanisms were extended to an institution-independent level in [ST88].

In Fig. 1, we present a kernel of specification-building operations and their semantics over an arbitrary institution, similar to the one introduced in [ST88]. The second and third columns of the table contain the model-level and the theory-level semantics for the corresponding **structured specification** SP, denoted $Mod[SP]$ and $Thm[SP]$ respectively. The signature of SP, denoted $Sig[SP]$ is defined as follows: (i) $Sig[(\Sigma, E)] = \Sigma$, (ii) $Sig[SP_1 \cup SP_2] = Sig[SP_1]$ $(= Sig[SP_2])$, (iii) $Sig[\sigma(SP)] = \Sigma'$, where Σ' is the target signature of σ and finally (iv) $Sig[\sigma^{-1}(SP)] = \Sigma$, where Σ is the source signature of σ. Note that we restrict attention to hiding against inclusion morphisms. Moreover, we will only consider basic specifications that are finite.

Without hiding, the two semantics can be regarded as dual because we have $Thm[SP] = Th_{Sig[SP]}(Mod[SP])$, which is called *soundness* and *completeness* in [ST11]. But completeness does not hold in general in the presence of hiding

SP	$Mod[SP]$	$Thm[SP]$
(Σ, E) $E \subseteq Sen(\Sigma)$	$Mod_\Sigma(E)$	$Cl_\Sigma(E)$
$SP_1 \cup SP_2$ $Sig[SP_1] = \Sigma$ $Sig[SP_2] = \Sigma$	$Mod(SP_1) \cap Mod(SP_2)$	$Cl_\Sigma(Thm[SP_1] \cup Thm[SP_2])$
$\sigma(SP)$ $\sigma : Sig[SP] \to \Sigma'$	$\{M \in Mod(\Sigma') \mid M\vert_\sigma \in Mod[SP]\}$	$Cl_\Sigma(\{\sigma(e) \mid e \in Thm[SP]\})$
$\sigma^{-1}(SP)$ $\sigma : \Sigma \hookrightarrow Sig[SP]$	$\{M\vert_\sigma \mid M \in Mod[SP]\}$	$\{e \in Sen(\Sigma) \mid \sigma(e) \in Thm[SP]\}$

Fig. 1. Semantics of Structured Specifications

[Bor02]. Moreover, in [ST11] it is proved that this choice for defining the theory level semantics is the strongest possible choice with good structural properties (e.g. compositionality). This shows that the mismatch between theory-level semantics and model-level semantics cannot be bridged in this way. We will argue below that this is not a failure of formalist methods in general; instead, we will pursue a different approach that takes model-level aspects into account while staying mechanizable.

The mismatch between model and theory-level semantics is particularly apparent when looking at **refinements**. For two Σ-specifications SP and SP', we write $SP \leadsto_\Sigma SP'$ if $Mod[SP'] \subseteq Mod[SP]$. Without hiding, this is equivalent to $Thm[SP] \subseteq Thm[SP']$, which can be seen as soundness and completeness properties for refinements. But in the presence of hiding, both soundness (if SP has hiding) and completeness (if SP' has hiding) for refinements may fail.

3 LF and MMT

The Edinburgh Logical Framework LF [HHP93] is a proof theoretic logical framework based on a dependent type theory related to Martin-Löf type theory [ML74]. Precisely, it is the corner of the λ-cube [Bar92] that extends simple type theory with dependent function types. We will also use the notion of LF signature morphisms as given in [HST94]. Moreover, in [RS09], LF was extended with a module system based on MMT. MMT [RK10] is a generic module system which structures signatures using named imports and signature morphisms. The expressivity of MMT is similar to that of ASL or development graphs [AHMS99] except for a lack of hiding. In [Rab10], LF is used as a logical framework to represent both proof and model theory of object logics.

We give a brief summary of basic LF signatures, MMT-structured LF signatures, and the representation of model theory in LF in Sect. 3.1, 3.2, and 3.3, respectively. Our approach is not restricted to LF and can be easily generalized to other frameworks such as Isabelle or Maude along the lines of [CHK+10].

3.1 LF

LF expressions E are grouped into kinds K, kinded type-families $A : K$, and typed terms $t : A$. The kinds are the base kind `type` and the dependent function kinds $\Pi x : A.\, K$. The type families are the constants a, applications $a\, t$, and the dependent function type $\Pi x : A.\, B$; type families of kind `type` are called types. The terms are constants c, applications $t\, t'$, and abstractions $\lambda x : A.\, t$. We write $A \to B$ instead of $\Pi x : A.\, B$ if x does not occur in B. An LF **signature** Σ is a list of kinded type family declarations $a : K$ and typed constant declarations $c : A$. Optionally, declarations may carry definitions. A grammar that subsumes LF is given in Sect. 3.2 below.

Given two signatures Σ and Σ', an LF **signature morphism** $\sigma : \Sigma \to \Sigma'$ is a typing- and kinding-preserving map of Σ-symbols to Σ'-expressions. Thus, σ maps every constant $c : A$ of Σ to a term $\sigma(c) : \overline{\sigma}(A)$ and every type family symbol $a : K$ to a type family $\sigma(a) : \overline{\sigma}(K)$. Here, $\overline{\sigma}$ is the homomorphic extension of σ to Σ-expressions, and we will write σ instead of $\overline{\sigma}$ from now on. Signature morphisms preserve typing and kinding: if $\vdash_\Sigma E : E'$, then $\vdash_{\Sigma'} \sigma(E) : \sigma(E')$.

Composition and identity of signature morphisms are straightforward, and we obtain a category $\mathbb{LF}$ of LF signatures and morphisms. This category has **inclusion** morphisms by taking inclusions between sets of declarations. Moreover, it has **pushouts** along inclusions [HST94]. Finally, a **partial morphism** from Σ to Σ' is a signature morphism from a subsignature of Σ to Σ'. Partiality will only be used in MMT structures below, and we do not need to define a composition of partial morphisms.

LF uses the Curry-Howard correspondence to represent axioms as constants and theorem as defined constants (whose definiens is the proof). Then the typing-preservation of signature morphisms corresponds to the theorem preservation of theory morphisms.

3.2 LF+MMT

The motivation behind the MMT structuring operations is to give a flattenable, concrete syntax for a module system on top of a declarative language. Signature morphisms are used as the main concept to relate and form modular signatures, and signature morphisms can themselves be given in a structured way. Moreover, signature morphisms are always named and can be composed into morphism expressions.

The grammar for the LF+MMT language is given below where $[-]$ denotes optional parts. Object level expressions E unify LF terms, type families, and kinds, and morphism level expressions are composed morphisms:

Signature graph	$G ::= \cdot \mid G,\ \%\mathsf{sig}\,T = \{\Sigma\} \mid \%\mathsf{view}\,v : S \to T = \{\sigma\}$
Signatures	$\Sigma ::= \cdot \mid \Sigma,\ \%\mathsf{struct}\,s : S = \{\sigma\} \mid \Sigma,\ c : E[= E']$
Morphisms	$\sigma ::= \cdot \mid \sigma,\ \%\mathsf{struct}\,s := \mu \mid \sigma,\ c := E$
Object level expr.	$E ::= \mathsf{type} \mid c \mid x \mid E\,E \mid \lambda x : E.\,E \mid \Pi x : E.\,E \mid E \to E$
Morphism level expr.	$\mu ::= \cdot \mid T.s \mid v \mid \mu\,\mu'$

The $\mathbb{LF}$ signatures and signature morphisms are those without the keyword %struct. Those are called **flat**[2].

Syntax. The module level declarations consist of named signatures R, S, T and two kinds of signature morphism declarations. Firstly, **views** $\%\mathsf{view}\,v : S \to T = \{\sigma\}$ occur on toplevel and declare an explicit morphism from S to T given by σ. Secondly, **structures** $\%\mathsf{struct}\,s : S = \{\sigma\}$ occur in the body of a signature T and declare an import from S into T. Structures carry a partial morphism σ from S to T, i.e., σ maps some symbols of S to expressions over T. Views and structures correspond to refinements and inclusion of subspecifications in unions in ASL and CASL.

MMT differs from ASL-like structuring languages in that it uses named imports. Consequently, the syntax of MMT can refer to all paths in the signature

---

[2] Note that in the grammars presented in this paper · stands for the empty entity.

graph using composed morphisms; these morphism level expressions μ are formed from structure names $T.s$, view names v, and diagram-order composition $\mu\,\mu'$.

MMT considers morphisms μ from S to T as expressions on the module level. Such a morphism μ has type S and is valid over T. Most importantly, MMT permits structured morphisms: The morphisms σ occurring in views and structures from S to T may map a structure %struct$\,r : R = \{\sigma\}$ declared in S (i.e., a morphism level constant of type R over S) to a morphism $\mu : R \to T$ (i.e., a morphism level expression of type R over T). These are called structure maps %struct$\,r := \mu$.

Semantics. The semantics of LF+MMT is given by **flattening**. Every well-formed LF+MMT signature graph G is flattened into a diagram $\overline{G}$ over $\mathbb{LF}$. Every signature S in G produces a node $\overline{S}$ in $\overline{G}$; every structure %struct$\,s : S = \{\sigma\}$ occurring in T produces an edge $\overline{T.s}$ from $\overline{S}$ to $\overline{T}$; and every view %view$\,v : S \to T = \{\sigma\}$ produces an edge $\overline{v}$ from $\overline{S}$ to $\overline{T}$. Accordingly, every morphism expression μ yields a morphism $\overline{\mu}$. These results can be found in [RS09], and we will only sketch the central aspects here.

The flattening is defined by recursively replacing all structure declarations and structure maps with lists of flat declarations. To flatten a structure declaration %struct$\,s : S = \{\sigma\}$ in a signature T, assume that S and σ have been flattened already. For every declaration $c : E[= E']$ in $\overline{S}$, we have in $\overline{T}$

- a declaration $s.c : \overline{T.s}(E) = E''$ in $\overline{S}$ if σ contains $c := E''$,
- a declaration $s.c : \overline{T.s}(E)\,[= \overline{T.s}(E')]$ in $\overline{S}$ otherwise.

The morphism $\overline{T.s}$ from $\overline{S}$ to $\overline{T}$ maps every $\overline{S}$-symbol c to the $\overline{T}$ symbol $s.c$.

For a view %view$\,v : S \to T = \{\sigma\}$, the morphism $\overline{v}$ from $\overline{S}$ to $\overline{T}$ is given by the flattening of σ. $\overline{\cdot}$ is the identity morphism in $\mathbb{LF}$, and $\overline{\mu\,\mu'}$ is the composition $\overline{\mu'} \circ \overline{\mu}$.

Finally, morphisms σ from S to T are flattened as follows. To flatten a structure map %struct$\,r := \mu$ where r is a structure from R to S, assume that R has been flattened already. Then the flattening of σ contains $s.c := \overline{\mu}(c)$ for every constant c in $\overline{R}$.

In particular, if %sig$\,T = \{\Sigma, $ %struct$\,s : S = \{\sigma\}\}$ the semantics of signature graphs is such that the left diagram below is a pushout. Here S_0 is a subsignature of $\overline{S}$ such that $\overline{\sigma} : S_0 \to \overline{\Sigma}$. Moreover, if S declares a structure r of type R, then the semantics of a structure map %struct$\,r := \mu$ occurring in σ is that the diagram on the right commutes.

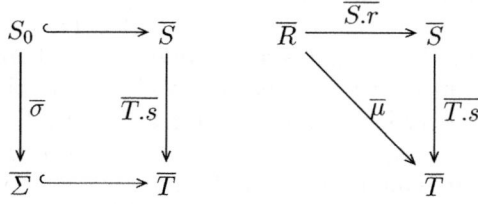

3.3   Representing Logics in LF

LF has been designed for the representation of the proof theory of logics. Recently we showed how this can be extended to representations of the model theory [Rab10]. The key idea is to use a signature Σ^{Mod} that represents models of a logical signature Σ and a signature $\mathcal{F}$ that represents the foundation of mathematics. Then individual Σ-models can be represented as morphisms $\Sigma^{Mod} \to \mathcal{F}$.

The feasibility of this approach has been demonstrated in [IR11], where we give detailed encodings of ZFC set theory, Mizar's set theory, and Isabelle's higher-order logic. Thus, we can choose the right foundation for every individual logic. In the following, it is sufficient to assume a fixed arbitrary signature $\mathcal{F}$. A comprehensive example has been given in [HR11] where we represent first-order logic with a set theoretical foundation. We will use a simplified variant of this methodology in the sequel and give a summary below, using first-order logic as a running example.

The commuting LF diagram on the right presents the representation of a logic L as a tuple $(L^{Syn}, L^{Mod}, L^{mod}, \mathcal{F})$.

L^{Syn} represents the syntax of the logic. We assume that L^{Syn} contains two distinguished declarations o : type and ded : $o \to$ type. For example, for first-order logic, L^{Syn} contains a constant $\wedge : o \to o \to o$ for conjunction along with proof rules for it. Then Σ-sentences are represented as closed $\beta\eta$-normal terms of type o over Σ^{Syn}, and correspondingly Σ-proofs of F as terms of type ded F. Theorems are represented as sentences F for which the type ded F is inhabited over Σ^{Syn}.

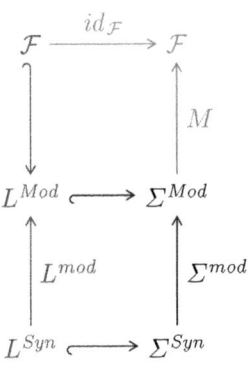

$\mathcal{F}$ represents the foundation of mathematics. In the case of set theory, $\mathcal{F}$ contains in particular symbols set : type, $prop$: type and $true$: $prop \to$ type for sets, propositions, and proofs. Moreover, it contains predicates $\in: set \to set \to prop$ and $eq : set \to set \to prop$ for elementhood and equality between sets.

L^{Mod} extends $\mathcal{F}$ with declarations for all components present in any model. For example, to represent set theoretical first-order models, L^{Mod} declares two symbols: a set $univ$: set for the universe and an axiom making $univ$ non-empty.

L^{mod} interprets the syntax in a model. It represents the fixed part of the interpretation function assigning semantics to the logical symbols. For example, in the case where $\mathcal{F}$ is set theory, $L^{mod}(o)$ could be an $\mathcal{F}$-type representing the set $\{0, 1\}$ of booleans, and $L^{mod}(\text{ded})$ the type family $\lambda x. \, true \, (x \, eq \, 1)$. Then L^{mod} would map $\wedge$ to the $\mathcal{F}$-expression representing the boolean AND function.

Individual signatures Σ are represented as inclusion morphisms $L^{syn} \hookrightarrow \Sigma^{Syn}$. In the case of first-order logic, Σ^{Syn} extends L^{Syn} with declarations for all the function and predicate symbols. Due to the Curry-Howard representation of proofs as terms, there is no conceptual difference between representing signatures and theories of the underlying logic. Therefore, Σ^{Syn} could also declare axioms.

From these, we obtain Σ^{Mod} and Σ^{mod} via a pushout in the category of LF signatures. Thus, Σ^{Mod} arises as the extension of L^{Mod} with declarations for all components present in a model of Σ. In our running example, Σ^{Mod} would declare, e.g., an n-ary function/relation on $univ$ for every n-ary function/predicate symbol declared in Σ^{Syn}.

Then a model M of Σ can finally be represented as a morphism $M : \Sigma^{Mod} \to \mathcal{F}$ such that $M|_{\mathcal{F}} = id_{\mathcal{F}}$. In our running example, $M(univ)$ is the universe of the model, and M maps every non-logical symbol declared in Σ^{Mod} to its interpretation.

Then finally, for a sentence $F : o$ over Σ^{Syn}, the homomorphic extension $M(\Sigma^{mod}(F))$ yields the truth value of F in M. In particular, the satisfaction of F in M is represented as the inhabitation of the type $M(\Sigma^{mod}(\mathsf{ded}\ F))$ over $\mathcal{F}$. If Σ^{Syn} contains axioms, then so does Σ^{Mod}. In that case, M must map each axiom to a proof in $\mathcal{F}$. Thus, the type-preservation property of LF signature morphisms guarantees that all such morphisms indeed yield models.

In [Rab10], the proof theory of the logic is represented in parallel to the model theory as a morphism $L^{pf} : L^{Syn} \to L^{Pf}$ where L^{Pf} adds the proof rules that populate the types $\mathsf{ded}\ F$. Here, we assume for simplicity that $L^{Syn} = L^{Pf}$, and our results easily extend to the general case.

4   Hiding in LF and MMT

In a proof theoretic setting as in LF+MMT, flattening is not a theorem but rather the way to assign meaning to a modular signature. Since hiding precludes flattening, it is a particularly difficult operation to add to systems like LF+MMT.

In this section, we develop an extension of LF+MMT with hiding. The basic idea is to represent signatures with hidden information as inclusions $\Sigma_v \hookrightarrow \Sigma_h$. Intuitively, Σ_v contains all declarations making up the visible interface, and $\Sigma_h \setminus \Sigma_v$ contains declarations for all the hidden operations. Thus, the hidden operations are never removed; instead, they are recorded in Σ_h. That way the hidden operations are still available when defining models. Intuitively, models will be represented as morphisms out of Σ_v that factor through Σ_h.

In the following, we will abstractly introduce LF signatures with hidden declarations and morphisms between such signatures in Sect. 4.1. They will be plain LF-signatures with a distinguished subsignature for the visible declarations. Intuitively, an LF signature with hiding is a plain LF-signature in which some declarations are flagged as hidden. These LF signatures with hiding do not use the module system yet, and we will extend the MMT module system to LF signatures with hiding in Sect. 4.2. Intuitively, LF+MMT with hiding works in exactly the same way as LF+MMT except for keeping track of which declarations are hidden.

4.1   LF with Hiding

We will not only introduce LF signatures with hidden declarations but also LF morphisms that hide constants. It is important to realize that we need hiding morphisms in addition to partial morphisms. Therefore, we introduce H-morphisms, which may have the two orthogonal properties of total/partial and revealing/hiding. Here *revealing* is used for H-morphisms that do not use hiding: The (partial) revealing H-morphisms will be exactly the (partial) $\mathbb{LF}$-morphisms from above.

Given two $\mathbb{LF}$-signatures Σ and Σ', an **H-morphism** from Σ to Σ' consists of two subsignatures $\Sigma_0 \hookrightarrow \Sigma_1 \hookrightarrow \Sigma$ and an $\mathbb{LF}$ signature morphism $\sigma : \Sigma_0 \to \Sigma'$. The intuition is that σ maps all constants in Σ_0 to Σ'-expressions and hides all constants in $\Sigma \setminus \Sigma_1$; for the intermediate declarations in $\Sigma_1 \setminus \Sigma_0$, σ is left undefined, i.e., partial. We call Σ_0 the **revealed domain** and Σ_1 the **non-hidden domain** of σ. We call σ **total** if $\Sigma_1 = \Sigma_0$ and otherwise **partial**; and we call σ **revealing** if $\Sigma = \Sigma_1$ and otherwise **hiding**.

For a Σ-expression E, we say that σ **maps** E if E is a Σ_0-expression and that σ **hides** E if E is not a Σ_1-expression. Then we can define a **composition** of total H-morphisms as follows: The revealed domain of $\sigma' \circ \sigma$ is the largest subsignature of the revealed domain of σ that comprises only constants c such that σ' maps $\sigma(c)$; then we can put $(\sigma' \circ \sigma)(c) = \sigma'(\sigma(c))$. We omit the technical proof that the revealed domain of $\sigma' \circ \sigma$ well-defined.

An **H-signature** is a pair $\Sigma = (\Sigma_v, \Sigma_h)$ such that Σ_v is a subsignature of Σ_h. We call Σ_h the **domain** and Σ_v the **visible domain** of Σ.

Finally, we define the category $\mathbb{LFH}$ whose objects are H-signatures and whose morphisms $(\Sigma_v, \Sigma_h) \to (\Sigma'_v, \Sigma'_h)$ are total H-morphisms from Σ_v to Σ'_v. Note that these morphisms are exactly the total morphisms from Σ_h to Σ'_v whose revealed domain is at most Σ_v. The $\mathbb{LFH}$ identity of (Σ_v, Σ_h) is the $\mathbb{LF}$ identity of Σ_v. Associativity follows after observing that $\sigma'' \circ (\sigma' \circ \sigma)$ hides c iff $(\sigma'' \circ \sigma') \circ \sigma$ hides c.

$\mathbb{LFH}$-morphisms only translate between the visible domains and may even use hiding in doing so. We are often interested in whether the hidden information could also be translated. Therefore, we define:

Definition 2. *For an $\mathbb{LFH}$-morphism $\sigma_0 : \Sigma \to \Sigma'$ with revealed domain Σ_0, we write $\sigma_0 : \Sigma \overset{!}{\to} \Sigma'$ if σ_0 can be extended to a total revealing morphism $\sigma : \Sigma_h \to \Sigma'_h$, i.e., if there is an $\mathbb{LF}$ morphism $\sigma : \Sigma_h \to \Sigma'_h$ that agrees with σ_0 on Σ_0.*

4.2   LF+MMT with Hiding

We can now extend the MMT structuring to $\mathbb{LFH}$, i.e., to a base language with hiding. The flattening of signature graphs with hiding will produce $\mathbb{LFH}$-diagrams.

We avoid using pairs (Σ_v, Σ_h) in the concrete syntax for H-signatures and instead extend the grammar of LF+MMT as follows:

Signatures $\Sigma ::= \cdot \mid \Sigma, [\%\text{hide}] \%\text{struct}\, s : S = \{\sigma\} \mid \Sigma, [\%\text{hide}]\, c : E[= E]$
Morphisms $\sigma ::= \cdot \mid \sigma, \%\text{struct}\, s := \mu \mid \sigma, c := E \mid \%\text{hide}\, c \mid \%\text{hide}\, \%\text{struct}\, s$

If a declaration in Σ has the %hide modifier, we call it **hidden**, otherwise **visible**. Hidden declarations are necessary to keep track of the hidden information. From a proof theoretical perspective, it may appear more natural to delete them, but this would not be adequate to represent ASL specifications with hiding.

If σ contains $c := E$ (or %hide c), we say that σ **maps** (or **hides**) c, and accordingly for structures. As before, we call signatures or morphisms **flat** if they do not contain the %struct keyword.

The semantics of a well-formed signature graph G is given in two steps: first G is flattened into a flat signature graph $\widetilde{G}$, second the semantics of a flat signature graph G is given by an $\mathbb{LFH}$-diagram $\overline{G}$. In particular, every composite μ from S to T occurring in G induces a total H-morphism $\overline{\mu} : \overline{S}_v \to \overline{T}_v$.

Well-formedness and semantics are defined in a joint induction on the structure of G, and only minor adjustments to the definition of $\overline{G}$ for LF+MMT are needed. We begin with the flat syntax.

Firstly, a flat signature %sig $T = \{\Sigma\}$ induces a hiding signature $\overline{T} = (\overline{T}_v, \overline{T}_h)$ as follows: $\overline{T}_h$ contains all declarations in Σ, and $\overline{T}_v$ is the largest subsignature of $\overline{T}_h$ that contains only visible declarations. Σ is well-formed if this is indeed a well-formed $\mathbb{LFH}$-object.

Secondly, consider a flat morphism σ and two flat signatures S and T in G. σ induces an H-morphism from $\overline{S}_v$ to $\overline{T}_v$ as follows: Its revealed domain is the smallest subsignature of $\overline{S}_v$ that contains all constants mapped by σ; its non-hidden domain is the largest subsignature of $\overline{S}_v$ that contains no constants hidden by σ. σ is well-formed if this is indeed a well-formed H-morphism from $\overline{S}_v$ to $\overline{T}_v$.

Next we define the semantics of the full syntax by flattening an arbitrary signature graph G to $\widetilde{G}$. We use the same definition as in [RS09] except for additionally keeping track of which declarations are hidden. In particular, the semantics is unchanged if no declarations are hidden.

Firstly, consider a signature T with a structure %struct $s : S = \{\sigma\}$, and consider a declaration of c in $\widetilde{S}$. Then $\widetilde{T}$ contains a constant $s.c$ defined in the same way as for LF+MMT. Moreover, $s.c$ is hidden in $\widetilde{T}$ if s is hidden in T, c is hidden in S, or $\widetilde{\sigma}$ hides c.

Secondly, consider an occurrence of %struct $s := \mu$ in σ in a structure or view declaration with domain S. Since the semantics $\overline{\mu}$ of μ is a total H-morphism, we must consider two cases for every visible constant c in $\widetilde{S}$: if c is in the revealed domain of $\overline{\mu}$, then $\widetilde{\sigma}$ contains $s.c := \overline{\mu}(c)$ as for LF+MMT; otherwise, $\widetilde{\sigma}$ contains %hide $s.c$.

Thirdly, consider an occurrence of %hide %struct s in σ in a structure or view declaration with domain S. Then $\widetilde{\sigma}$ contains %hide $s.c$ for every visible constant c of $\widetilde{S}$.

Finally, to define well-formedness of signature graphs, we use the same inference system as in [RS09] with the following straightforward restriction for morphisms: In a structure declaration %struct $s : S = \{\sigma\}$ within T or in a view declaration %view $v : S \to T = \{\sigma\}$, $\widetilde{\overline{\sigma}}$ must be an H-morphism from

$\overline{\overline{S}}_v$ to $\overline{\overline{T}}_v$. $\overline{\overline{\sigma}}$ must be total for views and may be partial for structures. Such structures and views induce edges $\overline{T.s}$ and $\overline{v}$ in $\overline{G}$ in the obvious way.

It is easy to show that well-formedness of the flat syntax is decidable. More-over, the following result can be proved by a straightforward induction on the structure of G.

Theorem 3. $\overline{\overline{G}}$ *is a diagram over* $\mathbb{LFH}$ *for every well-formed signature graph* G.

The morphisms σ in structures and views may only map symbols of the visible domain. Moreover, they may hide some of these symbols. However, if we inspect the definition of the flattening of a structure %**struct** $s : S = \{\sigma\}$, we see that it imports all constants of $\overline{S}$ including the hidden ones and including those hidden by σ. Therefore, we have:

Lemma 4. *Assume a well-formed signature graph with hiding* G *containing a structure* %**struct** $s : S = \{\sigma\}$ *in* T. *Then* $\overline{T.s} : \overline{S} \xrightarrow{!} \overline{T}$.

Proof. The extension of $\overline{T.s}$ to $\overline{S}_h$ maps every constant c to s.c.

5   Interpreting ASL in LF+MMT

We now introduce the translation from ASL-style structured specifications into LF+MMT. We assume that there is a representation of an institution I in LF (see Sect. 5.1), such that when translating an ASL-style specification over I (see Sect. 5.2), the resulting MMT specification is based on this representation. The subsequent subsections deal with proving adequacy of the translation.

5.1   Logics

Consider an encoding as in Sect. 3 for an institution I. We make the following assumptions about the adequacy of the encoding.

Definition 5. *We say that a* **foundation** *is* **adequately** *represented by an LF signature* $\mathcal{F}$ *if there is (i) an* $\mathcal{F}$-*type prop* : **type** *such that there is a bijective representation* $\ulcorner - \urcorner$ *of formal statements* F *of the foundation as* $\beta\eta$-*normal* $\mathcal{F}$-*terms* $\ulcorner F \urcorner$: *prop, and (ii) an* $\mathcal{F}$-*type family true* : *prop* $\to$ **type** *such that there is a bijective representation of formal proofs of* F *in the foundation as* $\beta\eta$-*normal* $\mathcal{F}$-*terms of type true* $\ulcorner F \urcorner$.

In the following, we will assume a fixed signature $\mathcal{F}$ that adequately represents the foundation of mathematics, in which the models of our specifications are expressed. For example, in order to represent an institution whose models are defined in terms of Zermelo-Fraenkel set theory, $\mathcal{F}$ can be the signature given in [HR11]; in that case the terms of type *prop* represent first-order formulas over the binary predicate symbol $\in$.

Definition 6. *We say that an* **institution** *I of the form $(Sig, Sen, Mod, \models)$ is* **adequately** *represented as $(L^{Syn}, \mathcal{F}, L^{Mod}, L^{mod})$ if there is a functor $\Phi : Sig \to \mathbb{LF}/L^{Syn}$ such that for every signature Σ (i) $\Phi(\Sigma) = \Sigma^{Syn}$ is an extension of L^{Syn}, (ii) there is a bijection $\ulcorner - \urcorner$ mapping Σ-sentences to $\beta\eta$-normal Σ^{Syn}-terms of type o, and $\ulcorner - \urcorner$ is natural with respect to sentence translation $Sen(\sigma)$ and morphism application $\Phi(\sigma)$ (iii) there is a bijection $\ulcorner - \urcorner$ mapping Σ-models to $\mathbb{LF}$-morphisms $\Sigma^{Mod} \to \mathcal{F}$, and $\ulcorner - \urcorner$ is natural with respect to model reduction $Mod(\sigma)$ and precomposition with $\Phi(\Sigma)^{mod}$, (iv) satisfaction $M \models_\Sigma F$ holds iff $\ulcorner M \urcorner$ maps $\Sigma^{mod}(\ulcorner F \urcorner)$ to an inhabited $\mathcal{F}$-type.*

Using the definitions of [Rab10], this can be stated more concisely as an institution comorphism from I to an appropriate institution based on LF.

Our assumption of a bijection between I-models and $\mathbb{LF}$-morphisms is quite strong. In most cases, not all models will be representable as morphisms. However, using canonical models constructed in completeness proofs, in many cases it will be possible to represent all models up to elementary equivalence.

Moreover, note that the bijection between models also directly implies that the represented institution has amalgamation, provided that we are able to transfer pushouts into the representation:

Theorem 7. *If the functor Φ of an adequate representation of an institution preserves pushouts, the represented institution I has amalgamation.*

5.2   Specifications

We define a translation from ASL specifications to LF+MMT signatures with hiding. Since the ASL structuring is built over an arbitrary institution, we assume that the underlying institution has already been represented in LF and the representation is adequate in the sense of Sect. 3.3 and Def. 6.

For every ASL specification SP over a signature Σ, we define two LF+MMT signatures of the form

$$\%\mathtt{sig}\, N_\Sigma = \{\%\mathtt{struct}\, l : L^{Syn}, \ulcorner\Sigma\urcorner\} \quad \%\mathtt{sig}\, N_{SP} = \{\%\mathtt{struct}\, s : N_\Sigma, \ulcorner SP\urcorner\}$$

$\ulcorner\Sigma\urcorner$ is a list of declarations representing the visible signature symbols of SP, and similarly $\ulcorner SP\urcorner$ represents the hidden signature symbols and all the axioms. There is some flexibility regarding the treatment of axioms: Alternatively, we could distinguish visible and hidden axioms and put the visible ones in $\ulcorner\Sigma\urcorner$. Our choice makes the technical details simpler.

$\ulcorner\Sigma\urcorner$ and $\ulcorner SP\urcorner$ need to refer to the logical symbols of the underlying logic. Therefore, N_Σ starts with an import from L^{Syn}.

It is important to realize that LF+MMT does not use signature expressions in the way ASL uses specification-building operations. In LF+MMT, imports are achieved by declaring structures within the body of a signature, and for efficiency reasons, they may import only named signatures. Therefore, the translation from ASL to LF+MMT must translate nested specification-building operations into multiple named LF+MMT signatures; consequently, it results in an LF+MMT signature graph. We will use N_Σ and N_{SP} to denote the fresh names generated during the translation for the nodes and edges of this signature graph. Note that this leads to an increase in size but not to the exponential blow-up incurred when flattening.

In the following, we will define $\ulcorner\Sigma\urcorner$ and $\ulcorner SP\urcorner$ inductively on the structure of the specification SP. The cases of the translation are given in Fig. 2.

We will describe the translation step by step visualizing the involved objects using diagrams in $\mathbb{LF}$. First we introduce one simplification of the notation. Recall that technically, the semantics $\overline{N_{SP}}$ of N_{SP} is an $\mathbb{LFH}$ object $(\overline{N_{SP_v}}, \overline{N_{SPh}})$ and similarly for $\overline{N_\Sigma} = (\overline{N_{\Sigma v}}, \overline{N_{\Sigma h}})$. A simple induction will show that N_Σ never contains hiding and that $\overline{N_{SP}.s} : \overline{N_{\Sigma v}} = \overline{N_{\Sigma h}} \to \overline{N_{SP_v}}$ is an isomorphism in $\mathbb{LF}$. Therefore, we will always write N_Σ instead of $\overline{N_{\Sigma v}}$, N_{SP} instead of $\overline{N_{SPh}}$, and $N_{SP}.s$ instead of $\overline{N_{SP}.s}$.

The rule *Basic* translates basic specification $SP = (\Sigma, E)$ using the LF representation of the underlying institution. $\ulcorner\Sigma\urcorner$ contains one declaration for every non-logical symbol declared in Σ. For example, if L^{Syn} encodes first-order logic and has a declaration $i : \texttt{type}$ for the universe, a binary predicate symbol p in Σ leads to a declaration $p : l.i \to l.i \to l.o$ in $\ulcorner\Sigma\urcorner$. All axioms $F \in E$, lead to a declaration $a : \texttt{ded}\ulcorner F\urcorner$ where a is a fresh name. This has the effect that axioms are always hidden, which simplifies the notation significantly; it is not harmful because the semantics of ASL does not depend on whether an axiom is hidden or visible.

$$\begin{array}{c} N_\Sigma \\ \Big\downarrow {\scriptstyle N_{SP}.s} \\ N_{SP} \end{array}$$

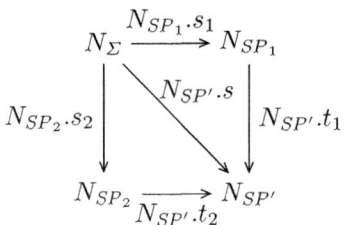

The rule *Union* assumes translations of Σ, SP_1, and SP_2 and creates the translation of $SP' = SP_1 \cup SP_2$ by instantiating N_{SP_1} and N_{SP_2} in such a way that they share N_Σ. The semantics of LF+MMT guarantees that the resulting diagram on the left is a pushout in $\mathbb{LF}$.

The rule *Transl* translates $SP' = \sigma(SP)$ assuming that σ and SP have been translated already. The signature morphism σ is translated to a view in a straightforward way. Recall that N_Σ and $N_{\Sigma'}$ contain no hidden declarations or axioms so that N_σ is a (total) morphism in $\mathbb{LF}$. The resulting diagram is the left diagram below; it is again a pushout in $\mathbb{LF}$.

Similarly, the rule *Hide* translates $SP = \sigma^{-1}(SP')$ assuming that SP has been translated already. As $\sigma : \Sigma \hookrightarrow \Sigma'$ is an inclusion, we only need to know the names c_i of the symbols in Σ and the names h_j of the symbols in $\Sigma' \setminus \Sigma$, which are to be hidden. Then we can form N_{SP} by importing from $N_{SP'}$ and mapping all symbols that remain visible to their counterparts in N_{SP} and hiding the remaining symbols. The resulting diagram is the right diagram below. Note that by Lem. 4, $N_{SP}.t$ extends to a total $\mathbb{LF}$ morphism $N_{SP}.t^*$; moreover, it is easy to verify that $N_{SP}.t^*$ is an isomorphism.

$$SP := (\Sigma, \{F_1, \ldots, F_n\})$$
$$\text{--} \text{ Basic}$$

$\%\text{sig}\, N_\Sigma = \{\%\text{struct}\, l : L^{Syn}, \ulcorner \Sigma \urcorner\}$

$\%\text{sig}\, N_{SP} = \{\%\text{struct}\, s : N_\Sigma,$

$\qquad\qquad \%\text{hide}\, a_1 : \text{ded}\, \ulcorner F_1 \urcorner, \ldots, \%\text{hide}\, a_n : \text{ded}\, \ulcorner F_n \urcorner\}$

$\Sigma = Sig[SP_1] = Sig[SP_2] \quad \%\text{sig}\, N_\Sigma = \{\%\text{struct}\, l : L^{Syn}, \ulcorner \Sigma \urcorner\}$

$SP' := SP_1 \cup SP_2 \qquad\quad \%\text{sig}\, N_{SP_1} = \{\%\text{struct}\, s_1 : N_\Sigma, \ulcorner SP_1 \urcorner\}$

$\qquad\qquad\qquad\qquad\quad\ \%\text{sig}\, N_{SP_2} = \{\%\text{struct}\, s_2 : N_\Sigma, \ulcorner SP_2 \urcorner\}$
$$\text{--} \text{ Union}$$

$\%\text{sig}\, N_{SP'} = \{\%\text{struct}\, s : N_\Sigma,$

$\qquad\qquad\ \%\text{struct}\, t_1 : N_{SP_1} = \{\%\text{struct}\, s_1 := s\},$

$\qquad\qquad\ \%\text{struct}\, t_2 : N_{SP_2} = \{\%\text{struct}\, s_2 := s\}\}$

$\qquad\qquad\ \%\text{sig}\, N_\Sigma = \{\%\text{struct}\, l : L^{Syn}, \ulcorner \Sigma \urcorner\}$

$\sigma : \Sigma \to \Sigma' \qquad \%\text{sig}\, N_{\Sigma'} = \{\%\text{struct}\, l' : L^{Syn}, \ulcorner \Sigma' \urcorner\}$

$SP' := \sigma(SP) \qquad \%\text{sig}\, N_{SP} = \{\%\text{struct}\, s : N_\Sigma, \ulcorner SP \urcorner\}$

$\qquad\qquad\ \%\text{view}\, N_\sigma : N_\Sigma \to N_{\Sigma'} = \{\%\text{struct}\, l := l', \ulcorner \sigma \urcorner\}$
$$\text{--} \text{ Transl}$$

$\%\text{sig}\, N_{SP'} = \{\%\text{struct}\, s' : N_{\Sigma'},$

$\qquad\qquad\ \%\text{struct}\, t : N_{SP} = \{\%\text{struct}\, s := N_\sigma s'\}\}$

$\sigma : \Sigma \hookrightarrow \Sigma'$

$dom(\Sigma) = \{c_1, \ldots, c_m\} \qquad\qquad \%\text{sig}\, N_\Sigma = \{\%\text{struct}\, l : L^{Syn}, \ulcorner \Sigma \urcorner\}$

$dom(\Sigma') \setminus dom(\Sigma) = \{h_1, \ldots, h_n\} \quad \%\text{sig}\, N_{\Sigma'} = \{\%\text{struct}\, l' : L^{Syn}, \ulcorner \Sigma' \urcorner\}$

$SP := \sigma^{-1}(SP') \qquad\qquad\qquad \%\text{sig}\, N_{SP'} = \{\%\text{struct}\, s' : N_{\Sigma'}, \ulcorner SP' \urcorner\}$
$$\text{--} \text{ Hide}$$

$\%\text{sig}\, N_{SP} = \{\%\text{struct}\, s : N_\Sigma,$

$\qquad\qquad\ \%\text{struct}\, t : N_{SP'} = \{\%\text{struct}\, s'.l' := s.l,$

$\qquad\qquad\qquad\qquad\qquad\quad s'.c_1 := s.c_1, \ldots, s'.c_m := s.c_m,$

$\qquad\qquad\qquad\qquad\ \%\text{hide}\, s'.h_1, \ldots, \%\text{hide}\, s'.h_n\}\}$

Fig. 2. Translation of ASL specifications to LF+MMT with Hiding

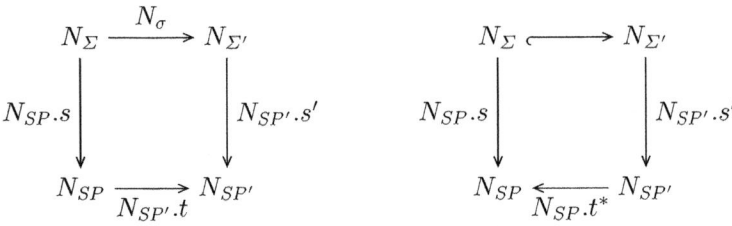

5.3   Adequacy for Specifications

The general idea of the encoding of models is given in Fig. 3. The diagram corresponds to the one from Sect. 3.3 except that we have two extensions of L^{Syn}, namely N_Σ and N_{SP}. Consequently, we obtain two pushouts N_Σ^{Mod} and N_{SP}^{Mod} as shown in the left diagram. Then $(N_{SP}.s)^{mod}$ arises as the unique factorization through the pushout N_Σ^{Mod}.

Our central result will be that models $M \in Mod^I[SP] \subseteq Mod^I(\Sigma)$ can be represented as $\mathbb{LF}$ morphisms $m : N_\Sigma^{Mod} \to \mathcal{F}$ that factor through N_{SP}^{Mod}, i.e., such that there is an m^* with $m^* \circ (N_{SP}.s)^{mod} = m$.

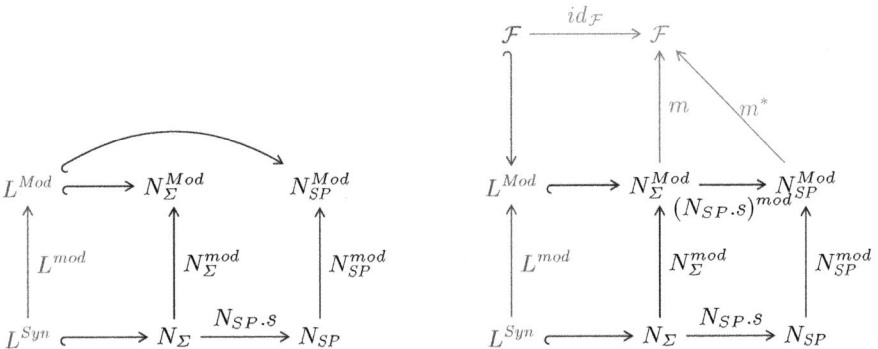

Fig. 3. Representation of Models in the Presence of Hiding

The translation of ASL to MMT yields pushouts between $\mathbb{LF}$ signatures extending L^{Syn}, but models are stated in terms of signatures extending L^{Mod}. Therefore, we use the following simple lemma:

Lemma 8. *Consider the left diagram below and assume that the rectangle is a pushout. For a morphism $L^{mod} : L \to L^{Mod}$, we form the signatures Σ_i^{Mod} as pushouts of Σ_i along L^{mod}; and we form the morphisms σ_i^{mod} as unique factorizations through the respective pushout. Then the rectangle in the resulting diagram on the right is also a pushout.*

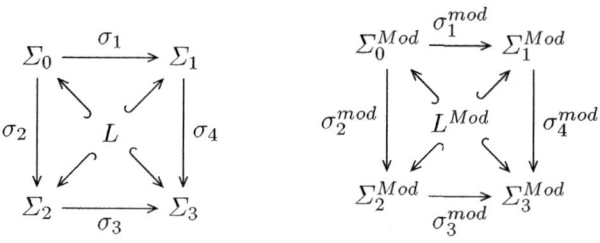

Proof. This is shown with a straightforward diagram chase.

Then we are ready to state our main result:

Theorem 9. *Let I be an institution that is adequately represented in LF. Then for any signature Σ, any ASL-structured specification SP with $Sig[SP] = \Sigma$, and any Σ-model M*

$$M \in Mod^I[SP] \quad \text{iff} \quad \text{exists } m^* : N_{SP}^{Mod} \to \mathcal{F} \text{ such that } (N_{SP}.s)^{mod}; m^* = \ulcorner M \urcorner$$

Proof. The proof is done by induction on the structure of SP. All cases will refer to diagrams that correspond to those in Sect. 5.2, but which refer to the model theory rather than to the syntax – i.e., every node N_X becomes N_X^{Mod}. Lem. 8 ensures that when making this transformation, the pushout properties of the diagrams are kept.

Case $SP = (\Sigma, E)$:
For the base case, the conclusion follows directly from the assumption that the representation of I in LF is adequate.

Case $SP = SP_1 \cup SP_2$:
Let $M \in Mod[SP]$ and $m := \ulcorner M \urcorner : N_{\Sigma}^{Mod} \to \mathcal{F}$. We want to factor m through N_{SP}^{Mod}. By definition, we have that $M \in Mod[SP_1]$ and $M \in Mod[SP_2]$. By the induction hypothesis for SP_1 and SP_2, we get that there are morphisms $m_i : N_{SP_i}^{Mod} \to \mathcal{F}$ such that $m = (N_{SP_i}.s)^{mod}; m_i$. Using the pushout property we get a unique morphism $m^* : N_{SP}^{Mod} \to \mathcal{F}$ such that $(N_{SP_i}.s)^{mod}; (N_{SP'}.t_i)^{mod}; m^* = m$ which gives us the needed factorization.

For the reverse inclusion, let $m := \ulcorner M \urcorner : N_{\Sigma}^{Mod} \to \mathcal{F}$ represent a Σ-model M and factor as $(N_{SP}.s)^{mod}; m^*$. Notice that by composing $(N_{SP}.t_i)^{mod}$ with m^* we get morphisms $m_i : N_{SP_i}^{Mod} \to \mathcal{F}$. By using the induction hypothesis, M is then a model of both SP_1 and SP_2 and by definition M is a model of SP.

Case $SP' = \sigma(SP)$:
Let $M' \in Mod[SP']$ and $m' := \ulcorner M' \urcorner : N_{\Sigma'}^{Mod} \to \mathcal{F}$. We want to prove that there is $m'^* : N_{SP'}^{Mod} \to \mathcal{F}$ such that $m' = (N_{SP'}.s')^{mod}; m'^*$. By definition $M'|_\sigma \in Mod[SP]$. By induction hypothesis for SP' we get a morphism $m := \ulcorner M'|_\sigma \urcorner : N_{\Sigma}^{Mod} \to \mathcal{F}$ and a morphism $m^* : N_{SP}^{Mod} \to \mathcal{F}$ such that $(N_{SP}.s)^{mod}; m^* = m = (N_\sigma)^{mod}; m'$, where the latter equality holds due to the definition of model reduct. Using the pushout property we get the desired m'^*.

For the reverse inclusion, assume $m' := \ulcorner M'^{\urcorner} : N_{\Sigma'}^{Mod} \to \mathcal{F}$ that factors as $(N_{SP'}.s')^{mod}; m'^*$. Then $(N_{SP}.s)^{mod}; (N_{SP'}.t)^{mod}; m'^*$ factors through N_{SP}^{Mod} and thus by induction hypothesis the reduct of M' is an SP-model, which by definition means that M' is an SP'-model.

Case $SP = \sigma^{-1}(SP')$:
Let M be an SP-model and let $m := \ulcorner M^{\urcorner} : N_{\Sigma} \to \mathcal{F}$. We want to prove that m factors through N_{SP}^{Mod}. By definition M has an expansion M' to an SP'-model. By induction hypothesis, there are morphisms $m' := \ulcorner M'^{\urcorner} : N_{\Sigma'}^{Mod} \to \mathcal{F}$ and $m'^* : N_{SP'}^{Mod} \to \mathcal{F}$ such that $(N_{SP'}.s')^{mod}; m'^* = m'$. Then $m = (N_{SP}.s)^{mod}; (N_{SP}.t^{*-1})^{mod}; m'^*$.

For the reverse inclusion, let $m := \ulcorner M^{\urcorner}$ be a morphism that factors as $(N_{SP}.s)^{mod}; m^*$. We need to prove that M has an expansion to a SP'-model. We obtain it by applying the induction hypothesis to $m' := (N_{SP'}.s')^{mod}; (N_{SP}.t^*)^{mod}; m^*$.

Corresponding to the adequacy for models, we can prove the adequacy for theorems by induction on SP, by observing that due to Lem. 4 structures always translate (possibly hidden) theorems to (possibly hidden) theorems.

Theorem 10. *Let I be an institution and assume that I has been represented in LF in an adequate way. Then for any signature Σ, any ASL-structured specification SP with $Sig[SP] = \Sigma$, and any Σ-sentence F*

$$F \in Thm^I[SP] \quad \text{iff} \quad N_{SP}.s(l.\mathsf{ded}\ulcorner F^{\urcorner}) \text{ inhabited over } N_{SP}$$

Note that both in Theorem 9 and Theorem 10 we make use of the fact that LF has model amalgamation for pushouts along injections. The results are thus valid also in the case when the institution I does not have model amalgamation or when the functor Φ of an adequate representation of I does not preserve pushouts.

5.4   Adequacy for Refinements

We want to give a syntactical criterion for refinement $SP \rightsquigarrow_\Sigma SP'$. Consider the diagram on the right. $SP \rightsquigarrow_\Sigma SP'$ states that for all m, if m'^* exists, then some m^* exists such that the diagram commutes. Clearly, this holds if there is an $\mathbb{LF}$ morphism $\rho : N_{SP}^{Mod} \to N_{SP'}^{Mod}$.

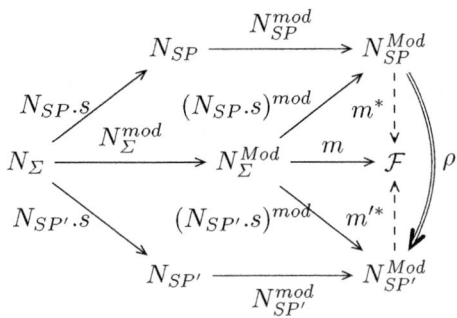

We can also prove the opposite implication if $\mathcal{F}$ has some additional technical properties. Intuitively, $\mathcal{F}$ must be able to represent I-models as $\mathcal{F}$-terms so that we can switch back and forth between models represented as morphisms and models represented as terms. This is the case if $\mathcal{F}$ declares an operation of tupling so that the components of a morphism into $\mathcal{F}$ can be collected in a tuple.

Theorem 11. *Let I be an institution that is adequately encoded in LF. More-over, assume that (i) $\mathcal{F}$ declares an operation of tupling and all I-models of a finite Σ can be represented as $\mathcal{F}$-tuples whose components correspond to the declarations in $\Sigma^{Mod} \setminus L^{Mod}$, and (ii) whenever $\mathcal{F}$ can prove the existence of such a tuple, there is an $\mathcal{F}$-term for such a model.*

Then for ASL-specifications SP and SP' over the signature Σ, we have that $SP \rightsquigarrow_\Sigma SP'$ iff there is an $\mathbb{LF}$ morphism $\rho : N_{SP}^{Mod} \to N_{SP'}^{Mod}$ such that $(N_{SP}.s)^{mod}; \rho = (N_{SP'}.s)^{mod}$.

Proof. The right-to-left implication follows immediately using Thm. 9.

For the left-to-right implication, we assume $SP \rightsquigarrow_\Sigma SP'$ and work within $N_{SP'}^{Mod}$. Due to the adequacy of $\mathcal{F}$, $\ulcorner SP \rightsquigarrow_\Sigma SP' \urcorner$ is a provable statement of $\mathcal{F}$ and thus of $N_{SP'}^{Mod}$ (iii). Using (i), we can tuple the declarations in $N_{SP'}^{Mod} \setminus \mathcal{F}$ (excluding the axioms) and obtain a Σ' model m as a term over $\mathcal{F}$; using the axioms in $N_{SP'}^{Mod} \setminus \mathcal{F}$, we can prove that m' is an SP' model. Using (iii), we show that m is also an SP model. Then using (ii), we obtain an expression m^* over $N_{SP'}^{Mod}$ that expresses a model of $N_{SP'}$. Finally, using (i), we can project out the components of m^*. The morphism ρ maps every symbol of $N_{SP}^{Mod} \setminus \mathcal{F}$ (excluding the axioms) to the corresponding components; it maps all axioms to proofs about m^*.

The assumption (i) of this theorem, while very specific, is mild in practice. In fact, if (i) did not hold, it would be dubious how the foundation can express institutions and models at all. The assumption (ii) is more restricting. For example, it does not hold necessarily for ZF set theory with a choice axiom but without a choice *operator*: without a choice operator, we may be able to prove the existence of a model in $\mathcal{F}$ without being able to turn it into a morphism into $\mathcal{F}$. Alternatively, we can establish (ii) for individual institutions by giving a constructive model existence proof.

6   Related Work

Our use of MMT is akin to the SML-style structure declarations that can occur within Extended ML signatures [KST97]. In fact, Extended ML can be extended with hiding in the same way as we extend LF+MMT, and a similar representation of ASL in Extended ML could be defined. The key advantage of LF+MMT is that LF is strong enough to encode to foundation as well and thus models as morphisms.

Our definition of hiding in LF+MMT is motivated by the approach taken in [GR04]. They also use a pair of two signatures one of which gives the visible interface. The flattening of LF+MMT with hiding corresponds to their semantics.

[BHK90] introduces *normal forms* for ASL-like structured specifications over many sorted classical first-order logic, where the normal form of a structured specification SP has the form $\sigma^{-1}(\Sigma, E)$, such that the normal form has the same model class as the original specification. This has been generalized to an arbitrary institution with weak amalgamation property in [Bor02]. The connection between our work and normal forms is done by the following result, that can be proved by structural induction.

Theorem 12. *Let SP be a structured specification and let us denote $nf(SP)$ its normal form. Then $N_{nf(SP)}$ is a flat H-signature that is isomorphic to the flattening of N_{SP}.*

The intuitive idea is that both the normal form and the hidden signature of N_{SP} hide the same symbols.

7   Conclusion

With the translation presented in this paper, it is possible to encode ASL- and CASL-style structured specifications with hiding in proof theoretic logical frameworks. This provides a new perspective on structured specifications that emphasizes constructive and mechanizable notions. Our translation is given for MMT-structured LF, but it easily generalizes to other MMT-structured logical frameworks.

Our work does not resolve the controversy between the formalistic (or proof-theoretic) and the semantic (or model-theoretic) approach. This, of course, could not be expected as certain aspects of a Platonic foundation are necessarily out of reach for a formalistic treatment. However, we have shown that substantial aspects of the semantic intuitions – here in particular, hiding – can be captured adequately in a formalistic framework.

Our encoding can be generalized to specifications represented as development graphs. In this context, our representation theorem for refinements can be strengthened to represent the hiding theorem links of [MAH06]. Even heterogeneous specifications [MT09, MML07] can be covered: as LF+MMT uses the same structuring operations for logics as for theories, this requires only the representation of the involved logics and logic translations in LF.

A theorem very similar to our representation theorem for refinements can be obtained for conservative extensions. This permits the interpretation of the proof calculus for refinement given in [Bor02]. In particular, the rules using an oracle for conservative extensions can be represented elegantly as the composition of LF signature morphisms.

The translation to MMT also has the benefit that we can re-use the infrastructure provided by languages like OMDoc [Koh06] (an XML-based markup format for mathematical documents) and tools like TNTBase [ZK09] (a versioned XML database for OMDoc documents that supports complex searches and queries, e.g., via XQuery). Further tools developed along these lines are the JOBAD framework (a JavaScript library for interactive mathematical documents), which will provide a web-based frontend for the Heterogeneous Tool Set, GMoc (a change management system), DocTip (a document and tool integration platform) and integration with the Eclipse framework (an integrated development environment).

Acknowledgement. This work has been supported by the German Research Center (DFG) under grant KO-2428/9-1.

References

[AHMS99]   Autexier, S., Hutter, D., Mantel, H., Schairer, A.: Towards an Evolutionary Formal Software-Development Using CASL. In: Bert, D., Choppy, C., Mosses, P.D. (eds.) WADT 1999. LNCS, vol. 1827, pp. 73–88. Springer, Heidelberg (2000)

[AKKB99]   Astesiano, E., Kreowski, H.-J., Krieg-Brückner, B.: Algebraic Foundations of Systems Specification. Springer, Heidelberg (1999)

[Bar92]   Barendregt, H.: Lambda calculi with types. In: Abramsky, S., Gabbay, D., Maibaum, T. (eds.) Handbook of Logic in Computer Science, vol. 2, Oxford University Press (1992)

[BG80]   Burstall, R., Goguen, J.: The semantics of Clear, a specification language. In: Bjorner, D. (ed.) Abstract Software Specifications. LNCS, vol. 86, pp. 292–332. Springer, Heidelberg (1980)

[BHK90]   Bergstra, J.A., Heering, J., Klint, P.: Module algebra. J. ACM 37(2), 335–372 (1990)

[Bor02]   Borzyszkowski, T.: Logical systems for structured specifications. Theor. Comput. Sci. 286(2), 197–245 (2002)

[CHK+10]   Codescu, M., Horozal, F., Kohlhase, M., Mossakowski, T., Rabe, F., Sojakova, K.: Towards Logical Frameworks in the Heterogeneous Tool Set Hets. In: Workshop on Abstract Development Techniques (2010)

[GB92]   Goguen, J., Burstall, R.: Institutions: Abstract model theory for specification and programming. Journal of the Association for Computing Machinery 39(1), 95–146 (1992)

[GR04]   Goguen, J., Rosu, G.: Composing Hidden Information Modules over Inclusive Institutions. In: Owe, O., Krogdahl, S., Lyche, T. (eds.) From Object-Orientation to Formal Methods. LNCS, vol. 2635, pp. 96–123. Springer, Heidelberg (2004)

[HHP93]   Harper, R., Honsell, F., Plotkin, G.: A framework for defining logics. Journal of the Association for Computing Machinery 40(1), 143–184 (1993)

[HR11]   Horozal, F., Rabe, F.: Representing Model Theory in a Type-Theoretical Logical Framework. Theoretical Computer Science (to appear, 2011), http://kwarc.info/frabe/Research/HR_folsound_10.pdf

[HST94]   Harper, R., Sannella, D., Tarlecki, A.: Structured presentations and logic representations. Annals of Pure and Applied Logic 67, 113–160 (1994)

[IR11]   Iancu, M., Rabe, F.: Formalizing Foundations of Mathematics. Mathematical Structures in Computer Science (to appear, 2011), http://kwarc.info/frabe/Research/IR_foundations_10.pdf

[KMR09]   Kohlhase, M., Mossakowski, T., Rabe, F.: The LATIN Project (2009), https://trac.omdoc.org/LATIN/

[Koh06]   Kohlhase, M.: OMDoc – An Open Markup Format for Mathematical Documents (version 1.2). LNCS (LNAI), vol. 4180. Springer, Heidelberg (2006)

[KST97]   Kahrs, S., Sannella, D., Tarlecki, A.: The definition of extended ML: A gentle introduction. Theoretical Computer Science 173(2), 445–484 (1997)

[MAH06]   Mossakowski, T., Autexier, S., Hutter, D.: Development graphs - Proof management for structured specifications. J. Log. Algebr. Program. 67(1-2), 114–145 (2006)

[ML74]   Martin-Löf, P.: An Intuitionistic Theory of Types: Predicative Part. In: Proceedings of the 1973 Logic Colloquium, pp. 73–118. North-Holland (1974)

[MML07]   Mossakowski, T., Maeder, C., Lüttich, K.: The Heterogeneous Tool Set, HETS. In: Grumberg, O., Huth, M. (eds.) TACAS 2007. LNCS, vol. 4424, pp. 519–522. Springer, Heidelberg (2007)

[Mos04]   Mosses, P.D. (ed.): CASL Reference Manual. LNCS, vol. 2960. Springer, Heidelberg (2004)

[MT09]   Mossakowski, T., Tarlecki, A.: Heterogeneous Logical Environments for Distributed Specifications. In: Corradini, A., Montanari, U. (eds.) WADT 2008. LNCS, vol. 5486, pp. 266–289. Springer, Heidelberg (2009)

[Pau94]   Paulson, L.C.: Isabelle: A Generic Theorem Prover. LNCS, vol. 828. Springer, Heidelberg (1994)

[Rab10]   Rabe, F.: A Logical Framework Combining Model and Proof Theory. Submitted to Mathematical Structures in Computer Science (2010), http://kwarc.info/frabe/Research/rabe_combining_09.pdf

[RK10]   Rabe, F., Kohlhase, M.: A Scalable Module System (2010), http://kwarc.info/frabe/Research/mmt.pdf

[RS09]   Rabe, F., Schürmann, C.: A Practical Module System for LF. In: Cheney, J., Felty, A. (eds.) Proceedings of the Workshop on Logical Frameworks: Meta-Theory and Practice (LFMTP), pp. 40–48. ACM Press (2009)

[ST88]   Sannella, D., Tarlecki, A.: Specifications in an arbitrary institution. Information and Computation 76, 165–210 (1988)

[ST11]   Sannella, D., Tarlecki, A.: Foundations of Algebraic Specification and Formal Program Development. Springer, Heidelberg (2011)

[SW83]   Sannella, D., Wirsing, M.: A kernel language for algebraic specification and implementation. In: ADT (1983)

[Wir86]   Wirsing, M.: Structured algebraic specifications: A kernel language. Theor. Comput. Sci. 42, 123–249 (1986)

[ZK09]   Zholudev, V., Kohlhase, M.: TNTBase: a Versioned Storage for XML. In: Proceedings of Balisage: The Markup Conference 2009. Balisage Series on Markup Technologies, vol. 3. Mulberry Technologies, Inc. (2009)

Towards Logical Frameworks
in the Heterogeneous Tool Set Hets

Mihai Codescu[1], Fulya Horozal[2], Michael Kohlhase[2], Till Mossakowski[1],
Florian Rabe[2], and Kristina Sojakova[3]

[1] DFKI GmbH, Bremen, Germany
[2] Computer Science, Jacobs University, Bremen, Germany
[3] Carnegie Mellon University, Pittsburgh, USA

Abstract. LF is a meta-logical framework that has become a standard
tool for representing logics and studying their properties. Its focus is
proof theoretic, employing the Curry-Howard isomorphism: propositions
are represented as types, and proofs as terms.

Hets is an integration tool for logics, logic translations and provers,
with a model theoretic focus, based on the meta-framework of institutions, a formalisation of the notion of logical system.

In this work, we combine these two worlds. The benefit for LF is that
logics represented in LF can be (via Hets) easily connected to various interactive and automated theorem provers, model finders, model checkers,
and conservativity checkers - thus providing much more efficient proof
support than mere proof checking as is done by systems like Twelf. The
benefit for Hets is that (via LF) logics become represented formally, and
hence trustworthiness of the implementation of logics is increased, and
correctness of logic translations can be mechanically verified. Moreover,
since logics and logic translations are now represented declaratively, the
effort of adding new logics or translations to Hets is greatly reduced.

This work is part of a larger effort of building an atlas of logics and
translations used in computer science and mathematics.

1   Introduction

There is a large manifold of different logical systems used in computer science,
such as propositional, first-order, higher-order, modal, description, temporal logics, and many more. These logical systems are supported by software, like e.g.
(semi-)automated theorem provers, model checkers, computer algebra systems,
constraint solvers, or concept classifiers, and each of these software systems
comes with different foundational assumptions and input languages, which makes
them non-interoperable and difficult to compare and evaluate in practice.

There are two main approaches to remedy this situation. The model theoretic approach of *institutions* [GB92, Mes89] provides a formalisation of the
notion of logical system. The benefit is that a large body of meta-theory can
be developed independent of the specific logical system, including specification
languages for structuring large logical theories. Recently, even a good part of

T. Mossakowski and H.-J. Kreowski (Eds.): WADT 2010, LNCS 7137, pp. 139–159, 2012.

model theory has been generalised to this setting [Dia08]. Moreover, the Heterogeneous Tool Set (Hets, [MML07]) provides an institution-independent software interface, such that a heterogeneous proof management involving different tools (as listed above) is practically realised. In Hets, logic translations, formalized as so-called institution comorphisms, become first-class citizens. Heterogeneous specification and proof management is done relative to a graph of logics and translations.

The proof theoretic approach of logical frameworks starts with one "universal" logic that is used as a *logical framework*. This is used for representing logics as theories (in the "universal" logic of the framework). For instance, the Edinburgh Logical Framework LF [HHP93] has been used extensively to represent logics [HST94, PSK+03, AHMP98], many of them included in the Twelf distribution [PS99]. Logic representations in Isabelle [Pau94] are notable for the size of the libraries in the encoded logics, especially for HOL [NPW02]. Logic representations in rewriting logic [MOM94] using the Maude system [CELM96] include the examples of equational logic, Horn logic and linear logic. A notable property of rewriting logic is *reflection* i.e. one can represent rewriting logic within itself. Zermelo-Fraenkel and related set theories were encoded in a number of systems, see, e.g., [PC93] or [TB85]. Other systems employed to encode logics include Coq [BC04], Agda [Nor05], and Nuprl [CAB+86]. Only few logic *translations* have been formalized systematically in this setting. Important translations represented using the logic programming interpretation of LF include cut elimination [Pfe00] and the HOL-Nurpl translation [SS04]. The latter guided the design of the Delphin system [PS08] for logic translations.

Both approaches provide the theoretical and practical infrastructure to define logics. However, there are two major differences. Firstly, Hets is based on model theory – the semantics of implemented logics and the correctness of translations are determined by model theoretic arguments. Proof theory is only used as a tool to discharge proof obligations and is not represented explicitly.

Secondly, the logics of Hets are specified on the meta-level rather than within the system itself. Each logic or logic translation has to be specified by implementing a Haskell interface that is part of the Hets code, and tools for parsing and static analysis have to be provided. Consequently, only Hets developers but not users can add them. Besides the obvious disadvantage of the cost involved when adding logics, this representation does not provide us with a way to reason about the logics or their translations themselves. In particular, each logic's static analysis is part of the trusted code base, and the translations cannot be automatically verified for correctness.

The work reported here is part of the ongoing project LATIN (Logic Atlas and Integrator, [KMR09]). LATIN has two main goals: to *fully integrate proof and model theoretic frameworks* described above preserving their respective advantages, and to create *modular formalizations of commonly used logics* together with *logic morphisms interrelating them*: the **Logic Atlas**. To this end, we develop general definition of a logical framework (the **LATIN metaframework** that covers logical frameworks such as LF, Isabelle, and rewriting logic and implement it in Hets. The LATIN metaframework follows a "logics as theories/translations as morphisms" approach such that a theory graph in a logical framework leads to a graph of institutions and comorphisms via a general construction. This means that new logics can now be added to Hets in a purely declarative way. Moreover, the declarative nature means that logics themselves are no longer only formulated in the semi-formal language of mathematics, but now are fully formal objects, such that one can reason about them (e.g. prove soundness of proof systems or logic translations) within proof systems like Twelf.

This paper is organized as follows. We give introductions to the model and proof theoretic approaches and the LATIN Atlas in Sect. 2. We introduce the LATIN metaframework in Sect. 3 and describe its integration into the Hets system in Sect. 4. We will use an encoding of first-order logic in the logical framework LF as a running example.

2 Preliminaries

2.1 The Heterogeneous Tool Set

The Heterogeneous Tool Set (Hets, [MML07]) is a set of tools for multi-logic specifications, which combines parsers, static analyzers, and theorem provers. Hets provides a heterogeneous specification language built on top of CASL [ABK+02] and uses the development graph calculus [MAH06] as a proof management component. The graph of logics supported by Hets and their translations is presented in Fig. 1.

Hets formalizes the logics and their translations using the abstract model theory notions of institutions and institution comorphisms (see [GB92]).

Definition 1. *An institution is a quadruple $I = (\mathbf{Sig}, \mathbf{Sen}, \mathbf{Mod}, \models)$ where:*

- **Sig** *is a category of* signatures;
- **Sen** : **Sig** $\rightarrow$ *Set is a functor to the category Set of small sets and functions, giving for each signature Σ its set of* sentences **Sen**(Σ) *and for any signature morphism $\varphi : \Sigma \rightarrow \Sigma'$ the sentence translation function* **Sen**$(\varphi) :$ **Sen**$(\Sigma) \rightarrow$ **Sen**(Σ') *(denoted by a slight abuse also φ);*
- **Mod** : **Sig**$^{op} \rightarrow$ *Cat is a functor to the category of categories and functors Cat [1] giving for any signature Σ its category of models* **Mod**(Σ) *and for any signature morphism $\varphi : \Sigma \rightarrow \Sigma'$ the model reduct functor* **Mod**$(\varphi) :$ **Mod**$(\Sigma') \rightarrow$ **Mod**(Σ) *(denoted $\_|_\varphi$);*
- *a satisfaction relation $\models_\Sigma \subseteq |\mathbf{Mod}(\Sigma)| \times \mathbf{Sen}(\Sigma)$ for each signature Σ*

such that the following satisfaction condition holds:

$$M'|_\varphi \models_{\Sigma'} e \Leftrightarrow M' \models_\Sigma \varphi(e)$$

for each $M' \in |\mathbf{Mod}(\Sigma')|$ and $e \in \mathbf{Sen}(\Sigma)$, expressing that truth is invariant under change of notation and context.

---

[1] We disregard here the foundational issues, but notice however that *Cat* is actually a so-called quasi-category.

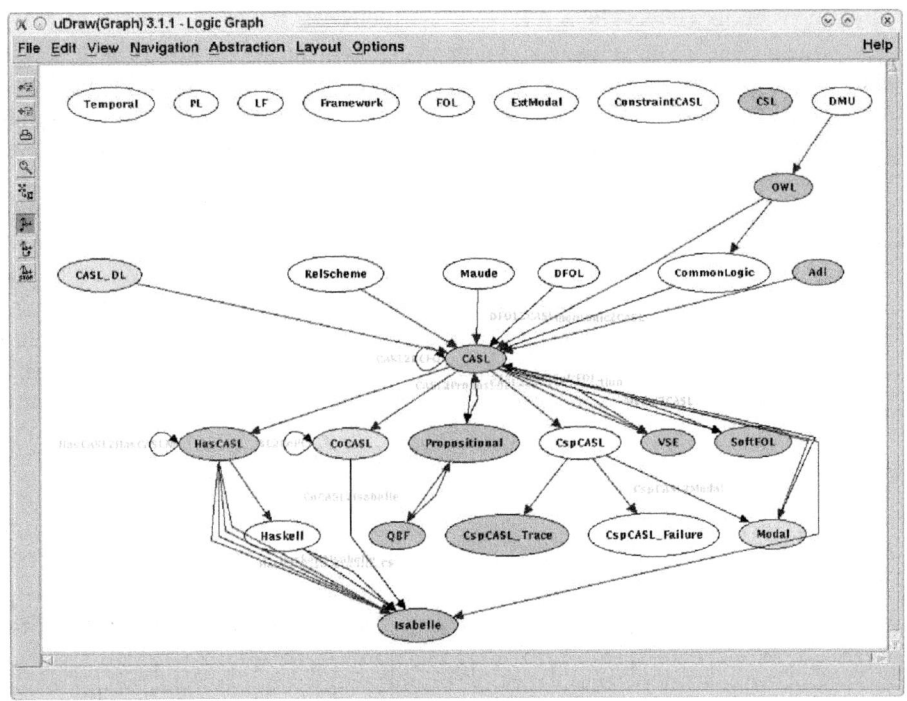

Fig. 1. Hets logic graph

For example, the institution of unsorted first-order logic $\mathbb{FOL}$ has signatures consisting of a set of function symbols and a set of predicate symbols, with their arities. Signature morphisms map symbols such that their arities are preserved. Models are first-order structures, and sentences are first-order formulas. Sentence translation means replacement of the translated symbols. Model reduct means reassembling the model's components according to the signature morphism. Satisfaction is the usual satisfaction of a first-order sentence in a first-order structure.

Definition 2. *Given two institutions I_1, I_2 with $I_i = (\mathbf{Sig}_i, \mathbf{Sen}_i, \mathbf{Mod}_i, \models^i)$, an institution comorphism from I_1 to I_2 consists of a functor $\Phi : \mathbf{Sig}_1 \to \mathbf{Sig}_2$ and natural transformations $\beta : \mathbf{Mod}_2 \circ \Phi \Rightarrow \mathbf{Mod}_1$ and $\alpha : \mathbf{Sen}_1 \Rightarrow \mathbf{Sen}_2 \circ \Phi$, such that the following satisfaction condition holds:*

$$M' \models^2_{\Phi(\Sigma)} \alpha_\Sigma(e) \iff \beta_\Sigma(M') \models^1_\Sigma e,$$

where Σ is an I_1 signature, e is a Σ-sentence in I_1 and M' is a $\Phi(\Sigma)$-model in I_2.

The process of extending Hets with a new logic can be summarized as follows. First, we need to provide Haskell datatypes for the constituents of the logic, e.g. signatures, morphisms and sentences. This is done via instantiating various Haskell type classes, namely *Category* (for the signature category of the institution), *Sentences* (for the sentences), *Syntax* (for abstract syntax of basic specifications, and a parser transforming input text into this abstract syntax), *StaticAnalysis* (for the static analysis, turning basic specifications into theories, where a theory is a signature and a set of sentences). All this is assembled in the type class *Logic*, which additionally provides logic-specific tools like provers and model finders. For displaying the output of model finders, also (finite) models are represented in Hets, and these can even be translated against comorphisms. The model theoretic foundation of Hets also is apparent from the fact that *StaticAnalysis* contains methods for checking amalgamability properties that are defined model theoretically (and therefore not available in purely proof theoretic logical frameworks). The type class *Logic* is used to represent logics in Hets internally. Finally, the new logic is made available by adding it to the list of Hets' known logics. Similarly, Hets represents comorphisms as instances of a type class *Comorphism*, which provides an interface for translating constituents of the source logic to the target logic of the comorphism. Notice that the domain of the translation can be restricted to a certain sublogic of the source using its sublogics hierarchy. Moreover, the methods of the class *Comorphism* include translation of theories, signature morphisms or sentences to the target logic.

The input language of Hets is HetCASL. It combines logic-specific syntax of basic specifications (as specified by an instance of *Syntax*) with the logic-independent structuring constructs of CASL (like extension, union, translation of specifications, or hiding parts). Moreover, there are constructs for choosing a particular logic, as well as for translating a specification along an institution comorphism.

2.2   Proof Theoretic Logical Frameworks

We use the term *proof theoretic* to refer to logical frameworks whose semantics is or can be given in a formal and thus mechanizable way without reference to a Platonic universe. These frameworks are declarative formal languages with an inference system defining a consequence relation between judgments. They come with a notion of language extensions called signatures or theories, which admits the structure of a category. Logic encodings represent the syntax and proof theory of a logic as a theory of the logical framework, and logical consequence is represented in terms of the consequence relation of the framework.

The most important logical frameworks are LF, Isabelle, and rewriting logic. LF [HHP93] is based on dependent type theory; logics are encoded as LF signatures, proofs as terms using the Curry-Howard correspondences, and consequence between formulas as type inhabitation. The main implementation is Twelf [PS99]. The Isabelle system [Pau94] implements higher-order logic [Chu40]; logics are represented as HOL theories, and consequence between formulas as HOL propositions. The Maude system [CELM96] is related to rewriting logic [MOM94]; logics are represented as rewrite theories, and consequence between formulas as rewrite judgments. Other languages such as Coq [BC04] or Agda

[Nor05] can be used as logical frameworks as well, but this is not the primary application encountered in practice.

In the following, we give an overview of **LF**, which we will use as a running example. LF extends simple type theory with dependent function types and is related to Martin-Löf type theory [ML74]. The following grammar is a simplified version of the LF grammar where we write · for the empty list. It includes LF signature morphisms, which were added to LF in [HST94] and added to Twelf in [RS09]:

$$\text{Signatures} \quad \Sigma ::= \cdot \mid \Sigma,\, c : E \mid \Sigma,\, c : E = E$$
$$\text{Morphisms} \quad \sigma ::= \cdot \mid \sigma,\, c := E$$
$$\text{Expressions} \, E ::= \textbf{type} \mid c \mid x \mid E\,E \mid \lambda_{x:E}\,E \mid \Pi_{x:E}\,E \mid E \to E$$

LF **expressions** E are grouped into kinds K, kinded type-families $A : K$, and typed terms $t : A$. The kinds are the base kind **type** and the dependent function kinds $\Pi_{x:A}\,K$. The type families are the constants a, applications $a\,t$, and the dependent function type $\Pi_{x:A}\,B$; type families of kind **type** are called types. The terms are constants c, applications $t\,t'$, and abstractions $\lambda_{x:A}\,t$. We write $A \to B$ instead of $\Pi_{x:A}\,B$ if x does not occur in B.

An LF **signature** Σ is a list of kinded type family declarations $a : K$ and typed constant declarations $c : A$. Both may carry definitions, i.e., $c : A = t$ and $a : K = A$, respectively. Due to the Curry-Howard representation, propositions are encoded as types as well; hence a constant declaration $c : A$ may be regarded as an axiom A, while $c : A = t$ additionally provides a proof t for A. Hence, an LF signature corresponds to what usually is called a logical *theory*.

Relative to a signature Σ, closed expressions are related by the judgments $\vdash_{\Sigma} E : E'$ and $\vdash_{\Sigma} E = E'$. Equality of terms, type families, and kinds are defined by $\alpha\beta\eta$-equality. All judgments for typing, kinding, and equality are decidable.

Given two signatures Σ and Σ', an LF **signature morphism** $\sigma : \Sigma \to \Sigma'$ is a typing- and kinding-preserving map of Σ-symbols to Σ'-expressions. Thus, σ maps every constant $c : A$ of Σ to a term $\sigma(c) : \overline{\sigma}(A)$ and every type family symbol $a : K$ to a type family $\sigma(a) : \overline{\sigma}(K)$. Here, $\overline{\sigma}$ is the homomorphic extension of σ to Σ-expressions, and we will write σ instead of $\overline{\sigma}$ from now on.

Signature morphisms preserve typing, i.e., if $\vdash_{\Sigma} E : E'$, then $\vdash_{\Sigma'} \sigma(E) : \sigma(E')$, and correspondingly for kinding and equality. Due to the Curry-Howard encoding of axioms, this corresponds to theorem preservation of theory morphisms. Composition and identity are defined in the obvious way, and we obtain a category $\mathbb{LF}$.

In [RS09], a **module system** was given for LF and implemented in Twelf. The module system permits to build both signatures and signature morphisms in a structured way. Its expressivity is similar to that of development graphs [AHMS99].

2.3   A Logic Atlas in LF

In the LATIN project [KMR09], we aim at the creation of a logic atlas based on LF. The Logic Atlas is a multi-graph of LF signatures and morphisms between them. Currently it contains formalizations of various logics, type theories, foundations of mathematics, algebra, and category theory.

Among the logics formalized in the Atlas are propositional (PL), first (FOL) and higher-order logic (HOL), sorted ($SFOL$) and dependent first-order logic ($DFOL$), description logics (DL), modal (ML) and common logic (CL) as illustrated in the diagram below. Single arrows ($\rightarrow$) in this diagram denote translations between formalizations and hooked arrows ($\hookrightarrow$) denote imports. Among the foundations are encodings of Zermelo-Fraenkel set theory, Isabelle's higher-order logic, and Mizar's Set theory [IR11].

Note that a logical framework leaves the choice of the foundation deliberately open. In this way, we can use one logical framework (e.g. LF) with several foundations (e.g. ZFC, as well as category theory). Only the representation of a logic includes the choice of a foundation.

Actually the graph is significantly more complex as we use the LF module system to obtain a maximally modular design of logics. For example, first-order, modal, and description logics are formed from orthogonal modules for the individual connectives, quantifiers, and axioms. For example, the $\wedge$ connective is only declared once in the whole Atlas and imported into the various logics and foundations and related to the type theoretic product via the Curry-Howard correspondence.

Moreover, we use individual modules for syntax, proof theory and model theory so that the same syntax can be combined with different interpretations. For example, our formalization of first-order logic (presented in [HR11]) consists of the signatures $Base$ and FOL^{Syn} for syntax, FOL^{Pf} for proof theory, and FOL^{Mod} for model theory as illustrated in the diagram on the right. $Base$ contains declarations $o :$ type and $i :$ type for the type of formulas and first-order individuals, and a truth judgment for formulas. FOL^{Syn} contains declarations for all logical connectives and quantifiers (see Fig. 4). FOL^{truth} is an inclusion morphism from $Base$ to FOL^{Syn}. FOL^{Pf} consists of declarations for judgments and inference rules associated with each logical symbol declared in FOL^{Syn}. FOL^{pf} is simply an inclusion morphism from FOL^{Syn} to FOL^{Pf}.

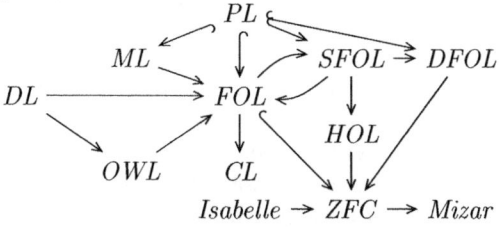

For the representation of FOL model theory, LF is not a suitable metalanguage because its type theory is minimalistic and the use of higher-order abstract syntax is incompatible with the natural way of adding computational support needed to express models. However, LF can serve as a minimal, neutral framework to formalize the metalanguage itself. We choose ZFC set theory as the appropriate metalanguage because it is the standard foundation of mathematics, and formalize it in LF (in the signature ZFC) and use it as the metalanguage to define models.

The ZFC encoding includes the type of sets, the membership predicate as a primitive non-logical symbol, and the usual ZFC set operations and axioms defined in a first-order language with description operator. Additionally, ZFC contains a type judgment $elem$ for the elements of a set as well as a binary operation $\implies$ on sets that returns the set of functions. This is important for being able to represent models as signature morphisms (see below): signature morphisms map types to types, and via $elem$, (carrier) sets can be turned into types.

FOL^{Mod} includes ZFC as a metalanguage and uses it to axiomatize the properties of FOL-models. More precisely, FOL^{Mod} declares a set $bool$ for the boolean values axiomatizing it to get the desired 2-element set $\{0, 1\}$, declares a fixed set $univ$ of individuals, along with an axiom stating that the universe is nonempty. For each logical symbol s^{Syn} in FOL^{Syn}, FOL^{Mod} declares a symbol s^{Mod} that represents the semantic operation used to interpret s^{Syn} along with axioms specifying its truth values. For instance, for disjunction, which is declared as $or : o \to o \to o$ in FOL^{Syn}, FOL^{Mod} declares the symbol $\vee$ as a ZFC-function from $bool^2$ to $bool$ and axiomatizes it to be the binary supremum in the boolean 2-element lattice. This corresponds to the case-based definition of the semantics of a formula.

FOL^{Syn}	FOL^{Mod}	ZFC
i : type	$univ : set$	set : type
o : type	$bool : set$	$prop$: type
$or : o \to o \to o$	$\vee : elem\,(bool \implies bool \implies bool)$	$\vee : prop \to prop \to prop$
$forall : (i \to o) \to o$	$\forall\ :\ elem\,((univ \implies bool) \implies bool)$	$\forall : (set \to prop) \to prop$
		$\in : set \to set \to prop$
		$elem : set \to$ type
		$\implies : set \to set \to set$

The morphism FOL^{mod} interprets the syntax of FOL in the semantic realm specified by FOL^{Mod}: It maps the type i of individuals to the type of elements of $univ$, the type o of formulas to the type of elements of $bool$, and the logical operations to the corresponding operations on booleans.

The individual FOL-models are represented as LF signature morphisms from FOL^{Mod} to ZFC that are the identity on ZFC. In other words, a model M maps $univ$ to a nonempty set expressed by using the set operations of ZFC. M interprets the boolean operations in FOL^{Mod} in terms of the usual set operations in ZFC. For instance, the universal quantification for the booleans is mapped to the intersection of a family of subsets. Given such a morphism M, the composition FOL^{mod} ; M then yields the interpretation of FOL^{Syn} in ZFC.

A particular aspect of our formalization is that soundness of FOL can be represented naturally as an LF signature morphism from FOL^{Pf} to FOL^{Mod} making the diagram above commute. Note that a morphisms in the opposite direction, i.e., from FOL^{Mod} to FOL^{Pf}, does not yield completeness.

3 The LATIN Metaframework

In this section we describe the theoretical background of our LATIN metaframework (LMF) based on the approach taken in [Rab10]. The LMF is an abstract framework that allows to represent logical frameworks as declarative languages given by categories of theories. The LMF is generic in the sense that it can be instantiated with specific logical frameworks such as LF, Isabelle or rewriting logic, thus allowing Hets to be flexible in the choice of the logical framework in which logics should be represented.

In Sect. 3.1, we show that our abstract representation of logical frameworks complies with the notion of institutions and institution comorphisms. Here we deliberately restrict attention to a special case of [Rab10] that makes the ideas clearest and discuss generalizations in Sect. 3.2.

3.1 Main Definition

Definition 3 (Inclusions). *A category with inclusions consists of a category together with a broad subcategory that is a partial order. We write $B \hookrightarrow C$ for the inclusion morphism from B to C.*

Definition 4 (Logical Framework). *A tuple $(\mathbb{C}, Base, \mathbf{Sen}, \vdash)$ is a logical framework if*

- *$\mathbb{C}$ is a category that has inclusions and pushouts along inclusions,*
- *$Base$ is an object of $\mathbb{C}$,*
- *$\mathbf{Sen} : \mathbb{C} \backslash Base \to Set$ is a functor, where $\mathbb{C} \backslash Base$ is the so-called slice category of $\mathbb{C}$ over $Base$, whose objects are arrows in $\mathbb{C}$ of source $Base$ and morphisms make triangles commute,*
- *for $t \in \mathbb{C} \backslash Base$, $\vdash_t$ is a unary predicate on $\mathbf{Sen}(t)$,*
- *$\vdash$ is preserved under signature morphisms: if $\vdash_t F$ then $\vdash_{t'} \mathbf{Sen}(\sigma)(F)$ for any morphism $\sigma : t \to t'$ in $\mathbb{C} \backslash Base$.*

$\mathbb{C}$ is the category of theories of the logical framework. Our focus is on declarative frameworks where theories are lists of named declarations. Typically these have inclusions and pushouts along them in a natural way.

Logics are encoded as theories Σ of the framework, but not all theories can be naturally regarded as logic encodings. Logic encodings must additionally distinguish certain objects over Σ that encode logical notions. Therefore, we consider $\mathbb{C}$-morphisms $t : Base \to \Sigma$ where $Base$ makes precise what objects must be distinguished.

We leave the structure of *Base* abstract, but we require that slices $t : Base \rightarrow \Sigma$ provide at least a notion of sentences and truth for the logic encoded by Σ. Therefore, $\mathbf{Sen}(t)$ gives the set of sentences, and the predicate $\vdash_t F$ expresses the truth of F.

Example 1 (LF). We define a logical framework $\mathbb{F}^{LF}$ based on the category $\mathbb{C} = $ $\mathbb{LF}$. $\mathbb{LF}$ has inclusions by taking the subset relation between sets of declarations. Given $\sigma : \Sigma \rightarrow \Sigma'$ and an inclusion $\Sigma \hookrightarrow \Sigma, c : A$, a pushout is given by

$$(\sigma, \ c := c) \ : \ (\Sigma, \ c : A) \ \rightarrow \ (\Sigma', \ c : \sigma(A))$$

(except for possibly renaming c if it is not fresh for Σ'). The pushouts for other inclusions are obtained accordingly.

   Base is the signature with the declarations o : type and $ded : o \rightarrow$ type. For every slice $t : Base \rightarrow \Sigma$, we define $\mathbf{Sen}(t)$ as the set of closed $\beta\eta$-normal LF-terms of type $t(o)$ over the signature Σ. Moreover, $\vdash_t F$ holds iff the Σ-type $t(ded)$ F is inhabited.

   Given $t : Base \rightarrow \Sigma$ and $t' : Base \rightarrow \Sigma'$ and $\sigma : \Sigma \rightarrow \Sigma'$ such that $\sigma \circ t = t'$, we define the sentence translation by $\mathbf{Sen}(\sigma)(F) = \sigma(F)$. Truth is preserved: assume $\vdash_t F$; thus $t(ded)$ F is inhabited over Σ; then $\sigma(t(ded)$ $F) = t'(ded)$ $\sigma(F)$ is inhabited over Σ'; thus $\vdash_{t'} \mathbf{Sen}(\sigma)(F)$.

Example 2 (Isabelle). A logical framework based on Isabelle is defined similarly. $\mathbb{C}$ is the category of Isabelle theories and theory morphisms (for the latter, see [BJL06]). *Base* consists of the declarations $bool$: type and $\mathtt{trueprop} : bool \rightarrow$ prop where prop is the type of Isabelle propositions. Given $t : Base \rightarrow \Sigma$, we define $\mathbf{Sen}(t)$ as the set of Σ-terms of type $t(bool)$, and $\vdash_t F$ holds if $t(\mathtt{trueprop})$ F is an Isabelle theorem over Σ.

Example 3 (Rewriting logic). A logical framework based on rewriting logic can be defined along the lines of [MOM94]. $\mathbb{C}$ is the category of rewriting logic theories and theory morphisms. *Base* consists of the following declarations:

```
sorts Prop FormList Sequent .
subsorts Prop < FormList .
op empty : -> FormList .
op tt : -> Prop .
op __⊢__ : FormList FormList -> Sequent .
```

where Prop stands for the type of propositions, tt for the formula *true*, and $\vdash$ turns two lists of formulas into a sequent. Given $t : Base \rightarrow \Sigma$, we define $\mathbf{Sen}(t)$ as the set of Σ-terms of type $t(\mathtt{Prop})$, and $\vdash_t F$ holds for some term F of type $t(\mathtt{Prop})$ if $\mathtt{empty} \vdash F \Rightarrow_\Sigma \mathtt{empty} \vdash \mathtt{tt}$. $\vdash_t$ is preserved by rewriting logic theory morphisms because rewriting must be preserved.

We use logical frameworks to define institutions. The basic idea is that slices $t : Base \rightarrow L^{Syn}$ define logics (L^{Syn} specifies the syntax of the logic), signatures of that logic are extensions $L^{Syn} \hookrightarrow \Sigma^{Syn}$, and sentences and truth are given by $\mathbf{Sen}$ and $\vdash$. We could represent the logic's models in terms of the models of the logical framework, but that would complicate the mechanizable representation of models. Therefore, we represent models as $\mathbb{C}$ morphisms into a fixed theory that represents the foundation of mathematics. We need one auxiliary definition to state this precisely:

Definition 5. *Fix a logical framework, and assume* $L^{mod} : L^{Syn} \rightarrow L^{Mod}$ *in* $\mathbb{C}$ *as in the diagram below.*

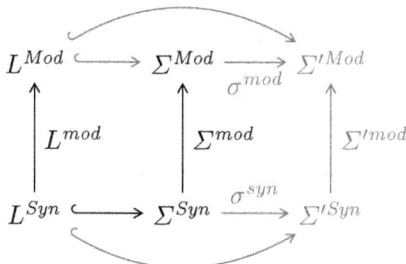

Firstly, for every inclusion $L^{Syn} \hookrightarrow \Sigma^{Syn}$*, we define* Σ^{Mod} *and* Σ^{mod} *such that* Σ^{Mod} *is a pushout. Secondly, for every* $\sigma^{syn} : \Sigma^{Syn} \rightarrow \Sigma'^{Syn}$*, we define* $\sigma^{mod} : \Sigma^{Mod} \rightarrow \Sigma'^{Mod}$ *as the unique morphism such that the above diagram commutes.*

Then we are ready for our main definition:

Definition 6 (Institutions in LMF). *Let* $\mathbb{F} = (\mathbb{C}, Base, \mathbf{Sen}, \vdash)$ *be a logical framework. Assume* $L = (L^{Syn}, L^{truth}, L^{Mod}, \mathcal{F}, L^{mod})$ *as in the following diagram:*

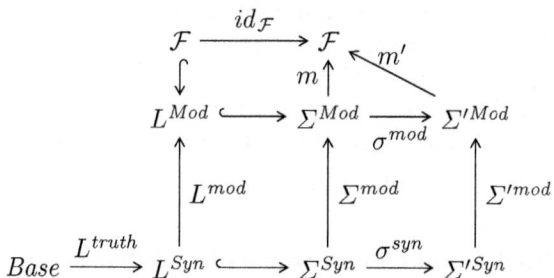

Then we define the institution $\mathbb{F}(L) = (\mathbf{Sig}^L, \mathbf{Sen}^L, \mathbf{Mod}^L, \models^L)$ *as follows:*

- $\mathbf{Sig}^L$ *is the full subcategory of* $\mathbb{C}\backslash L^{Syn}$ *whose objects are inclusions. To simplify the notation, we will write* Σ^{Syn} *for an inclusion* $L^{Syn} \hookrightarrow \Sigma^{Syn}$ *below.*
- $\mathbf{Sen}^L$ *is defined by*

$$\mathbf{Sen}^L(\Sigma^{Syn}) = \mathbf{Sen}((L^{Syn} \hookrightarrow \Sigma^{Syn}) \circ L^{truth}) \quad \text{and} \quad \mathbf{Sen}^L(\sigma) = \mathbf{Sen}(\sigma).$$

- $\mathbf{Mod}^L$ *is defined by*

$$\mathbf{Mod}^L(\Sigma^{Syn}) = \{m : \Sigma^{Mod} \rightarrow \mathcal{F} \mid m \circ (\mathcal{F} \hookrightarrow \Sigma^{Mod}) = id_\mathcal{F}\}$$
$$\mathbf{Mod}^L(\sigma^{syn})(m') = m' \circ \sigma^{mod}.$$

All model categories are discrete.

- We make the following abbreviation: For a model $m \in \mathbf{Mod}^L(\Sigma^{Syn})$, we write $\overline{m}$ for $m \circ \Sigma^{mod} \circ (L^{Syn} \hookrightarrow \Sigma^{Syn}) \circ L^{truth} : Base \to \mathcal{F}$. Then we define satisfaction by

$$m \models^L_{\Sigma^{Syn}} F \quad \text{iff} \quad \vdash_{\overline{m}} \mathbf{Sen}(m \circ \Sigma^{mod})(F).$$

Theorem 1 (Institutions in LMF). *In the situation of Def. 6, $\mathbb{F}(L)$ is an institution.*

Proof. We need to show the satisfaction condition. So assume $\sigma^{syn} : \Sigma^{Syn} \to \Sigma'^{Syn}$, $F \in \mathbf{Sen}^L(\Sigma^{Syn})$, and $m' \in \mathbf{Mod}^L(\Sigma'^{Syn})$. First observe that $\overline{m'} = m' \circ \Sigma'^{mod} \circ (L^{Syn} \hookrightarrow \Sigma'^{Syn}) \circ L^{truth} = (m' \circ \sigma^{mod}) \circ \Sigma^{mod} \circ (L^{Syn} \hookrightarrow \Sigma^{Syn}) \circ L^{truth} = \overline{m' \circ \sigma^{mod}}$. Then $\mathbf{Mod}^L(\sigma)(m') \models^L_{\Sigma^{Syn}} F$ iff $\vdash_{\overline{m' \circ \sigma^{mod}}} \mathbf{Sen}((m' \circ \sigma^{mod}) \circ \Sigma^{mod})(F)$ iff $\vdash_{\overline{m'}} \mathbf{Sen}(m' \circ \Sigma'^{mod})(\mathbf{Sen}(\sigma^{syn})(F))$ iff $m' \models^L_{\Sigma'^{Syn}} \mathbf{Sen}^L(\sigma^{syn})(F)$.

Example 4 (FOL). We can now obtain an institution from the encoding of first-order logic in Sect. 2.3 based on the logical framework $\mathbb{F}^{LF}$. First-order logic is encoded as the tuple $FOL = (FOL^{Syn}, FOL^{truth}, FOL^{Mod}, ZFC, FOL^{mod})$ as in Sect. 2.3.

We obtain an institution comorphism $\mathbb{FOL} \to \mathbb{F}^{LF}(FOL)$ as follows. Signatures of $\mathbb{FOL}$ are mapped to the extension of FOL^{Syn} with declarations $f : i \to \ldots \to i \to i$ for function symbols f, $p : i \to \ldots \to i \to o$ for predicate symbols p. If we want to map $\mathbb{FOL}$ theories as well, we add declarations $ax : ded\ F$ for every axiom F. Signature morphisms are mapped in the obvious way. The sentence translation is an obvious bijection. The model translation maps every $m : \Sigma^{Mod} \to \mathcal{F}$ to the model whose universe is given by $m(univ)$ and which interprets symbols f and p according to $m(f)$ and $m(p)$. The model translation is not surjective as there are only countably many morphisms m in $\mathbb{F}^{LF}(FOL)$. However, since $\mathbb{FOL}$ has a constructive existence proof of canonical models, these models can be represented as ZFC terms and are in the image of the model translation. The satisfaction condition can be proved by an easy induction. $\mathbb{F}^{LF}(FOL)$ is complete thus $\mathbb{FOL}$ and $\mathbb{F}^{LF}(FOL)$ have the same consequence relation.

Logical frameworks can also be used to encode institution comorphisms in an intuitive way:

Theorem 2 (Institution Comorphisms in LMF). *Fix a logical framework $\mathbb{F} = (\mathbb{C}, Base, \mathbf{Sen}, \vdash)$. Assume two logics $L = (L^{Syn}, L^{truth}, L^{Mod}, \mathcal{F}, L^{mod})$ and $L' = (L'^{Syn}, L'^{truth}, L'^{Mod}, \mathcal{F}, L'^{mod})$. Then a comorphism $\mathbb{F}(L) \to \mathbb{F}(L')$ is induced by morphisms (l^{syn}, l^{mod}) if the following diagram commutes*

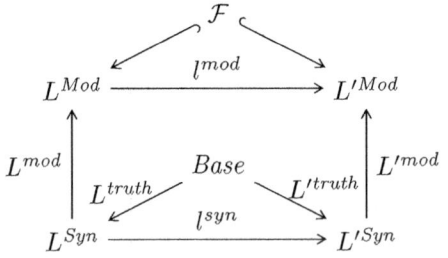

Proof. A signature $L^{Syn} \hookrightarrow \Sigma^{Syn}$ is translated to $L'^{Syn} \hookrightarrow \Sigma'^{Syn}$ by pushout along l^{syn} yielding $\sigma^{syn} : \Sigma^{Syn} \to \Sigma'^{Syn}$. Sentences are translated by applying σ^{syn}. We obtain $\sigma^{mod} : \Sigma^{Mod} \to \Sigma'^{Mod}$ as the unique morphism through the pushout Σ^{Mod}. Then models are translated by composition with σ^{mod}. We omit the details.

It is easy to see that comorphisms that are embeddings can be elegantly represented in this way, as well as many inductively defined encodings. However, the assumptions of this theorem are too strong to permit the encoding of some less trivial comorphisms. For example, non-compositional sentence translations, which come up when translating modal logic to first-order logic, cannot be represented as signature morphisms. Or signature translations that do not preserve the number of non-logical symbols, which come up when translating partial to total function symbols, often cannot be represented as pushouts. More general constructions for the special case of LF are given in [Rab10] and [Soj10].

3.2   Generalizations

In Ex. 4, we do not obtain a comorphism in the opposite direction. There are three reasons for that. Firstly, $\mathbb{F}^{LF}(FOL)$ contains a lot more signatures than needed because the definition of $\mathbf{Sig}^L$ permits any extension of L^{Syn}, not just the ones corresponding to function and predicate symbols. Secondly, the discrete model categories of $\mathbb{F}^{LF}(FOL)$ cannot represent the model morphisms of $\mathbb{FOL}$. Thirdly, only a (countable) subclass of the models of $\mathbb{FOL}$ can be represented as $\mathbb{LF}$ morphisms. Moreover, Def. 4 and 6 are restricted to institutions, i.e., the syntax and model theory of a logic, and exclude the proof theory. We look at these problems below.

Signatures. In order to solve the first problem we need to restrict $\mathbb{F}(L)$ to a subcategory of $\mathbf{Sig}^L$. However, it is difficult to single out the needed subcategory in a mechanizable way. Therefore, we restrict attention to those logical frameworks where $\mathbb{C}$ is the category of theories of a declarative language.

In a declarative language, the theories are given by a list of typed symbol declarations. In order to formalize this definition without committing to a type system, we use MMT expressions ([Rab08]) as the types. MMT expressions are formed from variables, constants, applications $@(E, l)$ of an expression E to a list of expressions l, bindings $\beta(E, l, E')$ of a binder E with scope E' binding a list of variables typed by the elements of l. To that we add jokers $*$, which matches an arbitrary expressions, and $\overline{E}$, which matches a list of expressions each of which matches E.

Such MMT expression patterns give us a generic way to pattern-match declarations of the logical framework. If a concrete logic definition contains a set P of patterns, we represent its logical signatures as $\mathbb{C}$-objects Σ^{Syn} that extend L^{Syn} only with declarations matching one of the patterns in P. For example, the patterns for first-order logic from Ex. 4 would be $@(\to, \overline{i}, i)$ and $@(\to, \overline{i}, o)$ for function and predicate symbols of arbitrary arity, and $@(ded, *)$ for axioms.

Here $*$ stands for an arbitrary expression, which in this case must be a sentence to be well-typed.

Model Morphisms. Regarding the second problem, if $\mathbb{C}$ is a 2-category, we can define the model morphisms of $\mathbb{F}(L)$ as 2-cells in $\mathbb{C}$. However it is difficult in practice to obtain 2-categories for type theories such as LF or Isabelle. In [Soj10], we give a syntactical account of logical relations that behave like 2-cells in sufficiently many ways to yield model morphisms.

Undefinable Models. The third problem is the most fundamental one because no formal logical framework can ever encode all models of a Platonic universe. Our encoding of ZFC is strong enough to encode any **definable** model. We call a model M definable if it arises as the solution to a formula $\exists^! M.F(M)$ for some parameter-free formula $F(x)$ of the first-order language of ZFC. This restriction is philosophically serious but in our experience not harmful in practice. Indeed, if infinite LF signatures are allowed, using canonical models constructed in completeness proofs, in many cases *all* models can be represented up to elementary equivalence.

Proof Theory. Our examples from Sect. 2.3 already encoded the proof theory of first-order logic in a way that treats proof theory and model theory in a balanced way. Our definitions can be easily generalized to this setting.

Logic encodings in a logical framework become 6-tuples $(L^{Syn}, L^{truth}, L^{Mod}, \mathcal{F}, L^{mod}, L^{Pf}, L^{pf})$ for $L^{pf} : L^{Syn} \to L^{Pf}$. L^{Pf} encodes the proof theory of a logic, which typically means to add auxiliary syntax, judgments, and proof rules to L^{Syn}. Def. 5 can be extended to obtain $\Sigma^{pf} : \Sigma^{Syn} \to \Sigma^{Pf}$ as a pushout in the same way as Σ^{mod}. Finally the logical framework must be extended with a component that yields a data structure of proofs (such as entailment systems or proof trees) for every slice out of *Base*.

For example, for the framework $\mathbb{F}^{LF}$, the proof trees for proofs of F using assumptions $F_1, \ldots, F_n$ can be defined as the $\beta\eta$-normal LF terms over Σ^{Pf} of type $\Sigma^{pf}\big(L^{truth}(ded)\,F_1 \to \ldots \to L^{truth}(ded)\,F_n \to L^{truth}(ded)\,F\big)$. A similar construction was given in [Rab10].

4   Logical Frameworks in Hets

The differences between LF and Hets mentioned in Sect. 2 exhibit complementary strengths, and a major goal of our work is to combine them. We have enhanced Hets with a component that allows the dynamic definition of new logics. The user specifies a logic by giving the representation of its constituents (syntax, model theory) in a logical framework and the combined system recognizes the new logic and integrates it into the Hets logic graph. The implementation follows the Hets principles of high abstraction and separation of concerns: we provide an implementation for the general concept of logical frameworks, which we describe in Sect. 4.1. This is further instantiated for the particular case of LF in Sect. 4.2. Finally, in Sect. 4.3 we present a complete description of the steps necessary to add a new logic in Hets using the framework of LF.

4.1   Implementing the LMF in Hets

This section sketches how the concept of logical frameworks is integrated into Hets. The integration is done entirely on the developer's side and a user wishing to add a new logic to Hets only has to select one of the available logical frameworks, which will serve as a meta-logic for the new object logic he or she specifies. We will give here just a brief overview of how the implementation is done and refer the interested reader to [Mos05] for a presentation of the theoretical foundations of Hets and to the Hets developers documentation pages [2] for a more detailed presentation of how the coding is actually done.

The central part of the implementation is a Haskell type class *LogicalFramework*, which is instantiated by the logics which can be used as logical frameworks, i.e. in which object logics can be specified by the user. Such candidates are for example LF, rewriting logic and Isabelle [3]. The class provides a selector for the *Base* signature and a method *writeLogic*, which takes an object logic name as an argument and generates the instances of the classes *Syntax*, *Sentences*, *StaticAnalysis*, and *Logic* for the given object logic.

Each logic implementing *LogicalFramework* must likewise implement the class *Category*, from which we get the category $\mathbb{C}$ mentioned in Def. 4. The sentence functor **Sen** is specified implicitly by the *writeLogic* method: the instantiation of the *StaticAnalysis* class determines exactly which sentences are valid for a particular signature of L, thus giving **Sen** on objects. Since the current implementation of logics in Hets does not include satisfaction of sentences in models, the predicate $\vdash_t$ is currently not represented as its main purpose is to define the satisfaction relation for object logics.

At the syntactic level, we must provide a way to write down new logic definitions in HetCASL, the underlying heterogenous algebraic specification language of Hets. Since definitions of new logics have a different status than usual algebraic specifications, we extend the language at the library level.

Concrete Syntax. We add the following concrete syntax (on the right) to HetCASL in order to define new logics. Here L is the name of the newly defined logic and $\mathbb{F}$ is an identifier pointing to the logical framework used. The identifiers $L^{truth}, L^{mod}, L^{pf}, \mathcal{F}$ are the components of the new logic L. They refer to previously declared signature morphisms of $\mathbb{F}$ and the signatures representing L^{Syn}, L^{Mod}, L^{Pf} can be inferred from them. $\mathcal{F}$ is a signature which gives the foundation. The declaration of patterns is optional.

```
newlogic L =
  meta F
  syntax L^truth
  models L^mod
  foundation F
  proofs L^pf
  patterns P
```

After encountering a `newlogic` declaration, Hets invokes a static analyzer, which retrieves the signatures and morphisms constituting the components of the logic L. The analyzer verifies the correct shape of the induced diagram and

---

[2] See `http://www.informatik.uni-bremen.de/agbkb/forschung/formal_methods/CoFI/hets/src-distribution/daily/Hets/docs/Logic.html`.

[3] Currently only LF has a full implementation as a logical framework.

instantiates the *Logic* class for the logic L as specified by the *writeLogic* method of the framework $\mathbb{F}$.

The logic L arising from the above `newlogic L` declaration differs slightly from the one described in Def. 6 in that it uses signatures of $\mathbb{F}$ that extend L^{Syn} rather than $\mathbb{F}$-inclusion morphisms out of L^{Syn}. Accordingly, the morphisms of L are those morphisms of $\mathbb{F}$ which are the identity on L^{Syn}. This is essentially the same thing, but has the advantage that the data types representing the signatures and morphisms of $\mathbb{F}$ can be directly reused for L and no separate instantiation of the class *Category* is required[4].

4.2   LF as a Logical Framework in Hets

In this section we outline how to turn $\mathbb{LF}$ into a logical framework in Hets, i.e. how to instantiate the *LogicalFramework* class for $\mathbb{LF}$. In order to do so we will make use of the instance of the *Logic* class for $\mathbb{LF}$.[5]

The *Base* signature is specified to be the $\mathbb{LF}$ signature containing the symbols o and ded, as described in Sect. 3. The instantiations of the classes *Logic*, *Syntax*, etc. provided by the *writeLogic* method mostly inherit their $\mathbb{LF}$ implementations, with one exception being the *StaticAnalysis* class. While both $\mathbb{LF}$ and the $\mathbb{LF}$ object logics use Twelf to verify the well-formedness of input specifications, a specification in an object logic is assumed to have been given relative to the L^{Syn} signature supplied when defining the object logic.

After receiving the input file, Twelf performs parsing, static analysis and reconstruction of types and implicit arguments. If the analysis succeeds, the output is stored as an OMDoc version of the input file, and is subsequently imported into Hets using standard XML technologies. Hets reads the imported OMDoc file and transforms it into corresponding $\mathbb{LF}$ signatures and morphisms in their Hets internal representation.

4.3   Adding a New Logic in Hets: FOL

We will now illustrate the steps needed to add first-order logic as a new logic in Hets. The aim of this section is not to show how to encode a particular logic in Twelf, which for the case of first-order logic has been described in [HR11], but rather to show how an existing encoding can be used to add the logic in Hets.

Given a *FOL* encoding as in Section 2.3, all that is needed to be done is to collect the components of the encoding in a `newlogic` definition, as in Fig. 2. The first lines import the morphism FOL^{truth} from *Base* to FOL^{Syn}, the morphism FOL^{mod} from FOL^{Syn} to FOL^{Mod}, and the morphism FOL^{pf} from FOL^{Syn} to FOL^{Pf} as in Ex. 4, from their respective directories. $STTIFOLEQ$ is a fragment of ZFC used to represent model theory. It is composed of simple type theory equipped with external intuitionistic first-order logic. Notice that we assume for convenience that the file with the new logic definition is in the folder that contains

---

[4] The theory presented in Section 3 could thus have been formulated equivalently, albeit less elegantly, without referring to slice categories.

[5] An institution for $\mathbb{LF}$ can be defined as for example in [Rab08].

```
from logics/first-order/syntax/fol get FOL_truth    %%FOL^truth
from logics/first-order/model_theory/fol get FOL_mod %%FOL^mod
from logics/meta/sttifol get STTIFOLEQ %%F
from logics/first-order/proof_theory/fol get FOL_pf %%FOL^pf

newlogic FOL =
  meta LF
  syntax FOL_truth
  models FOL_mod
  foundation STTIFOLEQ
  proofs FOL_pf
end
```

Fig. 2. Defining FOL as a new object logic

the directory of logics as sub-folder; the paths need to be adjusted if that is not the case. [6] The directory structure mirrors the modular design of logics in the Logic Atlas. As a result of calling Hets on the above file, a new directory called FOL is added to the source folder of Hets. The directory contains automatically generated files with the instances needed for the logic FOL. Moreover, the Hets variable containing the list of available logics is updated to include FOL. After recompiling Hets, the new logic is added to the logic graph of Hets (the node FOL in Fig. 1 for the dynamically-added logic) and can be used in the same way as any of the built-in logics.

In particular, we can now use the new object logic to write specifications. For example, the specification in Fig. 3 uses FOL as a current logic and declares a constant symbol c and a predicate p, together with an axiom that the predicate p holds for the constant c. Notice that the syntax for logics specified in a logical framework $\mathbb{F}$ is inherited from the framework (in our case $\mathbb{LF}$), but it has been extended with support for sentences, in the usual CASL syntax i.e. prefixed by the '.' character.

Fig. 4 presents the theory of SP as displayed from within Hets; as mentioned in Section 4.2, the theory is automatically assumed to extend FOL^{Syn}. Since in Hets all imports are internally flattened, the theory of SP when displayed will include all the symbols from FOL^{Syn}.

5   Conclusion and Future Work

We have described a prototypical integration of the institution-based Heterogeneous Tool Sets (Hets) with logical frameworks in general and LF and the Twelf tool in particular. The structuring language used by Hets has a model theoretic semantics, which has been reflected in the proof theoretic logical framework LF by representing models as theory morphisms into some foundation. While LF is

---

[6] The complete specification of FOL in LF can be found at https://svn.omdoc.org/repos/latin/twelf-r1687/

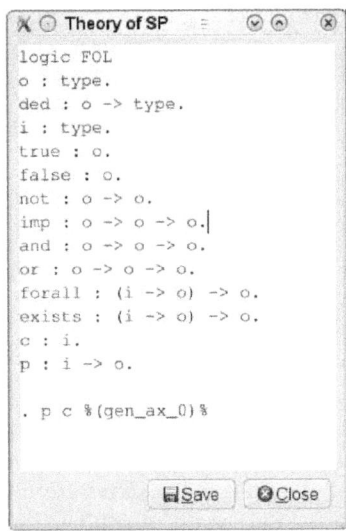

```
logic FOL
spec SP =
  c : i.
  p : i -> o.


  . p c
end
```

Fig. 3. Specification in
the new object logic

Fig. 4. Theory of SP

the logical framework of our current choice, both the theory and the implementation are so general that other frameworks like Isabelle can be used as well. We expect important synergy effects from this as Isabelle is already used as one of the main inference engines in Hets.

Proof theory of the represented logics has been treated only superficially in the present work, but in fact, we have represented proof calculi for all the LATIN logics within LF. Representing models in the system as well has enabled us to formally prove soundness of the calculi. It is straightforward to extend the construction of institutions out of logic representations in logical frameworks such that they deliver institutions with proofs. In the long run, we envision that the provers integrated in Hets also return proof terms, which Hets can then fill into the original file and rerun Twelf on it to validate the proof. Thus, Hets becomes the mediator that orchestrates the interaction between external theorem provers and Twelf as a trusted proof checker.

While the theory and implementation described in this paper make it possible to add logics to Hets in a purely declarative way, further work is needed to turn this into a scalable tool. Firstly, the logic translations-as-theory morphisms approach needs to be generalised in order to cover more practically useful examples. Secondly, the new LF generated logics present in Hets need to be connected (via institution comorphisms) to the existing hard-coded logics in order to share the connection of the latter to theorem provers and other tools. Thirdly, it will be desirable to have a declarative interface for specifying the syntax of new logics, such that one is not forced to use the syntax of the logical framework. We are currently examining whether Eclipse and Xtext are helpful here. Finally, also the various tool interfaces of Hets should be made more declarative, such that Hets logics specified in a logical framework can be directly connected to theorem provers and other tools, instead of using a comorphism into a hard-coded logic. Then, in the long run, it will be possible to entirely replace the hard-coded logics with declarative logic specifications in the LATIN metaframework — and only the latter needs to be hard-coded into Hets.

The Logic Atlas currently consists of a around 150 files containing some 700 signatures and views and producing over 10000 lines of Twelf output (including declarations that are generated by the module system). This is the result of roughly one year of development with substantial contributions from six different people, and due to the evolutionary improvement of our methodology, architecture, and expertise, growth has been exponential. Nevertheless, the representation and interconnection of logics is (and will remain) a task that requires a deep understanding of the respective logics, a good eye for the underlying primitives, and sound judgment in the design and layout of atlantes. We consider the current Logic Atlas to be a seed atlas that establishes best practices in these questions and provides a nucleus of logical primitives that can be extended to add particular logics by outside logic and system developers.

We explicitly invite researchers outside the LATIN project to contribute their logics. This should usually be a matter of importing the aspects that are provided by Logic Atlas theories, and LF-encoding the aspects that are not.

Acknowledgments. This paper mainly addresses the model theoretic side of the logic atlas developed in the LATIN project — funded by the German Research Council (DFG) under grant KO-2428/9-1.

References

[ABK+02]   Astesiano, E., Bidoit, M., Kirchner, H., Krieg-Brückner, B., Mosses, P., Sannella, D., Tarlecki, A.: CASL: The Common Algebraic Specification Language. Theoretical Computer Science 286(2), 153–196 (2002)

[AHMP98]   Avron, A., Honsell, F., Miculan, M., Paravano, C.: Encoding modal logics in logical frameworks. Studia Logica 60(1), 161–208 (1998)

[AHMS99]   Autexier, S., Hutter, D., Mantel, H., Schairer, A.: Towards an Evolutionary Formal Software-Development Using CASL. In: Bert, D., Choppy, C., Mosses, P.D. (eds.) WADT 1999. LNCS, vol. 1827, pp. 73–88. Springer, Heidelberg (2000)

[BC04]   Bertot, Y., Castéran, P.: Coq'Art: The Calculus of Inductive Constructions. Springer, Heidelberg (2004)

[BJL06]   Bortin, M., Broch Johnsen, E., Lüth, C.: Structured formal development in Isabelle. Nordic Journal of Computing 12, 1–20 (2006)

[CAB+86]   Constable, R., Allen, S., Bromley, H., Cleaveland, W., Cremer, J., Harper, R., Howe, D., Knoblock, T., Mendler, N., Panangaden, P., Sasaki, J., Smith, S.: Implementing Mathematics with the Nuprl Development System. Prentice-Hall (1986)

[CELM96]   Clavel, M., Eker, S., Lincoln, P., Meseguer, J.: Principles of Maude.
            In: Meseguer, J. (ed.) Proceedings of the First International Work-
            shop on Rewriting Logic, vol. 4, pp. 65–89 (1996)
[Chu40]   Church, A.: A Formulation of the Simple Theory of Types. Journal
            of Symbolic Logic 5(1), 56–68 (1940)
[Dia08]   Diaconescu, R.: Institution-independent Model Theory. Birkhäuser
            (2008)
[GB92]   Goguen, J., Burstall, R.: Institutions: Abstract model theory for spec-
            ification and programming. Journal of the Association for Computing
            Machinery 39(1), 95–146 (1992)
[HHP93]   Harper, R., Honsell, F., Plotkin, G.: A framework for defining logics.
            Journal of the Association for Computing Machinery 40(1), 143–184
            (1993)
[HR11]   Horozal, F., Rabe, F.: Representing Model Theory in a Type-
            Theoretical Logical Framework. Theoretical Computer Science (to ap-
            pear, 2011), http://kwarc.info/frabe/Research/HR_folsound_10.
            pdf
[HST94]   Harper, R., Sannella, D., Tarlecki, A.: Structured presentations and
            logic representations. Annals of Pure and Applied Logic 67, 113–160
            (1994)
[IR11]   Iancu, M., Rabe, F.: Formalizing Foundations of Mathematics. Math-
            ematical Structures in Computer Science (to appear, 2011), http://
            kwarc.info/frabe/Research/IR_foundations_10.pdf
[KMR09]   Kohlhase, M., Mossakowski, T., Rabe, F.: The LATIN Project (2009),
            https://trac.omdoc.org/LATIN/
[MAH06]   Mossakowski, T., Autexier, S., Hutter, D.: Development Graphs -
            Proof Management for Structured Specifications. Journal of Logic
            and Algebraic Programming 67(1-2), 114–145 (2006)
[Mes89]   Meseguer, J.: General logics. In: Ebbinghaus, H.-D., et al. (eds.)
            Proceedings of Logic Colloquium, 1987, pp. 275–329. North-Holland
            (1989)
[ML74]   Martin-Löf, P.: An Intuitionistic Theory of Types: Predicative Part.
            In: Proceedings of the 1973 Logic Colloquium, pp. 73–118. North-
            Holland (1974)
[MML07]   Mossakowski, T., Maeder, C., Lüttich, K.: The Heterogeneous Tool
            Set, HETS. In: Grumberg, O., Huth, M. (eds.) TACAS 2007. LNCS,
            vol. 4424, pp. 519–522. Springer, Heidelberg (2007)
[MOM94]   Martí-Oliet, N., Meseguer, J.: General logics and logical frameworks.
            In: What is a Logical System?, pp. 355–391. Oxford University Press,
            Inc., New York (1994)
[Mos05]   Mossakowski, T.: Heterogeneous Specification and the Heteroge-
            neous Tool Set. Habilitation thesis (2005), http://www.informatik.
            uni-bremen.de/~till/
[Nor05]   Norell, U.: The Agda WiKi (2005), http://wiki.portal.chalmers.
            se/agda
[NPW02]   Nipkow, T., Paulson, L., Wenzel, M.: Isabelle/HOL — A Proof As-
            sistant for Higher-Order Logic. Springer, Heidelberg (2002)
[Pau94]   Paulson, L.C.: Isabelle: A Generic Theorem Prover. LNCS, vol. 828.
            Springer, Heidelberg (1994)
[PC93]   Paulson, L., Coen, M.: Zermelo-Fraenkel Set Theory. Isabelle distri-
            bution, ZF/ZF.thy (1993)

[Pfe00]   Pfenning, F.: Structural cut elimination: I. intuitionistic and classical logic. Information and Computation 157(1-2), 84–141 (2000)

[PS99]   Pfenning, F., Schürmann, C.: System Description: Twelf - A Meta-Logical Framework for Deductive Systems. In: Ganzinger, H. (ed.) CADE 1999. LNCS (LNAI), vol. 1632, pp. 202–206. Springer, Heidelberg (1999)

[PS08]   Poswolsky, A., Schürmann, C.: System Description: Delphin - A Functional Programming Language for Deductive Systems. In: Abel, A., Urban, C. (eds.) International Workshop on Logical Frameworks and Metalanguages: Theory and Practice. ENTCS, pp. 135–141 (2008)

[PSK⁺03]   Pfenning, F., Schürmann, C., Kohlhase, M., Shankar, N., Owre, S.: The Logosphere Project (2003), http://www.logosphere.org/

[Rab08]   Rabe, F.: Representing Logics and Logic Translations. PhD thesis, Jacobs University Bremen (2008), http://kwarc.info/frabe/Research/phdthesis.pdf

[Rab10]   Rabe, F.: A Logical Framework Combining Model and Proof Theory. Submitted to Mathematical Structures in Computer Science (2010), http://kwarc.info/frabe/Research/rabe_combining_09.pdf

[RS09]   Rabe, F., Schürmann, C.: A Practical Module System for LF. In: Cheney, J., Felty, A. (eds.) Proceedings of the Workshop on Logical Frameworks: Meta-Theory and Practice (LFMTP), pp. 40–48. ACM Press (2009)

[Soj10]   Sojakova, K.: Mechanically Verifying Logic Translations. Master's thesis, Jacobs University Bremen (2010)

[SS04]   Schürmann, C., Stehr, M.: An Executable Formalization of the HOL/Nuprl Connection in the Metalogical Framework Twelf. In: 11th International Conference on Logic for Programming Artificial Intelligence and Reasoning (2004)

[TB85]   Trybulec, A., Blair, H.: Computer Assisted Reasoning with MIZAR. In: Joshi, A. (ed.) Proceedings of the 9th International Joint Conference on Artificial Intelligence, pp. 26–28 (1985)

An Institution for Graph Transformation[*]

Andrea Corradini[1], Fabio Gadducci[1], and Leila Ribeiro[2]

[1] Dipartimento di Informatica, Università di Pisa
[2] Instituto de Informática, Universidade Federal do Rio Grande do Sul

Abstract. The development of a denotational framework for graph transformation systems proved elusive so far. Despite the existence of many formalisms for modelling various notions of rewriting, the lack of an explicit, algebraic notion of "term" for describing a graph (thus different from the usual view of a graph as an algebra in itself) frustrated the efforts of the researchers. Resorting to the theory of institutions, the paper introduces a model for the operational semantics of graph transformation systems specified according to the so-called double-pullback approach.

Keywords: Graph transformation systems, institutions.

1 Introduction

Graph transformation [24] is a flexible framework for the specification and verification of distributed systems, whose strength lies in the visual nature of the specification process. The topology of the states traversed by a system is represented by a graph, and system evolution is described by a set of rewriting rules, specified according to some operational mechanism for graph manipulation.

The formalism has a rich set of theoretical tools available, such as a well-established theory of concurrency [8,1] (that proved pivotal in developing verification techniques [2]) and a growing family of methodologies for the modular specification of systems [13,17]. However, so far the development of a denotational framework for graph transformation proved elusive. There exist many formalisms that are able to properly capture various notions of term rewriting, while offering a proper denotational model [21]. Yet, the lack of an explicit notion of "term" for describing a graph (thus different from the usual view of a graph as an algebra in itself) and the relevance given in the theory to a structured notion of rewriting step among states (usually given by a partial morphism among graphs) contributed to frustrate so far the efforts of the researchers.

Institutions [12] are a powerful formalism relating to the *Abstract Model Theory for Specification and Programming*, as the title of the seminal work states. They provide a denotational framework that avoids resorting to the concrete presentation of a specification formalism, abstracting away the actual syntax

---

[*] Research partially supported by the MIUR project SisteR (PRIN 20088HXMYN) and by the FAPERGS/CNPq project 10/0043-0.

T. Mossakowski and H.-J. Kreowski (Eds.): WADT 2010, LNCS 7137, pp. 160–174, 2012.

in favor of a general notion of signature and related sentences. They intend to capture the essence of what a logical system is, at the same time equipping such systems with methodologies for their structural presentation that are independent of the language at hand. Their range of application is quite large, moving from algebraic specifications towards more operational formalisms such as process algebras [22], and they come equipped with a large array of tools for specification and verification [25].

The aim of the paper is to cast graph transformation into the abstract setting for system specification provided by institutions. The first two authors investigated the presentation of (term-)graph rewriting by using enriched categories [4,3]. The work presented here has a more foundational nature, since it requires the development of suitable notions of models for graph rewriting, while the categorical presentation aimed at recasting the operational semantics. Moreover, the development of an institution for graph transformation might enrich the family of methodologies for structured specification and verification of graph transformation systems (as well as representing a sanity check for those techniques adopted in the graph rewriting community).

One of the crucial issues in the definition of an institution for graph transformation systems is the notion of signature. So, before presenting the actual institution, recall that in current algebraic approaches to graph transformation all graphs are *typed*, e.g., system states are elements of a comma category: (total) graph morphisms $G \rightarrow T$. Rules (according e.g. to the SPO approach) are injective partial morphims $L \rightarrow R$ among typed graphs (over the same type graph): each rule specifies which items of the starting graph must be preserved, deleted, and also defines which items will be created. Indeed, rule application means to find a (total) morphism $L \rightarrow G$ among typed graphs such that the obvious pushout exists: the derivation is its leg $G \rightarrow H$. The intended semantics of a graph transformation system, considered as a set of rules over the same type graph, is then just the transition system whose elements are typed graphs and whose transitions are partial graph morphisms induced by the rules themselves.

As a consequence, we believe that a declarative view of graph transformation systems must put typed graphs at its core. Indeed, we consider (type) graphs as signatures, and (injective) graph morphisms as signature morphisms. Sentences are just rules, i.e., (injective) partial morphisms among typed graphs: the impact of a morphism between type graphs on a rule is a retyping of the rule, obtained by composition. Models are Kripke-like structures, i.e., diagrams interpreted over the category of typed graphs and partial graph (mono)morphisms; transition labels highlight which part of (the label of) a state is preserved along a transition.

So far, the notion of model we sketched above is independent of the chosen approach adopted for graph rewriting. The choice comes into play in the definition of the satisfaction relation, i.e., in determining when a graph transformation rule is satisfied by a model. We focus on a minimalistic solution: a rule is satisfied if, whenever it can be applied in a given state, it is indeed applied, and a transition labelled by the derivation corresponding to the rule application occurs. The previous definition makes sense for any algebraic approach to graph transformation. And, even if we introduce our case studies at a concrete level of graphs for the sake of presentation, all of our considerations could be formulated at the abstract level of *adhesive categories* [19], which allows to replace graphs and their morphisms with an arbitrary category satisfying suitable exactness conditions. However, we are going to show that only the choice of the *double pullback* mechanism [10] ensures that our formalism verifies the satisfaction condition, linking satisfiability between a model and its reduct.

The paper has the following structure. After recalling in Section 2 the basics of graphs (§ 2.1), graph transformation (§ 2.2) and institutions (§ 2.3), Section 3 presents our institution for graph transformation, defining signatures (§ 3.1), sentences (§ 3.2), models (§ 3.3) and the satisfiability relation (§ 3.4), and finally proving when the satisfiability condition holds (§ 3.5). Section 4 then presents in some details the double-pullback approach to graph transformation, indicating how it relates to the other approaches. Finally, Section 5 draws some conclusions and sketches some directions for further works.

2   Background

2.1   Graphs

Let us start presenting the definition of typed graphs.

Definition 1 (graphs). *A graph is a tuple $\langle V, E, in, out \rangle$ where V, E are the sets of nodes and edges and $in, out : E \to V$ are the input and output functions, mapping each edge to its source and target nodes.*

From now on we denote the components of a graph G by V_G, E_G, in_G and out_G, dropping the subscript whenever clear from the context.

Definition 2 (graph morphisms). *Let G and H be graphs. A graph morphism $f : G \to H$ is a pair of functions $\langle f_V : V_G \to V_H, f_E : E_G \to E_H \rangle$ preserving source and target, i.e., such that $f_V \circ in_G = in_H \circ f_E$ and $f_V \circ out_G = out_H \circ f_E$.*

The category of graphs is denoted by **Graph**. We now give the definition of typed graph [5], i.e., a graph labelled over a structure that is itself a graph.

Definition 3 (typed graphs). *Let Γ be a graph. A Γ-typed graph G (or typed over Γ) is a graph $|G|$ with a graph morphism $\tau_G : |G| \to \Gamma$.*

Thus, graphs typed over Γ are just the objects of the comma category **Graph** $\downarrow$ Γ: the arrows are defined accordingly.

Definition 4 (typed graph morphisms). *Let G and G' be Γ-typed graphs. A Γ-typed graph morphism $f : G \to G'$ is a graph morphism $f : |G| \to |G'|$ consistent with the typing, i.e., such that $\tau_G = \tau_{G'} \circ f$.*

The category of graphs typed over Γ (i.e., the comma category **Graph** $\downarrow$ Γ) is denoted by Γ-**Graph**.

2.2 Graph Rewriting

In this section we introduce some of the basic definitions concerning the algebraic approaches to the rewriting of typed graphs. Several mechanisms have been proposed in the literature: here we consider the widely known *double-pushout* (DPO) and *single-pushout* (SPO) approaches [24], as well as the less known *double-pullback* (DPB) approach [10,15] (we shall discuss the main differences among the three approaches in Section 4). A first thing to be noted is that the notion of *rule* is essentially the same for the three approaches.[1]

Definition 5 (rule). *A Γ-typed graph rule $p = (L \xleftarrow{l} K \xrightarrow{r} R)$ is a pair of Γ-typed graph monomorphisms $l : K \rightarrowtail L$, $r : K \rightarrowtail R$. A graph transformation system $\mathcal{G}$ is a pair $\langle \Gamma, P \rangle$, for Γ a type graph and P a set of Γ-typed graph rules.*

So, a rule is just a *span* of monomorphisms in the category Γ-**Graph**: the mono requirement asks for the arrows to be injective in **Graph**. Differently with respect to rules, the definition of *direct derivation*, i.e., of the effect of applying a rule to a graph, depends on the chosen approach. In the paper we use $\mathcal{A}$ as a metavariable ranging over the three algebraic approaches to graph transformation, i.e., $\mathcal{A} \in \{\text{DPO}, \text{SPO}, \text{DPB}\}$.

Definition 6 (derivation). *Let $p = (L \xleftarrow{l} K \xrightarrow{r} R)$ be a graph rule and G a graph, both over Γ. A match of p in G is a Γ-typed graph morphism $m_L : L \rightarrow G$.*

*A direct derivation from G to H via production p and match m_L is a diagram as depicted in Figure 1: it is called a DPO-derivation if (1) and (2) are pushouts in Γ-**Graph**; it is called a DPB-derivation if (1) and (2) are pullbacks in Γ-**Graph**; and it is called an SPO-derivation if (1) is a final pullback complement and (2) is a pushout.[2] We denote this direct derivation by $p/m : G \Rightarrow_{\mathcal{A}} H$, for $m = \langle m_L, m_K, m_R \rangle$, or simply by $G \Rightarrow H$.*

A match $m_L : L \rightarrow G$ of a rule $p = (L \xleftarrow{l} K \xrightarrow{r} R)$ is $\mathcal{A}$-valid, if it can be completed to an $\mathcal{A}$-derivation as in Fig. 1, with $\mathcal{A} \in \{\text{DPO}, \text{SPO}, \text{DPB}\}$.

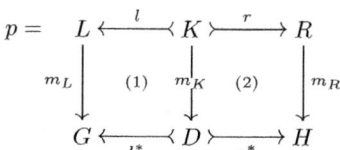

Fig. 1. A direct derivation

Explicit conditions for validity of matches have been studied in the literature. We just briefly remind, without reporting details that are irrelevant here, that

---

[1] Even if for the SPO approach a rule is usually defined in an alternative, yet equivalent way as a partial injective typed graph morphism.

[2] The reader is referred to [7] for the relationship between the standard definition of SPO derivation and the present one.

- a DPO-valid match must satisfy the *identification* and the *dangling conditions*, see [9];
- an SPO-valid or DPB-valid match must satisfy the *conflict-freeness condition*, see [10,7].

A match m is *conflict-free* if whenever $m(x) = m(y)$, then either both x and y are in $l(K)$, or none of them is in $l(K)$. As shown in [7], only for conflict-free matches an SPO-derivation can be defined as in Definition 6, and this is why we require SPO-valid matches to be conflict-free. However it is fair to remind that in the original definition of SPO-derivation (a pushout in the category of graphs and partial morphisms [20]) a rule can be applied to a non-conflict-free match as well: but in this case the resulting *co-match* $m_R : R \to H$ is a partial morphism, an effect which is often considered to be counterintuitive.

The following properties of the above conditions will be exploited later.

Proposition 1 (preservation and reflection of conditions). *Let $g : G \rightarrowtail G'$ be a Γ-typed monomorphism, let $p = (L \hookleftarrow K \rightarrowtail R)$ be a rule, and let $m : L \to G$, $m' : L \to G'$ be two matches such that $m' = g \circ m$. Then*

- *m satisfies the dangling condition if m' satisfies it;*
- *m satisfies the identification condition if and only if m' satisfies it;*
- *m satisfies the conflict-freeness condition if and only if m' satisfies it.*

2.3   Institutions

Finally, we recall what an institution is.

Definition 7 (institution). *An institution is a tuple $\langle \mathbf{Sign}, \mathbf{Sen}, \mathbf{Mod}, \models \rangle$ where*

- **Sign** *is a category of signatures;*
- **Sen** : **Sign** $\to$ **Set** *is a functor associating with each signature Σ the set of sentences $\mathbf{Sen}(\Sigma)$, and with each signature morphism $\sigma : \Sigma \to \Sigma'$ the sentence translation map $\mathbf{Sen}(\sigma) : \mathbf{Sen}(\Sigma) \to \mathbf{Sen}(\Sigma')$;*
- **Mod** : **Sign**^{op} $\to$ $\mathcal{CAT}$ *is a functor associating with each signature Σ, the category of models $\mathbf{Mod}(\Sigma)$ and with each signature morphism $\sigma : \Sigma \to \Sigma'$ the reduct functor $\mathbf{Mod}(\sigma) : \mathbf{Mod}(\Sigma') \to \mathbf{Mod}(\Sigma)$;*
- *$\models_\Sigma \subseteq |\mathbf{Mod}(\Sigma)| \times \mathbf{Sen}(\Sigma)$ is a relation for each $\Sigma \in |\mathbf{Sign}|$, such that given $\sigma : \Sigma \to \Sigma'$ the following satisfaction condition holds*

$$M' \models_{\Sigma'} \mathbf{Sen}(\sigma)(\phi) \quad \Longleftrightarrow \quad \mathbf{Mod}(\sigma)(M') \models_\Sigma \phi$$

for each $M' \in |\mathbf{Mod}(\Sigma')|$ and $\phi \in \mathbf{Sen}(\Sigma)$.

Most often, $\mathbf{Sen}(\sigma)(\phi)$ is written as $\sigma(\phi)$, and $\mathbf{Mod}(\sigma)(M')$ as $M'|_\sigma$.

3   An Institution for GTS

Given the basic preliminaries above, we can start filling in the picture.

3.1 Signatures

As anticipated in the introduction the type graphs play the role of signatures.

Definition 8 (signatures). *The category of* (graph transformation) *signatures is* **MGraph***, the sub-category of* **Graph** *of graphs and their injective morphisms.*

As this definition is one of the crucial ingredient of the design of our institution for graph transformation, it is worth providing a thorough justification. As explained in the introduction, typed graphs gained a central role in the algebraic approaches to graph rewriting. Actually, in the original presentations (like the seminal [9]), the classical theory of the double-pushout approach to graph transformation was developed for *colored graphs*, i.e., graphs where nodes and edges are labeled over corresponding *sets* of colors, and colors are preserved by morphisms. The idea of working instead with *typed graphs* arose in the early 1990s when trying to define "reasonable" notions of *morphisms* for graph transformation systems, as generalizations of morphisms for Petri nets; thus, in a sense, when moving from specifications "in the small" (a single system) to specifications "in the large" (relating several systems).

Even if this shift is, in essence, a minor syntactical change (simply adding a graphical structure to the sets of colors), it reflects a major conceptual achievement, namely, that graph rewriting can be seen as the result of lifting the theory of Petri nets from the category of sets to that of graphs (see the discussion in [6]). In fact, a morphism between two Petri nets can be conveniently described as a diagram in **Set**, looking at a marking as a function from the *set* of tokens to the *set* of places. Replacing **Set** with **Graph**, one can re-interpret the same diagram as defining a morphism between GTSs: now a state ("marking") is a *graph* (of "tokens") with a mapping (a homomorpism) to a *graph* (of "places"). Thus the type graph of a GTSs plays the role of the set of places of a Petri net. A graph $\Gamma \in$ **MGraph**, the category of signatures, is thus seen as a type graph.

Let us conclude this section by commenting briefly why we require signature morphisms to be injective. As the reduct functor (see Section 3.3) will be defined using a pullback construction along the signature morphism, the reduct functor induced by a non-injective morphism between type graphs could relate matches that satisfy the application condition of a rule to matches that do not satisfy it. This fact would not allow the satisfaction condition to hold (see Section 3.5). Similar considerations for the institutions of CSP are discussed in [22], leading to an analogous restriction on signature morphisms.

3.2 Sentences

Let us introduce now the sentences of our institution.

Definition 9 (sentences). *The functor of* (graph transformation) *sentences* **Rules** : **MGraph** $\rightarrow$ **Set** *maps a type graph Γ to the set of graph rules typed over Γ, i.e.,* **Rules**$(\Gamma) = \{ L \overset{l}{\hookleftarrow} K \overset{r}{\rightarrowtail} R \mid l, r \in \Gamma\text{-}\mathbf{Graph} \}$.

A sentence for a signature Γ is just a rule typed over it. This is reminiscent of the usual institution for term rewriting systems (or, for the matter, for pre-ordered algebras), even if the span of monomorphims is going to add discriminating power. If $\gamma : \Gamma \rightarrowtail \Gamma'$ is a morphism in **MGraph**, the action of functor **Rules** on γ consists of "retyping" the rule by post-composing in the obvious way the typing morphisms with γ

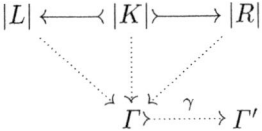

where for the sake of a clearer visual presentation the typing and the signature morphisms are depicted by dotted arrows.

3.3   Models

Let us now turn our attention to models. Let us denote as **Span**(Γ-**Graph**) the category whose objects are graphs typed over Γ, and whose arrows are *abstract spans* of Γ-typed graph morphisms, i.e., up to an isomorphic choice of the graph representing the common source of the span.

Definition 10 (graph transition system). *A graph transition system $\mathcal{M}$ over Γ is a diagram over* **Span**(Γ-**Graph**) *such that all arrows are spans of monomorphisms. More explicitly, $\mathcal{M}$ is a pair $\langle M, g \rangle$, where M is a graph and g a graph morphism from M to the graph underlying the category* **Span**(Γ-**Graph**), *such that all the spans in the image of g are made of monomorphisms.*

For the sake of verbal clarity, in the definition above M is going to be called a *transition system*, composed by a set S_M of *states* and a set T_M of *transitions*. Thus, the *labeling* g is a pair $\langle g^S, g^T \rangle$ of functions, where $g^S(s)$ is a graph typed over Γ for each state $s \in S_M$, and for each transition $t \in T_M$, $g^T(t)$ is an abstract span of Γ-typed graph monomorphisms between $g^S(in_M(t))$ and $g^S(out_M(t))$, as shown below (where, for the sake of visual clarity, the transition is represented via a double arrow and the labeling via a double-dotted arrow) for G_t any representative of the equivalence class.

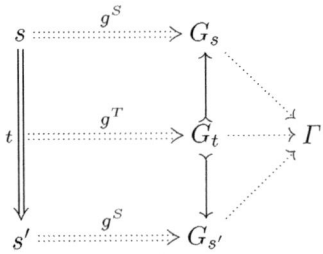

The models of a signature Γ are going to be graph transition systems over Γ. As it always occur, the notion of model morphism is trickier, and open to multiple choices. The next definition introduces the most obvious notion, namely, transition system morphisms that strictly preserve the labeling.

Definition 11 (morphisms). *Let $\langle M, g \rangle$ and $\langle M', g' \rangle$ be graph transition systems over Γ. A graph transition system morphism $\sigma : \langle M, g \rangle \rightarrow \langle M', g' \rangle$ is a graph morphism $\sigma : M \rightarrow M'$ consistent with the labelling, i.e., such that $g = g' \circ \sigma$.*

The category of graph transition systems over Γ and their morphisms is denoted $\mathbf{GTS}(\Gamma)$.

In order to complete the definition of the model functor $\mathbf{GTS} : \mathbf{MGraph}^{op} \rightarrow \mathcal{CAT}$ we need to extend the mapping $\Gamma \mapsto \mathbf{GTS}(\Gamma)$ to arrows, contravariantly.

Definition 12 (reduct). *Let $\langle M', g' \rangle$ be a graph transition system over Γ' and $\gamma : \Gamma \rightarrowtail \Gamma'$ a signature morphism. The reduct of $\langle M', g' \rangle$ along γ is the graph transition system $\langle M', g' \rangle{\restriction_\gamma} \overset{def}{=} \langle M', g'_\gamma \rangle$ over Γ, having the same states and transitions of the original system, and such that, for each state s of M', $g'^S_\gamma(s)$ is the sub-graph of $g'^S(s)$ obtained by removing those items that are not typed in the image of $\gamma(\Gamma)$; $g'^T_\gamma(t)$ is defined analogously for each transition in M'. Note that the graph $g'^S_\gamma(s)$ is characterized, as in the following diagram, as a pullback (a concretely chosen one) of the typing morphism $g'(s) \rightarrow \Gamma'$ along γ.*

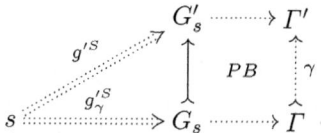

It should then be obvious how the reduct construction works for morphisms: given a signature morphism $\gamma : \Gamma \rightarrowtail \Gamma'$ and a graph transition system morphism $\sigma : \langle M, g \rangle \rightarrow \langle M', g' \rangle$ typed over Γ', the associated $\sigma{\restriction_\gamma} : \langle M, g_\gamma \rangle \rightarrow \langle M', g'_\gamma \rangle$ behaves exactly like σ on M: the consistency with the labeling trivially holds.

The canonical choice of the reduct allows to obtain a reduct functor, mapping each graph morphism $\gamma : \Gamma \rightarrowtail \Gamma'$ into a functor $\_{\restriction_\gamma} : \mathbf{GTS}(\Gamma') \rightarrow \mathbf{GTS}(\Gamma)$.

Definition 13 (model functor). *The (graph transformation) model functor $\mathbf{GTS} : \mathbf{MGraph}^{op} \rightarrow \mathcal{CAT}$ maps Γ to the category $\mathbf{GTS}(\Gamma)$ of graph transition systems typed over Γ and each signature morphim $\gamma : \Gamma \rightarrowtail \Gamma'$ to the reduct functor $\_{\restriction_\gamma} : \mathbf{GTS}(\Gamma') \rightarrow \mathbf{GTS}(\Gamma)$.*

The well-definedness of the functor $\mathbf{GTS}$ just introduced is based on the fact that the reduct is defined using a concrete choice of pullbacks along monomorphisms in the category of graphs, and thus it is functorial. In fact, this choice guarantees that the reduct $\langle M, g \rangle{\restriction_{id_\Gamma}}$ computed along an identity $id_\Gamma : \Gamma \rightarrowtail \Gamma$ is $\langle M, g \rangle$ itself, and similarly that $\langle M, g \rangle{\restriction_{\gamma_2 \circ \gamma_1}} = \langle M, g \rangle{\restriction_{\gamma_2}}{\restriction_{\gamma_1}}$.

Any other choice of pullbacks could be allowed, though, as long as it verifies the two properties above, in order to guarantee that $\mathbf{GTS}$ is a functor.

3.4 Satisfiability

We now define when a sentence is satisfied by a model, i.e., when a rule is satisfied by a graph transition system. This notion actually depends uniformly on the chosen graph transformation approach, which we will indicate as a parameter of the satisfaction relation $\models$. We shall consider the DPO, SPO and DPB approaches.

Definition 14 (satisfaction). *Let $\langle M, g \rangle$ be a graph transition system and let $p = (L \xleftarrow{l} K \xrightarrow{r} R)$ be a graph rule, both over Γ. Then, $\langle M, g \rangle$ satisfies p according to the approach $\mathcal{A} \in \{\text{DPO}, \text{SPO}, \text{DPB}\}$, written $\langle M, g \rangle \models^{\mathcal{A}}_{\Gamma} p$, if whenever there is an $\mathcal{A}$-valid match for p in the label $g(s)$ of a state s, as depicted in the left diagram below, then there is a transition witnessing the application of the rule p according to the approach $\mathcal{A}$, as depicted in the right diagram.*

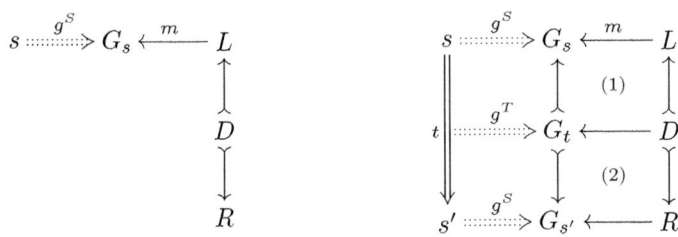

Here $t \in T_M$, and (1), (2) are both pushouts if $\mathcal{A} = \text{DPO}$, are both pullbacks if $\mathcal{A} = \text{DPB}$, and (1) is a final pullback complement, (2) is a pushout if $\mathcal{A} = \text{SPO}$.

In general, independently of the chosen approach, it is easy to check that a morphism $\sigma : \langle M, g \rangle \to \langle M', g' \rangle$ neither *preserves* nor *reflects* the satisfaction of a rule p. In fact, suppose that $\langle M, g \rangle \models^{\mathcal{A}} p$. If M' has a state s' such that p has an $\mathcal{A}$-valid match to $g'(s')$, if s' is not in the image of σ it is well possible that there is no transition with source s' witnessing the application of p, in which case $\langle M', g' \rangle \not\models^{\mathcal{A}} p$. Conversely, if $\langle M', g' \rangle \models^{\mathcal{A}} p$, M might still contain a state s with an $\mathcal{A}$-valid match for p in its label, but without a transition corresponding to the application of p leaving from s.

However, it is interesting to observe that a different notion of morphism, well-known in the literature and already used for standard transition systems (see e.g. [11]), does reflect satisfaction.

Definition 15 (tp-morphisms (aka *co-homomorphisms, open maps*)). *Let $\langle M, g \rangle$ and $\langle M', g' \rangle$ be graph transition systems over Γ. A transition preserving morphism (shortly, tp-morphisms) $\sigma : \langle M, g \rangle \to \langle M', g' \rangle$ is a graph transition system morphism, according to Definition 11, which additionally reflect transitions, i.e., such that it satisfies*

$$\forall s \in S_M . \forall t' \in T_{M'} . in_{M'}(t') = \sigma(s) \Rightarrow \exists t \in T_M . \sigma(t) = t' \wedge in_M(t) = s$$

Proposition 2 (tp-morphisms reflect satisfaction). *Let $\langle M, g \rangle$ and $\langle M', g' \rangle$ be graph transition systems over Γ and $\sigma : \langle M, g \rangle \to \langle M', g' \rangle$ a tp-morphims. Then for each rule p and approach $\mathcal{A}$, $\langle M', g' \rangle \models^{\mathcal{A}}_{\Gamma} p$ implies $\langle M, g \rangle \models^{\mathcal{A}}_{\Gamma} p$.*

Proof. Suppose that there is an $\mathcal{A}$-valid match m for p in graph $g^S(s)$, with $s \in M_S$. Then, since labels are preserved, m is also a match for p in $g'^S(\sigma(s))$; since $\langle M', g' \rangle \models^{\mathcal{A}}_\Gamma p$ there is a transition $t' : \sigma(s) \to s'$ in M' witnessing the application of p, and since σ is a tp-morphism there is a corresponding transition $t : s \to s''$ such that $\sigma(t) = t'$. As t and t' have the same label, t' witnesses the application of p to $g^S(s)$ according to the approach $\mathcal{A}$.

3.5   Satisfaction Condition

With the categories and functors introduced so far (Definitions 8, 9, 13 and 14), we can form three tuples like $\langle \mathbf{MGraph}, \mathbf{Rules}, \mathbf{GTS}, \models^{\mathcal{A}} \rangle$, with $\mathcal{A}$ varying in $\{\text{DPO}, \text{SPO}, \text{DPB}\}$. To see whether these are institutions or not, we must check if the *satisfaction condition* of Definition 7 holds for them. We will show that this condition holds for $\mathcal{A} = \text{DPB}$, and only for this approach.

Rephrasing the satisfaction condition for graph transformation systems, we must show that given a signature morphism $\gamma : \Gamma \to \Gamma'$

$$\langle M, g \rangle \models^{\mathcal{A}}_{\Gamma'} \mathbf{Rules}(\gamma)(p) \quad \Longleftrightarrow \quad \langle M, g \rangle |_\gamma \models^{\mathcal{A}}_\Gamma p \tag{1}$$

for each $\langle M, g \rangle \in |\mathbf{GTS}(\Gamma')|$ and $p \in \mathbf{Rules}(\Gamma)$.

Let us first analise Figure 2. In the left part, vertically, we depicted the Γ-typed rule $p = (L \hookleftarrow K \rightarrowtail R)$, drawing explicitly the typing morphisms to Γ. Notice that $\mathbf{Rules}(\gamma)(p)$ is basically the same rule, but typed over Γ' (by composing the typing morphisms with γ).

In the right part we have (also vertically) a generic transition $t : s \Rightarrow s'$ of M; in the graph transition system $\langle M, g \rangle \in |\mathbf{GTS}(\Gamma')|$ the transition t is labeled by

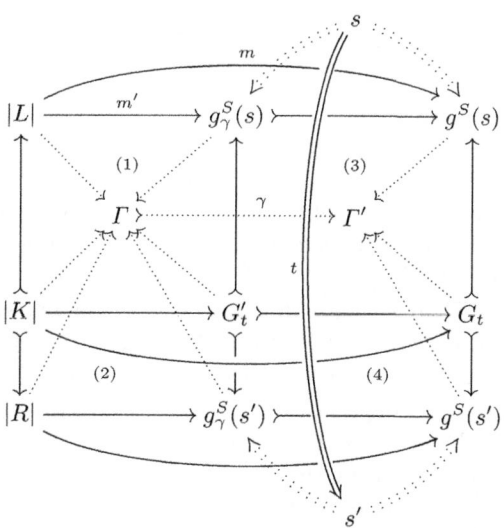

Fig. 2. Visualising the satisfaction condition

the span $g^S(s) \leftarrowtail G_t \rightarrowtail g^S(s')$ typed over Γ', while in the reduct $\langle M, g \rangle \!\upharpoonright_\gamma =$ $\langle M, g_\gamma \rangle \in |\mathbf{GTS}(\Gamma)|$ it is labeled by the span $g_\gamma^S(s) \leftarrowtail G'_t \rightarrowtail g_\gamma^S(s')$ typed over Γ. Note that squares (3) and (4) are pullbacks by standard decomposition properties, since obtained by taking suitable pullbacks along $\gamma : \Gamma \rightarrowtail \Gamma'$.

Now for the $\Leftarrow$-part of (1), suppose that $\langle M, g \rangle \!\upharpoonright_\gamma \models_\Gamma^{\mathcal{A}} p$, and that there is an $\mathcal{A}$-valid match m of $\mathbf{Rules}(\gamma)(p)$ in the label $g^S(s)$ of a state s of system $\langle M, g \rangle$. Thus m is as depicted in the top of Fig. 2, and (†) $|L| \xrightarrow{m} g^S(s) \to \Gamma' = |L| \to \Gamma \xrightarrow{\gamma} \Gamma'$ because m is a Γ'-typed morphism. As $g_\gamma^S(s)$ is the pullback of $g^S(s)$ along γ, by (†) there is a unique $m' : |L| \to g_\gamma^S(s)$ making the induced triangles commute. For any $\mathcal{A}$, match m' is $\mathcal{A}$-valid because so is m, by Prop. 1. Since $\langle M, g \rangle \!\upharpoonright_\gamma \models_\Gamma^{\mathcal{A}} p$, we know that there is a transition $t : s \Rightarrow s'$, labeled with span $g_\gamma^S(s) \leftarrowtail G'_t \rightarrowtail g_\gamma^S(s')$, witnessing the application of p to $g_\gamma^S(s)$, using match m' and the approach $\mathcal{A}$. Thus there are morphisms $|K| \to G'_t$ and $|R| \to g_\gamma^S(s')$ forming commutative squares (1) and (2), typed over Γ, whose nature depends on $\mathcal{A}$, as usual. Since transiton t is in the reduct $\langle M, g \rangle \!\upharpoonright_\gamma$, it is also in system $\langle M, g \rangle$, labeled with span $g^S(s) \leftarrowtail G_t \rightarrowtail g^S(s')$, and (3) and (4) are pullbacks. To conclude, let us make a simple case analysis on $\mathcal{A}$

- If $\mathcal{A} = \mathrm{DPB}$, then both (1) and (2) are pullbacks; therefore also the composed squares (1)+(3) and (2)+(4) are so, and thus transition t witnesses the application of $\mathbf{Rules}(\gamma)(p)$ to $g^S(s)$. Therefore (‡) $\langle M, g \rangle \models_{\Gamma'}^{\mathrm{DPB}} \mathbf{Rules}(\gamma)(p)$.
- If $\mathcal{A} \in \{\mathrm{DPO}, \mathrm{SPO}\}$, then (2) is a pushout, but the composed square (2)+(4) in general is not. The following diagram is a counter-example: the left square is a pushout, the right one is a pullback, but the composition is not a pushout. Starting from this, it is not difficult to build a counter-example showing that under the given hypotheses $\langle M, g \rangle \not\models_{\Gamma'}^{\mathcal{A}} \mathbf{Rules}(\gamma)(p)$, if $\mathcal{A} \in \{\mathrm{DPO}, \mathrm{SPO}\}$

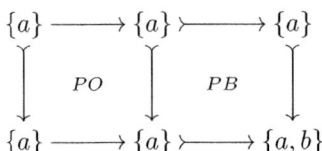

Let us consider now the $\Rightarrow$-part of (1) for the DPB approach only, assuming that $\langle M, g \rangle \models_{\Gamma'}^{\mathrm{DPB}} \mathbf{Rules}(\gamma)(p)$. Referring again to Fig. 2, suppose that there is a DPB-valid match m' of p in the label $g_\gamma^S(s)$ of a state s of system $\langle M, g \rangle \!\upharpoonright_\gamma$, the γ-reduct of $\langle M, g \rangle$. Then m' extends to a Γ'-typed match $m : L \to g^S(s)$ for $\mathbf{Rules}(\gamma)(p)$, which is DPB-valid by Prop. 1. Since $\langle M, g \rangle \models_{\Gamma'}^{\mathrm{DPB}} \mathbf{Rules}(\gamma)(p)$, there is a transition $t : s \Rightarrow s'$, labeled with span $g^S(s) \leftarrowtail G_t \rightarrowtail g^S(s')$, witnessing the application of p to $g^S(s)$: therefore the "big" squares $|K| \rightarrowtail |L| \xrightarrow{m} g^S(s) = |K| \to G_t \rightarrowtail g^S(s)$ and $|K| \rightarrowtail |R| \to g^S(s') = |K| \to G_t \rightarrowtail g^S(s')$ are pullbacks. Next, morphism $|K| \to G'_t$ is uniquely induced by $|K| \to \Gamma$ and $|K| \to G_t$, because G'_t is a pullback object, and similarly for morphism $|R| \to g_\gamma^S(s')$. Finally, since (3) and (4) are pullbacks, so are squares (1) and (2) by a standard decomposition property, showing that transition t witnesses the application of rule p to match m' using the DPB approach. Thus $\langle M, g \rangle \!\upharpoonright_\gamma \models_\Gamma^{\mathrm{DPB}} p$.

This concludes the proof that the satisfaction condition holds for the DPB-approach, and the discussion is summed up by the result below.

Theorem 1 (institution for double-pullback graph rewriting). *The tuple* $\langle \mathbf{MGraph}, \mathbf{Rules}, \mathbf{GTS}, \models^{\mathrm{DPB}} \rangle$ *is an institution.*

4   About the Double-Pullback Approach

The *double-pullback* approach to graph transformation has been introduced in [10,15] as a *loose* semantics of graph rules, in contraposition to the "strict" semantics of the single- and double-pushout approaches. In fact, one can observe that the SPO and DPO approaches assume an implicit *frame condition*: the application of a rule affects a graph only in the part matched by the left-hand side while the rest, the *context*, is preserved identically. On the contrary, in the DPB approach a rule can predicate only on the matched part of the system, while the rest may evolve on its own: in a derivation step, besides the effects described by the rule, arbitrary items can be deleted from the start graph and/or added to the target graph. This looser interpretation of the effect of a rule is suitable to describe the evolution of an open/reactive system, whose behaviour depends also on effects caused by the environment.

For the reader familiar with the DPO approach, the relationship with the DPB approach can be clarified by the following result [10]: if the match m and the co-match m' (see the next diagram) satisfy the identification condition, then there is a DPB-transformation $G \Rightarrow_{\mathrm{DPB}} H$ if and only if there is a DPO-transformation $G' \Rightarrow_{\mathrm{DPO}} H'$ and injective morphisms $G' \rightarrowtail G$ and $H' \rightarrowtail H$.

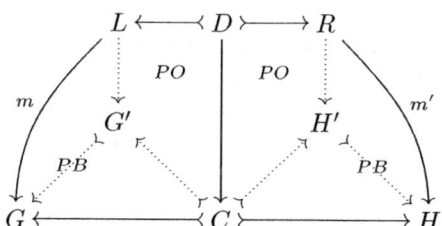

Classical results of the DPO approach have been studied for the DPB in [15], including independence, parallelism and local Church-Rosser properties, while in [14] the DPB approach has been exploited to define a loose semantics of graph transformation systems based on coalgebras.

Besides relating graph rewriting to other formalisms, one of the aims of an institutional view is to obtain structuring mechanisms, as briefly discussed in the final section. For the latter, a notion of loose semantics has already been recognized as needed by some attempts to define, for example, parameterized graph transformations [18,16]. Thus, it seems natural that an institution can be obtained only for the DPB approach, because structuring mechanisms are "built-in" in this formal framework and, specifically, the satisfaction property needs a loose interpretation of sentences to hold (see e.g. analogous considerations for the Logic Programming framework in [23]).

5   Conclusions

We proposed an institutional framework for modelling the algebraic approaches to graph transformation. After introducing the basics of (typed) graph and their transformation, we proved how the pulback/pushout mechanisms for graph rewriting can be modelled in terms of a suitable institution, based on graph-labelled transition systems. We remark the fact that in our proposal the models represent explicitly the "trace" formed by a derivation, denoting which elements are preserved along it. This represents the main difference with respect to adopting an institution based on (pre-)ordered algebras, where the rewriting step would be considered as atomic, even if possibly closed with respect to the operators of the algebra. The notion of trace lies at the basis of the concurrent semantics for graph transformation and we plan to exploit it in our future work.

As discussed in the previous sections, among the three algebraic approaches considered in the paper only the DPB gives rise to an institution. Nevertheless, it is easy to show that any DPO or SPO derivation step is also a DPB transformation. Thus given a rule p, the class of models that satisfy p according to the DPB approach contains all models satisfying p according to the DPO or SPO approaches. An interesting topic for future research is to identify which universal properties, if any, can characterize the latter models within the larger category of DPB models. However, preliminary investigations suggest that the strict preservation of labels required by the current notion of graph transition system morphisms does not allow to relate models that satisfy a rule "strictly" (i.e., with DPO or SPO) to models that satify it "loosely" (with DPB) via morphisms. Therefore we shall first investigate the possibility of defining a looser notion of morphism among models, one that does not preserve strictly the labels.

We thought of distilling an institution for graph transformation primarily as a sanity check, in order to lift and/or compare modularity concepts so far developed in the latter formalism. The current work is however just preliminary, since our chosen category of signatures (i.e., the category of graphs and their injective morphisms) lacks pushouts: this fact implies that the standard tools for model structuring (based on a suitable preservation of colimit diagrams by the model functor) are not available. In principle, the situation might be mitigated by adopting the solution proposed in [22]: basically, any diagram over **MGraph** is extended to one in **Graph**, and the colimit computed there. Now, it is easy to show that the resulting colimit diagram (the image of the diagram, plus the object and arrows induced by universal property) is also composed by monomorphisms. Thus, any colimit diagram in **Graph** out of a diagram factorizing through **MGraph** induces a commuting diagram in **MGraph**, for which a semi-exactness property can be stated (and hopefully proved).

Acknowledgements. We are indebt to Fernando Orejas who, based on his experience with institutions for Logic Programming [23], suggested us to consider a loose semantics of GTSs.

References

1. Baldan, P., Corradini, A., Montanari, U., Ribeiro, L.: Unfolding Semantics of Graph Transformation. Information and Computation 205(5), 733–782 (2007)
2. Baldan, P., Corradini, A., König, B.: A framework for the verification of infinite-state graph transformation systems. Information and Computation 206(7), 869–907 (2008)
3. Corradini, A., Gadducci, F.: A 2-Categorical Presentation of Term Graph Rewriting. In: Moggi, E., Rosolini, G. (eds.) CTCS 1997. LNCS, vol. 1290, pp. 87–105. Springer, Heidelberg (1997)
4. Corradini, A., Gadducci, F.: An algebraic presentation of term graphs, via gs-monoidal categories. Applied Categorical Structures 7(4), 299–331 (1999)
5. Corradini, A., Montanari, U., Rossi, F.: Graph processes. Fundamenta Informaticae 26(3/4), 241–265 (1996)
6. Corradini, A.: Concurrent Graph and Term Graph Rewriting. In: Montanari, U., Sassone, V. (eds.) CONCUR 1996. LNCS, vol. 1119, pp. 438–464. Springer, Heidelberg (1996)
7. Corradini, A., Heindel, T., Hermann, F., König, B.: Sesqui-Pushout Rewriting. In: Corradini, A., Ehrig, H., Montanari, U., Ribeiro, L., Rozenberg, G. (eds.) ICGT 2006. LNCS, vol. 4178, pp. 30–45. Springer, Heidelberg (2006)
8. Ehrig, H., Kreowski, H.J., Montanari, U., Rozenberg, G. (eds.): Handbook of Graph Grammars and Computing by Graph Transformation. Concurrency, Parallelism and Distribution, vol. 3. World Scientific (1999)
9. Ehrig, H.: Introduction to the Algebraic Theory of Graph Grammars (A Survey). In: Claus, V., Ehrig, H., Rozenberg, G. (eds.) Graph Grammars 1978. LNCS, vol. 73, pp. 1–69. Springer, Heidelberg (1979)
10. Ehrig, H., Heckel, R., Llabrés, M., Orejas, F., Padberg, J., Rozenberg, G.: Double-Pullback Graph Transitions: A Rule-Based Framework with Incomplete Information. In: Ehrig, H., Engels, G., Kreowski, H.-J., Rozenberg, G. (eds.) TAGT 1998. LNCS, vol. 1764, pp. 85–102. Springer, Heidelberg (2000)
11. Ferrari, G.L., Montanari, U.: Towards the unification of models for concurrency. In: Trees in Algebra and Programming, pp. 162–176 (1990)
12. Goguen, J.A., Burstall, R.M.: Institutions: Abstract model theory for specification and programming. Journal of ACM 39(1), 95–146 (1992)
13. Große-Rhode, M., Parisi-Presicce, F., Simeoni, M.: Formal software specification with refinements and modules of typed graph transformation systems. Journal of Computer and System Sciences 64(2), 171–218 (2002)
14. Heckel, R., Ehrig, H., Wolter, U., Corradini, A.: Double-pullback transitions and coalgebraic loose semantics for graph transformation systems. Applied Categorical Structures 9(1), 83–110 (2001)
15. Heckel, R., Llabrés, M., Ehrig, H., Orejas, F.: Concurrency and loose semantics of open graph transformation systems. Mathematical Structures in Computer Science 12(4), 349–376 (2002)
16. Kreowski, H.J., Kuske, S.: Approach-Independent Structuring Concepts for Rule-Based Systems. In: Wirsing, M., Pattinson, D., Hennicker, R. (eds.) WADT 2003. LNCS, vol. 2755, pp. 299–311. Springer, Heidelberg (2003)
17. Kreowski, H.-J., Kuske, S., Rozenberg, G.: Graph Transformation Units - An Overview. In: Degano, P., De Nicola, R., Meseguer, J. (eds.) Montanari Festschrift. LNCS, vol. 5065, pp. 57–75. Springer, Heidelberg (2008)

18. Kuske, S.: Parameterized transformation units. In: Bauderon, M., Corradini, A. (eds.) GETGRATS Closing Workshop. Electronic Notes in Theoretical Computer Science, vol. 51, pp. 246–257. Elsevier (2001)
19. Lack, S., Sobociński, P.: Adhesive and quasiadhesive categories. Informatique Théorique et Applications/Theoretical Informatics and Applications 39(3), 511–545 (2005)
20. Löwe, M.: Algebraic approach to single-pushout graph transformation. Theoretical Computer Science 109(1/2), 181–224 (1993)
21. Meseguer, J.: Conditional rewriting logic as a unified model of concurrency. Theoretical Computer Science 96(1), 73–155 (1992)
22. Mossakowski, T., Roggenbach, M.: Structured CSP - A Process Algebra as an Institution. In: Fiadeiro, J.L., Schobbens, P.-Y. (eds.) WADT 2006. LNCS, vol. 4409, pp. 92–110. Springer, Heidelberg (2007)
23. Orejas, F., Pino, E., Ehrig, H.: Institutions for logic programming. Theoretical Computer Science 173(2), 485–511 (1997)
24. Rozenberg, G. (ed.): Handbook of Graph Grammars and Computing by Graph Transformation. Foundations, vol. 1. World Scientific (1997)
25. Sannella, D., Tarlecki, A.: Specifications in an arbitrary institution. Information and Computation 76(2/3), 165–210 (1988)

New Results on Timed Specifications[*]

Timothy Bourke[2], Alexandre David[1], Kim G. Larsen[1], Axel Legay[2],
Didier Lime[3], Ulrik Nyman[1], and Andrzej Wąsowski[4]

[1] Computer Science, Aalborg University, Denmark
[2] INRIA/IRISA, Rennes Cedex, France
[3] IRCCyN/Ecole Centrale de Nantes, France
[4] IT University of Copenhagen, Denmark

Abstract. Recently, we have proposed a new design theory for timed systems. This theory, building on Timed I/O Automata with game semantics, includes classical operators like satisfaction, consistency, logical composition and structural composition. This paper presents a new efficient algorithm for checking Büchi objectives of timed games. This new algorithm can be used to enforce liveness in an interface, or to guarantee that the interface can indeed be implemented. We illustrate the framework with an infrared sensor case study.

1   Introduction and State of the Art

Several authors have proposed frameworks for reasoning about interfaces of independently developed components (e.g. [20,13,9,12]). Most of these works have, however, devoted little attention to real-time aspects. Recently, we proposed a new specification theory for Timed Systems (TS) [11]. Syntactically, our specifications are represented as Timed I/O Automata (TIOAs) [19], i.e., timed automata whose discrete transitions are labeled by *Input* and *Output* modalities. In contrast to most existing frameworks based on this model, we view TIOAs as games between two players: Input and Output, which allows for an optimistic treatment of operations on specifications [13].

Our theory is equipped with features typical of a compositional design framework: a *satisfaction relation* (to decide whether a TS is an implementation of a specification), a *consistency check* (whether the specification admits an implementation), and a refinement (to compare specifications in terms of inclusion of sets of implementations). Moreover, the model is also equipped with *logical composition* (to compute the intersection of sets of implementations), *structural composition* (to combine specifications) and its dual operator *quotient*. Our framework also supports incremental design [14].

Refinement, Satisfaction, and Consistency problems can be reduced to solving timed-games. For example, if inconsistent states are states that cannot be implemented, since they violate assumptions of the abstraction, then deciding whether an interface is consistent is equivalent to checking if a strategy that avoids inconsistent states exists.

Our theory is implemented in ECDAR [17], a tool that leverages the game engine UPPAAL-TIGA [4], as well as the model editor and the simulator of the UPPAAL model

---

[*] Work partially supported by VKR Centre of Excellence – MT-LAB and by an "Action de Recherche Collaborative" ARC (TP)I.

T. Mossakowski and H.-J. Kreowski (Eds.): WADT 2010, LNCS 7137, pp. 175–192, 2012.

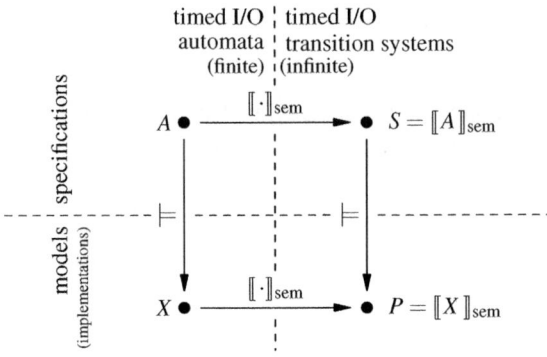

Fig. 1. Structure of our specification theory for real-time systems

checker [5]. The purpose of this paper is to describe enrichments to our theory, and to report on the evaluation of the tool on a concrete case study. Our contributions are:

1. *An on-the-fly algorithm for checking Büchi objectives of two-player timed games.* The algorithm builds on an existing, efficient method for solving reachability objectives [8,4], but it uses zones as a symbolic representation. We show how the method can be combined with a safety objective. This allows, for example, to guarantee that a player has a strategy to stay within a set of states without blocking the progress of time. Similar results were proposed by de Alfaro et al. [16] but for a restricted class of timed interfaces and without an implementation for the continuous case.

2. *A realistic case study.* Most existing interface theories have not been implemented and evaluated on concrete applications. We use ECDAR to show that our interface theory is indeed a feasible solution for the design of potentially complex timed systems. More precisely, we specify an infrared sensor for measuring short distances and for detecting obstructions. This extensive case study reveals both the advantages and disadvantages of our theory, which are summarized in this paper.

2   Background: Real Time Specifications as Games

Following [11], we now introduce the basic objects of this paper. Our specifications and models (implementations) are taken from the same class, timed games. They both exist in two flavors: infinite and finite. Fig. 1 summarizes this structure. The top–bottom division goes across the notion of satisfaction (models and specifications) and the left–right one across syntax-semantics (Timed I/O Transition Systems and Timed I/O Automata). This orthogonality is exploited to treat the intricacies of continuous time behaviour separately from those of algorithms. Roughly, the infinite models have been used to develop the theory, while the finite symbolic representations are used in the implementation.

Definition 1. *A Timed I/O Transition System (TIOTS) is a tuple $S = (St^S, s_0, \Sigma^S, \rightarrow^S)$, where St^S is an infinite set of states, $s_0 \in St$ is the initial state, $\Sigma^S = \Sigma_i^S \oplus \Sigma_o^S$ is a finite set of actions partitioned into inputs and outputs, and $\rightarrow^S : St^S \times (\Sigma^S \cup \mathcal{R}_{\geq 0}) \times St^S$ is a transition relation. We write $s \xrightarrow{a}^S s'$ instead of $(s, a, s') \in \rightarrow^S$ and use $i?$, $o!$ and d*

to range over inputs, outputs and $\mathcal{R}_{\geq 0}$ respectively. Also for any TIOTS we require:

[time determinism] whenever $s \xrightarrow{d}_S s'$ and $s \xrightarrow{d}_S s''$ then $s' = s''$,

[time reflexivity] $s \xrightarrow{0}_S s$ for all $s \in St^S$, and,

[time additivity] for all $s, s'' \in St^S$ and all $d_1, d_2 \in \mathcal{R}_{\geq 0}$ we have $s \xrightarrow{d_1 + d_2}_S s''$ iff $s \xrightarrow{d_1}_S s'$ and $s' \xrightarrow{d_2}_S s''$ for some $s' \in St^S$.

We write $s \xrightarrow{a}_S$ meaning that there exists a state s' such that $s \xrightarrow{a}_S s'$.

TIOTSs are abstract representations of real time behaviour. We use *Timed I/O Automata* (TIOAs) to represent them symbolically using finite syntax.

Let Clk be a finite set of *clocks*. $[Clk \mapsto \mathcal{R}_{\geq 0}]$ denotes the set of mappings from Clk to $\mathcal{R}_{\geq 0}$. A *valuation* over Clk is an element u of $[Clk \mapsto \mathcal{R}_{\geq 0}]$. Given $d \in \mathcal{R}_{\geq 0}$, we write $u + d$ to denote a valuation such that for any clock $r \in Clk$ we have $(u + d)(r) = x + d$ iff $u(r) = x$. We write $u[r \mapsto 0]_{r \in c}$ for a valuation which agrees with u on all values for clocks not in c, and gives 0 for all clocks in $c \subseteq Clk$. Let op be the set of relational operators: op $= \{<, \leq, >, \geq\}$. A *guard* over Clk is a finite conjunction of expressions of the form $x \prec n$, where $\prec \in$ op and $n \in \mathbb{N}$. We write $\mathcal{B}(Clk)$ for the set of guards over Clk using operators in the set op, and $\mathscr{P}(X)$ for the powerset of a set X.

Definition 2. *A Timed I/O Automaton (TIOA) is a tuple $A = (Loc, q_0, Clk, E, Act, Inv)$ where Loc is a finite set of locations, $q_0 \in Loc$ is the initial location, Clk is a finite set of clocks, $E \subseteq Loc \times Act \times \mathcal{B}(Clk) \times \mathscr{P}(Clk) \times Loc$ is a set of edges, Act is the action set $Act = Act_i \oplus Act_o$, partitioned into inputs and outputs respectively, and $Inv : Loc \mapsto \mathcal{B}(Clk)$ is a set of location invariants.*

If $(q, a, \varphi, c, q') \in E$ is an edge, then q is a source location, a is an action, φ is a constraint over clocks that must be satisfied when the edge is executed, c is a set of clocks to be reset, and q' is the target location. We will give examples of TIOAs in Sect. 4.

The expansion of the behaviour of a TIOA $A = (Loc, q_0, Clk, E, Act, Inv)$ is the following TIOTS $[\![A]\!]_{\text{sem}} = (Loc \times [Clk \mapsto \mathcal{R}_{\geq 0}], (q_0, \mathbf{0}), Act, \to)$, where $\mathbf{0}$ is a constant function mapping all clocks to zero, and $\to$ is generated by the two rules:

- Each $(q, a, \varphi, c, q') \in E$ gives rise to $(q, u) \xrightarrow{a} (q', u')$ for each clock valuation $u \in [Clk \mapsto \mathcal{R}_{\geq 0}]$ such that $u \models \varphi$ and $u' = u[r \mapsto 0]_{r \in c}$ and $u' \models Inv(q')$.
- Each location $q \in Loc$ with a valuation $u \in [Clk \mapsto \mathcal{R}_{\geq 0}]$ gives rise to a transition $(q, u) \xrightarrow{d} (q, u + d)$ for each delay $d \in \mathcal{R}_{\geq 0}$ such that $u + d \models Inv(q)$.

We refer to states and transitions of a TIOA, meaning the states and transitions of the underlying TIOTS. As stated above, these states are location–clock valuation pairs.

The TIOTSs induced by TIOAs conform to Def. 1. In addition, to guarantee determinism, for each action–location pair only one transition can be enabled at a time. This is a standard check. We assume that all TIOAs below are deterministic.

Implementations (models) are a subclass of specifications that are amenable to implementation. They have fixed timing behaviour (outputs occur at predictable times) and can always advance either by producing an output or delaying.

Definition 3. *A TIOA A is a specification if each state $s \in St^{[\![A]\!]_{\text{sem}}}$ is input-enabled:*

[input enabledness] $\forall i? \in \Sigma_i^{[\![A]\!]_{\text{sem}}}. s \xrightarrow{i?}_{[\![A]\!]_{\text{sem}}}$.

Definition 4. *An* implementation *A is a* specification *(so a suitable TIOA), where, in addition, for each state $p \in St^{[\![A]\!]_{sem}}$ the following two conditions hold:*

[output urgency] for each $o! \in \Sigma_o^{[\![A]\!]_{sem}}$ if $p \xrightarrow{o!}_{[\![A]\!]_{sem}}$ and $p \xrightarrow{d}_{[\![A]\!]_{sem}}$ then $d = 0$ and,

[independent progress] $(\forall d \geq 0. \; p \xrightarrow{d}_{[\![A]\!]_{sem}})$ or

$$(\exists d \in \mathcal{R}_{\geq 0}. \; \exists o! \in \Sigma_o^{[\![A]\!]_{sem}}. \; p \xrightarrow{d}_{[\![A]\!]_{sem}} p' \text{ and } p' \xrightarrow{o!}_{[\![A]\!]_{sem}})$$

Specifications are a subclass of TIOAs (the upper-left quadrant in Fig. 1) which induce TIOTSs that are input-enabled (the upper-right quadrant). Implementations are TIOAs (the lower-left quadrant) that induce both input-enabled and output-urgent TIOTSs able to progress independently (the lower-right quadrant). Although specifications and implementations are defined above by restricting their semantic properties, it is possible, although more clumsy, to rephrase these conditions syntactically and implement them in a tool. These are again standard checks.

A run ρ of a TIOTS S from its state s_1 is a sequence $s_1 \xrightarrow{a_1} s_2 \xrightarrow{a_2} \cdots \xrightarrow{a_{n-1}} s_n$ such that for all $i \in [1..n]$, $s_i \xrightarrow{a_i} s_{i+1}$ is a transition of S. We write $\mathsf{Runs}(s_1, S)$ for the set of runs of S starting in s_1, and $\mathsf{Runs}(S)$ for the set of runs starting from the initial state of S. We write $\mathsf{States}(\rho)$ for the set of states of S present in ρ and, if ρ is finite, $\mathsf{last}(\rho)$ for the last state occurring in ρ.

TIOAs are interepreted as two-player real-time games between the *output player* (the component) and the *input player* (the environment). The *input* plays with actions in Σ_i and the *output* plays with actions in Σ_o:

Definition 5. *A strategy f for the input (resp. output) player, $k \in \{i, o\}$, on the TIOA A is a partial function from $\mathsf{Runs}([\![A]\!]_{sem})$ to $Act_i \cup \{\mathsf{delay}\}$ (resp. $Act_o \cup \{\mathsf{delay}\}$) such that for every finite run ρ, if $f(\rho) \in Act_k$ then $\mathsf{last}(\rho) \xrightarrow{f(\rho)} s'$ for some state s' and if $f(\rho) = \mathsf{delay}$, then $\exists d > 0. \exists s''$ such that $\mathsf{last}(\rho) \xrightarrow{d} s''$.*

For a given strategy, we consider behaviors resulting from the application of the strategy to the TIOA, with respect to all possible strategies of the opponent:

Definition 6 (Outcome [15]). *Let A be a TIOA, f a strategy over A for the input player, and s a state of $[\![A]\!]_{sem}$. The outcome $\mathsf{Outcome}_i(s, f)$ of f from s is the subset of $\mathsf{Runs}(s, [\![A]\!]_{sem})$ defined inductively by:*

- *$s \in \mathsf{Outcome}_i(s, f)$,*
- *if $\rho \in \mathsf{Outcome}_i(s, f)$ then $\rho' = \rho \xrightarrow{e} s' \in \mathsf{Outcome}_i(s, f)$ if*
  $\rho' \in \mathsf{Runs}(s, [\![A]\!]_{sem})$ and one of the following three conditions hold:
  1. $e \in Act_o$,
  2. $e \in Act_i$ and $e = f(\rho)$,
  3. $e \in \mathcal{R}_{\geq 0}$ and $\forall 0 \leq e' < e. \exists s''. \mathsf{last}(\rho) \xrightarrow{e'} s''$ and $f(\rho \xrightarrow{e'} s'') = \mathsf{delay}$.
- *$\rho \in \mathsf{Outcome}_i(s, f)$ if ρ infinite and all its finite prefixes are in $\mathsf{Outcome}_i(s, f)$*

Let $\mathsf{MaxOutcome}_i(s, f)$ denote the maximal runs of $\mathsf{Outcome}_i(s, f)$, that is $\rho \in \mathsf{MaxOutcome}_i(s, f)$ iff $\rho \in \mathsf{Outcome}_i(s, f)$ and ρ has an infinite number of discrete actions, or ρ has a finite number of discrete actions, but there exist no $e \in Act \cup \mathcal{R}_{\geq 0}$ and no state s' with $\rho \xrightarrow{e} s' \in \mathsf{Outcome}_i(s, f)$, or the sum of the delays in ρ is infinite.

For a given TIOA A, a *winning condition* W for input is a subset of Runs($[\![A]\!]_{\text{sem}}$). We say that W does not depend on the progress of the opponent (here output) iff whenever $\rho \in W$ and $\rho = \rho' \xrightarrow{e} \rho''$, with $e \in Act_o$, then either there exists $e' \in Act_i$, $d \in \mathcal{R}_{\geq 0}$, a state s and a run ρ''' such that $\rho' \xrightarrow{d} s \xrightarrow{e'} \rho''' \in W$ or there exists $d \in \mathcal{R}_{\geq 0}$ and some state s such that $\rho' \xrightarrow{d} s \in W$. This restriction means that input should always be able to ensure progress by itself and that the actions of the opponent should not be abused to advance the game, since we cannot assume that the opponent will ever make use of them. For a winning condition W, we write Strip(W) to denote the subset of W in which the runs not satisfying this condition are removed.

A pair (A, W) is an *input timed game*. Given a winning condition W for input, a strategy f of input is *winning* from state s if MaxOutcome(s, f) $\subseteq W$. A state s is *winning* for input, if there exists a winning strategy for input from s. The game (A, W) is *winning* for input if the initial state of A is winning for it. For an input timed game (A, W), we write $\mathcal{W}_i(A, W)$ for the set of winning states for input and $\mathcal{F}_i(A, W, s)$ for all winning strategies for input from s. The winning conditions considered here are:

- Reachability objective: the input player must enforce a set Goal of "good" states. The corresponding winning condition is defined as

$$WR_i(\text{Goal}) = \text{Strip}\{\rho \in \text{Runs}([\![A]\!]_{\text{sem}}) \mid \text{States}(\rho) \cap \text{Goal} \neq \emptyset\} \quad (1)$$

- Safety objective: the player must avoid a set Bad of "bad" states. The corresponding winning condition is defined as:

$$WS_i(\text{Bad}) = \{\rho \in \text{Runs}([\![A]\!]_{\text{sem}}) \mid \text{States}(\rho) \cap \text{Bad} = \emptyset\} \quad (2)$$

- Büchi objective: the player must enforce visiting Goal, a set of "good" states, infinitely often. Let $|A|$ denote the cardinality of set A. The winning condition is:

$$WB_i(\text{Goal}) = \text{Strip}\{\rho \in \text{Runs}([\![A]\!]_{\text{sem}}) \mid |\text{States}(\rho) \cap \text{Goal}| = \infty\} \quad (3)$$

We define the outcomes Outcome$_o(s, f)$ and MaxOutcome$_o(s, f)$ of a strategy of the output player, as well as output timed games and all the related notions, by swapping 'i' and 'o' (for instance Act_i and Act_o) in the above definitions.

We now present discuss the *refinement relation*, which relates TIOTSs of two real time specifications, by determining which one allows more behaviour:

Definition 7. *A TIOTSs $S = (St^S, s_0, \Sigma, \rightarrow^S)$ refines a TIOTSs $T = (St^T, t_0, \Sigma, \rightarrow^T)$, written $S \leq T$, iff there exists a binary relation $R \subseteq St^S \times St^T$ containing (s_0, t_0) such that for each pair of states $(s, t) \in R$ we have:*

1. *if $t \xrightarrow{i?}^T t'$ for some $t' \in St^T$ then $s \xrightarrow{i?}^S s'$ and $(s', t') \in R$ for some $s' \in St^S$*
2. *if $s \xrightarrow{o!}^S s'$ for some $s' \in St^S$ then $t \xrightarrow{o!}^T t'$ and $(s', t') \in R$ for some $t' \in St^T$*
3. *if $s \xrightarrow{d}^S s'$ for $d \in \mathcal{R}_{\geq 0}$ then $t \xrightarrow{d}^T t'$ and $(s', t') \in R$ for some $t' \in St^T$*

A specification A_1 refines a specification A_2, written $A_1 \leq A_2$, iff $[\![A_1]\!]_{\text{sem}} \leq [\![A_2]\!]_{\text{sem}}$. If A_1 is an implementation then we also say that it satisfies A_2, written $A_1 \models A_2$.

Refinement between two automata may be checked by playing a safety game on the product of their two state spaces, avoiding the error states (where error states are pairs of states of S and T for which one of the above rules is violated). See details in [11,13]. Since the product can be expressed as a TIOA itself, the refinement can be checked using the safety game as defined above.

Consider two TIOTSs $S = (St^S, s_0^S, \Sigma^S, \to^S)$ and $T = (St^T, s_0^T, \Sigma^T, \to^T)$. We say that they are *composable* iff their output alphabets are disjoint $\Sigma_o^S \cap \Sigma_o^T = \emptyset$. The *product* of S and T is the TIOTS $S \otimes T = (St^S \otimes St^T, (s_0^S, s_0^S), \Sigma^{S \otimes T}, \to^{S \otimes T})$, where the alphabet $\Sigma^{S \otimes T} = \Sigma^S \cup \Sigma^T$ is partitioned into inputs and outputs in the following way: $\Sigma_i^{S \otimes T} = (\Sigma_i^S \setminus \Sigma_o^T) \cup (\Sigma_i^T \setminus \Sigma_o^S)$, $\Sigma_o^{S \otimes T} = \Sigma_o^S \cup \Sigma_o^T$. The transition relation is generated by the following rules:

$$\frac{s \xrightarrow{a}{}^S s' \quad a \in \Sigma^S \setminus \Sigma^T}{(s,t) \xrightarrow{a}{}^{S \otimes T} (s', t)} \text{[indep-l]} \qquad \frac{t \xrightarrow{a}{}^T t' \quad a \in \Sigma^T \setminus \Sigma^S}{(s,t) \xrightarrow{a}{}^{S \otimes T} (s, t')} \text{[indep-r]}$$

$$\frac{s \xrightarrow{a}{}^S s' \quad t \xrightarrow{a}{}^T t' \quad a \in \mathcal{R}_{\geq 0} \cup \Sigma_i^{S \otimes T} \cup (\Sigma_i^S \cap \Sigma_o^T) \cup (\Sigma_o^S \cap \Sigma_i^S)}{(s,t) \xrightarrow{a}{}^{S \otimes T} (s', t')} \text{[sync]}$$

Let undesirable be a set of error states that violate a safety property (for example, an elevator engine running while its door is open). Two specifications are *useful* with respect to one another if there is an environment that can avoid undesirable states in their product. The existence of such an environment is established by finding a winning strategy in the game formed by the product automaton and the objective WS_i(undesirable).

The parallel composition of S and T is defined as $S \mid T = prune(S \otimes T)$, where the prune operation removes from $S \otimes T$ all states which are not winning for the input player in the game $(S \otimes T, WS_i(\text{undesirable}))$. Parallel composition is defined for TIOTSs induced by both specifications and implementations. A similar construction can be given directly for specifications and implementations on the syntactic level [11].

In [11] we give constructions for two other operators computed as winning strategies in timed games. For TIOAs (TIOTSs) B and C we define conjunction $B \wedge C$, which computes an automaton representing shared implementations of B and C, and also quotient $B \setminus C$, which computes a specification describing implementations that when composed with C give a specification refining B. Rather than define these operations explicitly we characterize their essential properties, and refer the reader to [11] for precise details of the constructions. Let A be an implementation. Then:

$$A \models B \wedge C \quad \text{iff} \quad A \models B \text{ and } A \models C \tag{4}$$

$$A \models B \setminus C \quad \text{iff} \quad C \mid A \leq B \tag{5}$$

3   Büchi Objectives

Symbolic On-The-Fly Timed Reachability (SOFTR)[8] is an efficient algorithm for solving two-players reachability timed games used in UPPAAL-TIGA [4]. It operates on the simulation graph induced by a TIOA representing the game. It follows an established principle: begin with all reachable states and propagate the winning states backwards. Its major contribution is the use of zones rather than regions. Zones, which are unions of regions of Alur and Dill [3], are the most efficient representation of clock valuations known to date. In the following we recall SOTFTR, extend it to solve Büchi objectives, and provide a new algorithm to verify Büchi and safety objectives combined.

3.1 Solving Büchi Games with SOTFTR

For a TIOTS S and a set of states X, write $\mathsf{Pred}_a(X) = \{s \in St \mid \exists s' \in X.\ s \xrightarrow{a} s'\}$ for the set of all a-predecessors of states in X. We write $\mathsf{iPred}(X)$ for the set of all input predecessors, and $\mathsf{oPred}(X)$ for all the output predecessors of X, so $\mathsf{iPred}(X) = \bigcup_{a \in \Sigma_i^S} \mathsf{Pred}_a(X)$ and $\mathsf{oPred}(X) = \bigcup_{a \in \Sigma_o^S} \mathsf{Pred}_a(X)$. Also $\mathsf{post}_{[0,d_0]}(s)$ is the set of all time successors of a state s that can be reached by delays less than or equal to d_0: $\mathsf{post}_{[0,d_0]}(s) = \{s' \in St^S \mid \exists d \in [0,d_0].\ s \xrightarrow{d}_S s'\}$. The safe timed predecessors of a set X relative to an unsafe set Y are the states from which a state in X is reached after a delay while avoiding any of the states in Y (the subscript t in the definition of cPred_t below indicates that these are timed predecessors only):

$$\mathsf{cPred}_t(X,Y) = \{s \in St^S \mid \exists d_0 \in \mathcal{R}_{\geq 0}.\ \exists s' \in X.s \xrightarrow{d_0}_S s' \text{ and } \mathsf{post}^S_{[0,d_0]}(s) \subseteq \overline{Y}\}$$

Let A be a TIOA and G a set of "good" states in $[\![A]\!]_{\mathrm{sem}}$ that have to be reached, that is the objective is $WR_i(G)$. Consider the following computation [21,8]:

$$H_0 \leftarrow \emptyset$$
repeat $H_{k+1} \leftarrow H_k \cup \pi_i(H_k) \cup G$ **for** $k = 0, 1, \ldots$
until $H_{k+1} = H_k$

where $\pi_i(H) = \mathsf{cPred}_t(\mathsf{iPred}(H), \mathsf{oPred}(\mathsf{States}(\mathsf{Runs}([\![A]\!]_{\mathrm{sem}})) \setminus H))$. The π_i operator computes the predecessors of set H that can enforce H in one step, regardless of what the output player does. This is done by taking timed predecessors of input-predecessors of H, as long as we can avoid output predecessors of states outside H. The fixpoint of π_i is the set of states in which the input player can enforce reaching G eventually [21,8]. SOTFTR is a symbolic zone-based implementation of the above fixpoint.

The winning states of the output player can be computed by replacing π_i with $\pi_o(H) = \mathsf{cPred}_t(\mathsf{oPred}(H), \mathsf{iPred}(\mathsf{States}(\mathsf{Runs}([\![A]\!]_{\mathrm{sem}})) \setminus H))$. Thus, in the remainder, we focus on solving the game for the input player only.

The following algorithm for solving Büchi timed games is an adaptation of the above procedure given in [21], adjusted for a TIOA A and a Büchi objective. The set of "good" states, Goal, is to be enforced infinitely often:

$$W_0 \leftarrow \mathsf{States}(\mathsf{Runs}([\![A]\!]_{\mathrm{sem}}))$$
for $j = 0, 1, \ldots$ **repeat**
    $H_0 \leftarrow \emptyset$
    repeat $H_{k+1} \leftarrow H_k \cup \pi_i(H_k) \cup (\mathsf{Goal} \cap \pi_i(W_j))$ **for** $k = 0, 1, \ldots$
    until $H_{k+1} = H_k$
    $W_{j+1} \leftarrow H_k$
until $W_{j+1} = W_j$

Observe that a Büchi objective is essentially a closure of a reachability objective: it corresponds to finding a subset of "good" Goal states, from which reachability to the good subset again is guaranteed for the player, and then solving for reachability of that good subset. In the above computation, the inner loop finds states that can enforce a Goal state in at least one discrete step, and uses this information to determine which Goal states are actually "good" (the intersection with Goal). The outer loop removes the Goal states that are not "good" from the target set of the inner loop. In the fixpoint, we find both the subset of good Goal states and the states from which this subset can be reached regardless of what the opponent does.

SOTFTR itself computes the inner loop of this algorithm when $G = \text{Goal} \cap \pi_i(W_j)$, this observation leads to the *Symbolic Timed Büchi* games (STB) algorithm:

$$W_0 \leftarrow \text{States}(\text{Runs}(\llbracket A \rrbracket_{\text{sem}}))$$
$$\textbf{repeat } W_{j+1} \leftarrow \text{SOTFTR}(\text{Goal} \cap \pi_i(W_j)) \textbf{ for } j = 0, 1, \ldots$$
$$\textbf{until } W_{j+1} = W_j$$

Observe that STB uses exactly the same operations on zones as SOTFTR, which means that it can also be implemented in an efficient manner.

Theorem 1 ([8,21]). *For any input Büchi timed game* $(A, WB_i(\text{Goal}))$, *STB terminates and upon termination* $W_j = \mathcal{W}_i(A, WB_i(\text{Goal}))$.

The algorithm of [21] computes over infinite sets of states. Our algorithm is nothing more than a symbolic implementation of the original one. By construction and because of [8], the above correspondence is obtained directly. Termination is shown in [21].

3.2   Combining Safety and Büchi objectives

We now strengthen the Büchi objective so that not only the Goal states are visited infinitely often, but also the set of unsafe states Bad is avoided ($\text{Bad} \cap \text{Goal} = \emptyset$):

$$WBS(\text{Goal}, \text{Bad}) = \text{Strip}\{\rho \in \text{Runs}(\llbracket A \rrbracket_{\text{sem}}) \mid \text{States}(\rho) \cap \text{Bad} = \emptyset \text{ and}$$
$$|\text{States}(\rho) \cap \text{Goal}| = \infty\} \quad (6)$$

One application of such games is ensuring that the input player has a strategy to avoid Bad while ensuring that time is elapsing [16], eliminating the so called Zeno-behaviours.

If Bad can be expressed as a finite union of pairs of locations and finite unions of zones, then this objective can be reduced to the usual Büchi objective by transforming the game in the following way: (i) add a location $B \notin \text{Goal}$; (ii) add an *output* action $err \notin Act_i$; (iii) for each pair $(q, \bigcup_{i=1..n} Z_i) \in \text{Bad}$ such that q is a location of A and $\bigcup_{i=1..n} Z_i$ is a finite union of zones, add n edges E_i $(i = 0..n)$ labelled by err from q to B such that for all i, the guard of E_i is Z_i. Since location B has no outgoing edges and does not belong to Goal, entering B means losing the Büchi game. Suppose we want a winning strategy for the input player. Observe that the added edges belong to the opponent. By definition of outcomes, going through any state in Bad means that one of these edges can now be taken by the output player and, as $B \notin \text{Goal}$, the game is lost for the input player. The following theorem expresses the correctness of our transformation.

Theorem 2. *Let* $(A, WBS_i(\text{Goal}, \text{Bad}))$ *be a TIOA, and* A' *be its modification obtained by the above construction. Then* $\mathcal{F}_i(A, WBS_i(\text{Goal}, \text{Bad})) = \mathcal{F}_i(A', WB_i(\text{Goal}))$

Proof. Show that $\mathcal{F}_i(A, WBS_i(\mathsf{Goal}, \mathsf{Bad})) \subseteq \mathcal{F}_i(A', WB_i(\mathsf{Goal}))$. Let f be a strategy in $\mathcal{F}_i(A, WBS_i(\mathsf{Goal}, \mathsf{Bad}))$ and s_0 be the initial state of A and s_0' of A'. As f is winning, no run in $\mathsf{MaxOutcome}_i(s_0, f)$ goes through a state in Bad. By construction of A' we have that no run in $\mathsf{MaxOutcome}_i(s_0', f)$ goes through Bad and therefore the guards of the extra edges in A are never satisfied. Since, apart from these edges, A' is identical to A, and since f ensures infinite repetition of Goal in A, then it does also in A'.

Now, show $\mathcal{F}_i(A, WBS_i(\mathsf{Goal}, \mathsf{Bad})) \supseteq \mathcal{F}_i(A', WB_i(\mathsf{Goal}))$. Let f be a strategy in $\mathcal{F}_i(A, WBS_i(\mathsf{Goal}, \mathsf{Bad}))$ and s_0 be the initial state of A, and s_0' of A'. The runs of A' that go to location B are maximal and cannot belong to $WB_i(\mathsf{Goal})$ for B has no outgoing edge. Let ρ be a run in $\mathsf{MaxOutcome}_i(s_0', f)$ and $\rho = \rho' \to s \to \rho''$, and the guard of one of the *err* edges is satisfied in s. Then $\rho' \xrightarrow{err} (B, v)$ for some valuation v is a maximal run and thus belongs to $\mathsf{MaxOutcome}_i(s_0', f)$ and then $\mathsf{MaxOutcome}_i(s_0', f) \not\subseteq WB_i(\mathsf{Goal})$ which contradicts that f is winning. So the runs in $\mathsf{MaxOutcome}_i(s_0', f)$ never go through states in Bad. Furthermore, since A and A' are identical except for B and its incoming edges, it must then be that $\mathsf{MaxOutcome}_i(s_0, f) = \mathsf{MaxOutcome}_i(s_0', f)$ and so the runs in $\mathsf{MaxOutcome}_i(s_0, f)$ also repeat Goal infinitely often.   □

An Application: eliminating Zeno Strategies. Consider a TIOA A and a set Bad of bad states. Our objective is to find the set of states from which the input player (symmetrically the output player) has a strategy to avoid Bad while letting time elapse — as opposed to, for example, taking infinitely many discrete transitions without any delays.

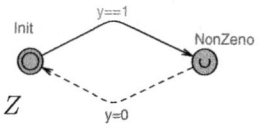

Fig. 2. Monitor for non-zeno strategies

In order to generate non-zeno strategies consider the product $A \times Z$ of A and the TIOA Z of Fig. 2. Then solve the timed game $(A \times Z, WBS_i(\mathsf{Goal}, \mathsf{Bad}))$, where Goal is the set of states of $A \times Z$ in which Z is in location $\mathsf{NonZeno}$. To fulfill this objective, the input player needs to avoid Bad *and* ensure that $\mathsf{NonZeno}$ is visited infinitely often: once in $\mathsf{NonZeno}$, the only way to revisit it is to pass through Init. This loop requires that 1 time unit elapses, so repeated visits to $\mathsf{NonZeno}$ ensure that time progresses.

Note that this does not prevent the opposing player from using a spoiling strategy producing zeno runs to prevent fulfillment of the objective.

Remark 1. One problem with the above setup is the effect of adding self-loops. Our interface theory requires TIOA to be input-enabled. This means that, in any state of the game, the input player should always be able to react on any of the input actions. This typically means that states have implicit loops on input actions when the designer does not specify any other transition for an input. Now, assume that the output player wants to win the game and guarantee that time elapses. The input player could always play such an input-loop and hence block time. This means that the potential addition of arbitrary inputs may corrupt the game. A solution to the above problem is to blame the input player each time it plays [16]. Then, the input player loses the game if there is a point of time after which it is blamed forever. De Alfaro et al. were the first to use blames [16]. We can also add a monitor for the blame situation. Another solution, in order to avoid adding an extra automaton, is to use a counter in ECDAR to bound the number of Inputs (Outputs) that can be played successively.

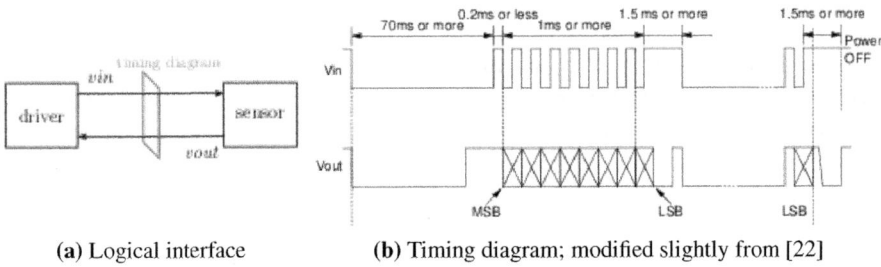

(a) Logical interface   (b) Timing diagram; modified slightly from [22]

Fig. 3. The driver/sensor system

4   Case Study

The ideas just presented have been implemented in the tool ECDAR [10], which supports graphical modeling of TIOAs, computing composition operators (including quotienting), and reachability analysis. For this paper, we have extended ECDAR with support for Büchi and Büchi with safety objectives. We apply it to the analysis of a simple but realistic example: a sensor component and the software required to interface with it.[1] The case study serves both to elucidate some of the technical definitions and to demonstrate their practicability.

4.1   Timing Diagram Model

The Sharp GP2D02 infrared sensor is a small component for measuring short distances and for detecting obstructions. Such sensors are incorporated into larger embedded systems through two communication wires which carry a protocol of rising and falling voltage levels. The four main components of a sensor subsystem are shown in Fig. 3a: an instance of the *sensor*, a *driver* component of a larger system, a *vin* wire controlled by the driver and read by the sensor, and a *vout* wire controlled by the sensor and read by the driver. The communication protocol between driver and sensor is described by the timing diagram of Fig. 3b.

The timing diagram describes the permissible interactions between a driver and a sensor. It represents a (partial) ordering of events and the timing constraints between them. With careful interpretation, against a background of engineering practice, the timing diagram can be modeled as the TIOA shown in Fig. 4 and henceforth called T. Note that constants are multiples of 0.1 ms, so the constant 0.2 ms in the timing diagram is represented by an integer constant 2 in the model. This model is the result of several choices and its fidelity can only be justified by informal argument [6, Chapter 4].

We now step through the timing diagram and the TIOA model in parallel describing the meaning of the former and justifying the latter. The interaction of driver and sensor is essentially quite simple: the driver requests a range reading, then after a brief delay the sensor signals that a reading has been made, the driver triggers the sensor to transmit the reading bit by bit, and, finally, the process is repeated or the sensor is powered off. The interaction takes place solely over the two communication wires.

---

[1] See http://www.tbrk.org/papers/wadt10.tar.gz for the implementation in ECDAR.

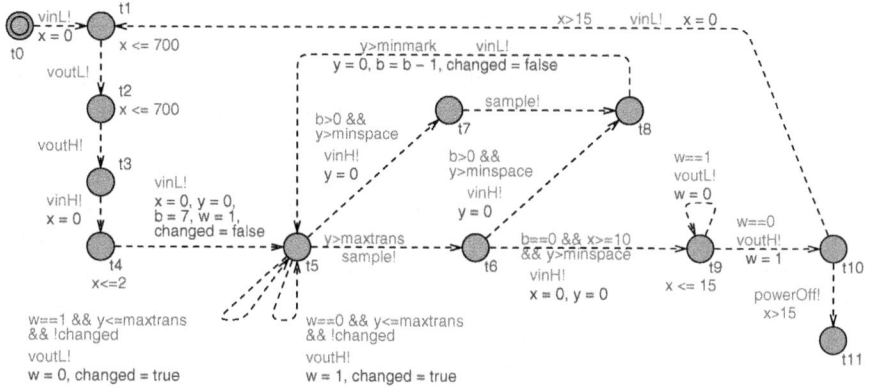

Fig. 4. TIOA model of the timing diagram: T

The signal controlled by the driver is shown in the top half of the timing diagram. Its most obvious features are the falling and rising transitions, these have been modeled in the TIOA as outputs called, respectively, vinL! and vinH!. The driver may also perform two other actions which are not entirely evident from the timing diagram. It may sample the *vout* signal to read a bit transmitted by the sensor, which we represent by an output called sample!, and it may stop using the sensor, which we represent by an output called powerOff!. The signal controlled by the sensor is shown in the bottom half of the timing diagram. The rising and falling transitions on this signal are modeled as outputs called, respectively, voutL! and voutH!. In fact, all of the actions in the model are outputs because the timing diagram describes a closed system. The model is thus trivially input-enabled and there is no need for self-looping input transitions on each state. Furthermore, the model can be simulated in isolation since all channels in ECDAR must be broadcast channels (i.e. outputs are non-blocking).

The driver requests a range reading with vinL!, i.e. by lowering the voltage level of *vin*. The sensor responds with voutL!, it then performs the necessary measurements before signaling completion with voutH!. The timing diagram guarantees that the sensor will complete a reading and respond after at least *70ms or more* have passed, after which the driver may perform a vinH!. This sequence can be seen in the model in the transitions linking states T_0–T_4. We model the timing constraint by resetting a clock x when the initial vinL! occurs, and adding the location invariant $x \leq 700$ to states T_1 and T_2. By rights this invariant should be strict, i.e. $x < 700$, but this is not currently permitted in ECDAR.For strict compliance with the timing diagram we should also add the guard $x > 700$ to the vinH! transition between T_3 and T_4, in practice, however, there are implementations that do not wait the full 700 ms but rather respond to voutH!. Both possible behaviors will be examined more closely in the next subsection.

After a reading has been made, the driver transfers the eight bits of the result from the sensor, from the most (MSB) to the least (LSB) significant bit. For each bit, the sensor sets the level of *vout* according to the value being transmitted, hence the 'crossed blocks' in Fig. 4. The timing diagram could be more precise about the details, but in our interpretation the driver triggers the next bit value with a vinL!, the sensor responds within a bounded time, and then the sensor may sample! the value and reset *vin* with a vinH!, in any order, before the next bit is requested. The triggering vinL! appears in the model from T_4 for the first bit and from T_8 for subsequent bits. The first action must occur in *0.2ms or less*, hence the invariant on T_4. The associated transition resets two clocks: x, for enforcing the *1ms or more* constraint across cycles, and y, for conditions on response times within each cycle. It also sets three variables: b, for counting the number of bits transmitted, w, for monitoring the level of *vout*, and changed, for limiting oscillations on *vout*. We use the w variable to ensure the strict alternation of voutL! and voutH!, an alternative approach is shown later. Two other constants appear around the loop T_5–T_8: maxtrans is a limit on the time it takes for *vout* to change after a triggering vinL!, and minspace is the minimum width of pulses on *vin*. We set both constants to zero for this case study.

Finally, after transmitting eight bits, the driver and sensor return their respective wires to a high level, and, after *1.5ms or more*, either another reading is requested, or the sensor is powered off. The timing constraint is expressed as an invariant on $T9$, i.e. a guarantee on the behavior of the sensor, and guards on the transitions from $T10$, i.e. a constraint on the behavior of the driver. The invariant is right-closed and the guards are left-open for the same reasons given above for the 700 ms constraint. Importantly, they do not overlap, so that time alone can be used to enforce the ordering between sensor and driver actions.

ECDAR is used to verify that the model is a valid (deterministic) specification, and also that it is consistent, i.e. that it has at least one valid implementation. We can also show two basic properties of the timing diagram model. The first, that vinL! and vinH! alternate strictly, is expressed using the automaton V^{in}, shown in Fig. 5b, and verified by the refinement $T \leq V^{in}$. The second, that voutL! and voutH! alternate strictly, is shown similarly using V^{out}, shown in Fig. 5c, and the refinement $T \leq V^{out}$. In fact, both properties can also be shown, using composition, by the single refinement $T \leq (V^{in} \mid V^{out})$.

4.2   Separate Driver and Sensor Models

While the single automaton model of the previous section is a suitable formalization of the timing diagram, there are at least two motivations for creating separate but interacting models for the roles of driver and sensor. First, this separation emphasizes the distinct behaviors of each and clarifies their points of synchronization; each of the two wires is, in effect, modeled separately. Second, each of the models may be used in isolation. This possibility is exploited in an appendix of the full version of this paper where a separate driver model serves as the specification for a model of an implementation in assembly language.

The components of the models are shown in Fig. 5. We discuss the driver models first, then the sensor, before relating them all to the timing diagram model.

The driver model. As previously mentioned, there are two ways for a driver to behave after it has requested a range reading: it can wait for a rising transition on the *vout* wire, or it can just wait 700 ms regardless. We model each possibility separately, both models shown in Fig. 5a. The model that responds to the sensor event is called D^{ev}, it comprises all locations except the one labeled *dd1*, which should be ignored together with all of its

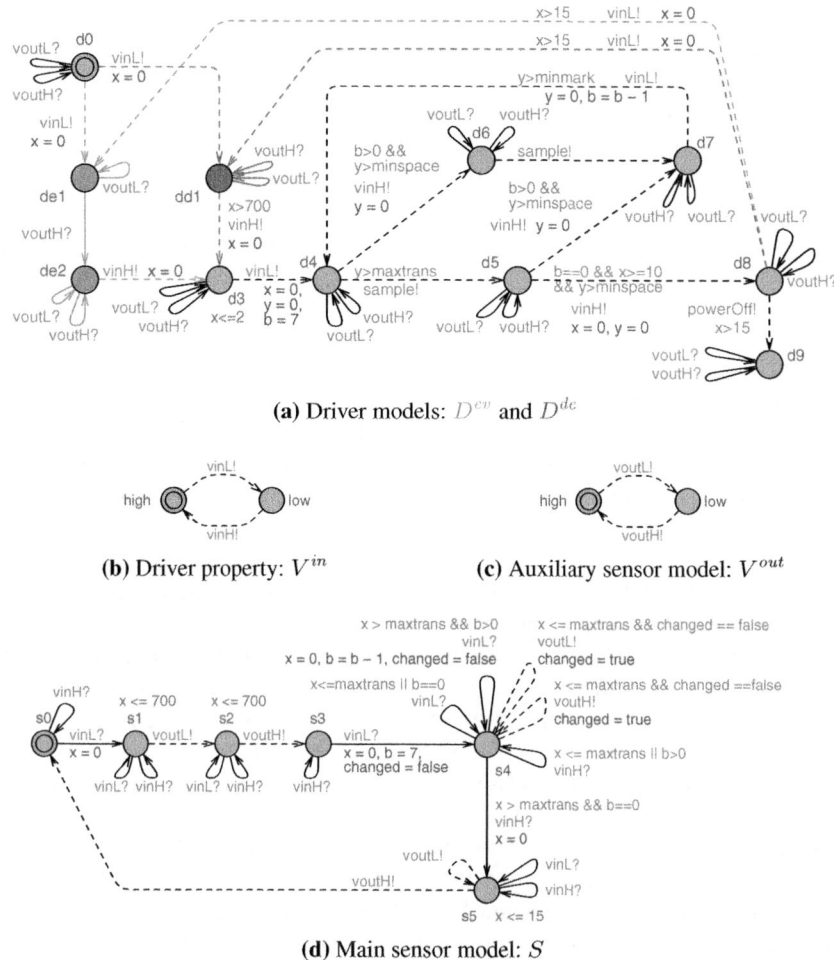

(a) Driver models: D^{ev} and D^{dc}

(b) Driver property: V^{in} **(c)** Auxiliary sensor model: V^{out}

(d) Main sensor model: S

Fig. 5. Sensor and driver models

incoming and outgoing transitions. The model that always delays is called D^{de}, it comprises all locations except those labeled *de1* and *de2* whose connected transitions are also excluded. The models cannot be combined without introducing non-determinism.

Aside from these initial differences the two models behave identically and their structures resemble that of the timing diagram model except that events on *vout* from the sensor are now modeled as the input actions voutL? and voutH?, and a counterpart for the state T_9 is not required. We explicitly model input-enabledness by adding self-loops, which, although not mandatory, since actions occur on broadcast channels, are necessary in ECDAR for verifying refinement. Note that both driver variants require little interaction with the sensor, relying instead on timing assumptions to ensure synchronization. In fact only D^{ev} reacts to sensor events directly, through the voutH? transition between D_1^{ev} and D_2^{ev}, though both models do sample the level of *vout*.

Refinement can be used to show a basic property of both driver models, that vinL! and vinH! alternate. This property is expressed as the automaton V^{in}, shown in Fig. 5b, and we use ECDAR to show $D^{ev} \leq V^{in}$ and $D^{de} \leq V^{in}$.

We would also like to claim that D^{de} refines D^{ev}, i.e. that $D^{de} \leq D^{ev}$, since D^{ev} can always wait after receiving voutH?, but ECDAR rejects this claim since D^{de} does not guarantee that voutH? will precede its initial vinH!. In fact, this type of refinement can only be shown in a conditional form where assumptions on the environment are made explicit. We revisit this idea after presenting a model for the sensor that embodies sufficient assumptions.

The sensor model. The sensor model S is shown in Fig. 5d. Events on the *vin* wire are now modeled as the inputs vinL? and vinH?, with additional self-loops on certain states, and the outputs sample! and powerOff! are not needed. The initial segment, S_0–S_3, mimics the corresponding part of the timing diagram model, but the clocking loop is reduced to a single location S_4 with five self-looping transitions and one outgoing transition.

In location S_4, the effect of the inputs, vinL? and voutL?, depends on the time elapsed since the last request for a bit, as measured by the clock x, and the number of bits remaining to transmit, as tracked by the counter b. The input vinL?, which requests the next bit, is ignored if it occurs (again) within the period given to the sensor to set the level of *vout*, and also when all bits have been transmitted, i.e. when $b = 0$. The input vinH! is ignored until all bits have been transmitted at which time, provided maxtrans units have elapsed since the last vinL?, it triggers an exit from S_4. The outputs voutL! and voutH! may only occur within maxtrans units of the last vinL?, and, furthermore, only at most one output may occur within any cycle, that is between any two successive and 'legal' vinL?s. The former constraint is expressed in the clause $x \leq$ maxtrans, and the latter using the variable changed.

Instead of a changed variable, an earlier model [6, Figure 4.16] has two states with three transitions from the first (changed = tt) to the second (changed = ff): one labeled with voutL!, another with voutH!, and the last unlabeled. This last τ-transition marks the possibility that the sensor decides not to change the voltage level, which occurs when two consecutive bits of a range reading are identical. Besides being more explicit, the two-state version is also more liberal since it is ready to accept vinH? and vinL? as soon as the value of *vout* has been set. Even with maxtrans = 0 there is a difference since in the current model there is always a non-zero delay after a triggering vinL! before subsequent vinL! or vinH! actions can influence the sensor. In any case, τ-steps are not permitted in TIOA and replacing them with an explicit output only makes modeling awkward, and, moreover, it is unnecessary since the driver models always wait and never respond immediately to vinL! or vinH! whose occurrence is a sufficient but not necessary indication of a stable value on *vout*.

The sensor model as it stands allows arbitrary interleaving of voutL! and voutH!. This is in contrast to the timing diagram model of Fig. 3b, where a variable, w, tracks the level of *vout*, or effectively which of voutL! or voutH! occurred most recently, and is used to constrain output events. The required alternating behavior is recovered using the conjunction operator and the TIOA V^{out}, depicted in Fig. 5c, giving the complete sensor specification: $(S \wedge V^{out})$. Here, the conjunction operator obviates the need to update and query a state variable on multiple transitions. A specific constraint is

Table 1. Counterexample for $D^{de} \leq D^{ev}$

D^{de}.vinL!		Attacker plays outputs on left of $\leq$
	D^{ev}.vinL!	Defender's response on right of $\leq$
D^{de} waits 701 ms		Attacker may delay on left of $\leq$
	D^{ev} waits 701 ms	Defender's response on right of $\leq$
D^{de}.vinH!		Attacker plays outputs on left of $\leq$
	no response	Defender loses!

expressed in a localized and obvious form and the rest of the model can be constructed under the assumption that it will hold. In ECDAR, the two automata, S and V^{out}, execute in parallel and must synchronize on voutL! and voutH!, neither of which may occur otherwise. Unlike for the timing diagram and the driver models, there is no need to separately verify the alternation of outputs—it is guaranteed by construction.

Relations between the models. Now that we have a few different models, we turn our attention to their interrelationships. It turns out that one of the driver models is more general than the other under certain assumptions. After verifying that fact, we turn our attention to validating the composition of the driver and sensor models against the timing diagram model. We also consider how the quotient operator might be applied.

The two driver models differ only in their initial interaction with the sensor, after requesting a range reading, D^{de} always waits 700 ms whereas D^{ev} may respond as soon as the sensor raises *vout*. One could thus suppose that D^{ev} is more general than D^{de}, since it can also refuse to act before 700 ms has passed even after receiving a voutH!. But, as described earlier, a first, naive attempt to show the refinement $D^{de} \leq D^{ev}$ fails! The counter-example strategy can be simulated in ECDAR, giving the results shown in Table 1. There is no guarantee that the inputs needed by D^{ev} will be provided. We must make these assumptions on the environment explicit by instead stating the relation as

$$\left(D^{de} \mid (S \wedge V^{out}) \right) \leq \left(D^{ev} \mid (S \wedge V^{out}) \right),$$

which is readily validated by ECDAR.[2] The verification fails if D^{de} and D^{ev} are swapped: D^{ev} can perform a vinH? when $x \leq 700$ while D^{de} cannot.

The compositions of the driver and sensor models have been proposed as alternatives to the timing diagram model. We state this, for the more general driver model, as two properties: $(D^{ev} \mid (S \wedge V^{out})) \leq T$, and $T \leq (D^{ev} \mid (S \wedge V^{out}))$. Both of which are verified almost instantaneously by ECDAR. For the similar properties with D^{de} instead of D^{ev}, only the version with T on the right of the refinement holds; as would be expected.

Even ignoring the conjunction operator, the possibility of verifying a refinement with a composition on the right-hand side is interesting, because it is not possible in any other existing tools for checking timed automata refinement. For instance, current implementations [7] of the usual construction for checking timed trace inclusion [18,23] require that the refined specification is an explicit automaton. The capability to address compositions is one advantage of incorporating the refinement verification into the model-checker itself.

---

[2] In the current version of ECDAR, S and V^{out} must be explicitly duplicated.

There are limited opportunities to apply the quotient operator in this case study, perhaps because there are only a small number of models and the operators are not nested in especially complicated ways. There are, though, two types of properties that may be attempted.

The first type of property uses the quotient on the right-hand side of a refinement instead of composition on the left-hand side. For instance, we can verify $D^{ev} \leq (T \setminus (S \wedge V^{out}))$ in ECDAR. The right hand side expresses the idea of the timing diagram modulo certain assumptions on the environment. Currently the tool requires the explicit definition of universal and inconsistent states when using the quotient operator, and simulations are not possible. These issues will be addressed in future versions.

Second, we could try the quotient on the left-hand side of a refinement. For instance, to propose the property $(T \setminus D^{ev}) \leq (S \wedge V^{out})$ as a means of finding out whether the sensor model is maximal with respect to the timing diagram and driver model. This cannot work in general, however, since as soon as D^{ev} cannot do an output from a state, like vinH! from the initial state for example, the quotient will have a transition to the universal state from which any output or delay can be chosen, at any time, to challenge the other side of the refinement.

Büchi objectives. Some aspects of specifying liveness are addressed by the algorithms presented earlier, and supported in ECDAR. It is possible, for example, to determine whether a given combination of a TIOA and a liveness constraint, expressed as a Büchi objective, are consistent; i.e. whether refinement is possible. But other important aspects are not yet addressed satisfactorily. Most notably, the interaction of Büchi constraints and refinement is limited.

Büchi objectives offer a way to further constrain specifications. For example, consider adding an additional requirement to the timing diagram model T: if an initial range reading is requested, the system must eventually be powered off. We will interpret this to mean that two behaviors are allowed: 1. resting forever in T_0, or, 2. terminating in T_{11}. Our first attempt is to simply try to solve a Büchi objective for the current model: $(T, WB(\{T_0, T_{11}\}))$. But this is not correct, and ECDAR reports that the model is inconsistent. While the model starts in T_0, and T_{11} is always reachable, the Büchi objective is only satisfied if either of T_0 or T_{11} is reentered infinitely often. Self-looping output transitions must be added to T_0 and T_{11} to allow 'resting' in these states. If we do this—choosing an arbitrary output that will not occur in any other models—and call the modified version T', ECDAR confirms that $(T', WB(\{T'_0, T'_{11}\}))$ is consistent.

The modified model is easily adapted to allow a system that never stops taking range readings: $(T', WB(\{T'_0, T'_{10}, T'_{11}\}))$. This model is obviously consistent since increasing the set of states in the Büchi objective cannot reduce the set of possible implementations. More information can be gained by verifying the consistency of $(T', WB(\{T'_{10}\}))$, which confirms that the model allows unbounded repetitions of the protocol. Compliance with the Büchi objective is achieved by pruning away the transition labelled powerOff!, so this verification does not show that the unadorned model T' does not allow termination, only that the model can choose to cycle continuously. Verifying the consistency of a model with a Büchi objective can be useful as a sanity check.

While Büchi objectives in ECDAR are quite useful for checking consistency properties, they work less well in combination with refinement. For instance, in ECDAR we can show $(T', WB(\{T'_0\})) \leq (T', WB(\{T'_{10}\}))$.

This is indeed correct, since any implementation of the left-hand side is also an implementation of the right-hand side, but it could be considered misleading, since the left-hand side specifies a system that never starts a range reading, while the right-hand side could be interpreted as specifying a system that never stops performing range readings whereas, in fact, it is a system where it is possible, but not strictly necessary, to keep performing range readings. The source of this mismatch is that the current refinement is based on partial observations rather than complete ones, which is adequate for safety but not for liveness.

The pruning of output transitions that can result from the combination of a TIOA and a Büchi objective gives models where a constraint that is supposedly on infinite behaviors also constrains finite behaviors, which, while not necessarily bad, is perhaps not completely reasonable [1]. The methodological implications for our theory are not yet clear, but we note here that this situation can be detected using refinement verification in ECDAR. The *machine closure* [2] of a TIOA A and a Büchi objective B can be checked by the refinement $A \leq (A, B)$, which will fail if a reachable output transition in A is not present in (A, B).

5 Summary and Future Work

We have shown that ECDAR and the underlying theory, are powerful enough to handle a small—in terms of the scale of systems developed by industry—but realistic case study. The input/output semantics of TIOA works well for open systems, and the game-based refinement semantics, i.e. the idea of challenging with inputs from the right-hand side and outputs or delays from the left-hand side, quickly comes to seem natural. Including refinement testing in the model checker itself is much more convenient than having to pass models through an external tool, and the concomitant feature of allowing composed models on either side of the relation is a powerful one. Finally, the conjunction operator is a very convenient modeling feature.

Still, several elements could be improved. While Büchi objectives are currently not without use, a different notion and implementation of refinement is needed to support more sophisticated applications. The quotient operator is supported by ECDAR, but its effect is not easily visualized or simulated. More work is needed to determine how it can be usefully applied to system development and verification; the sensor case study is too limited in this regard. ECDAR takes advantage of the mature UPPAAL user interface, but strategies, goals, and the effect of pruning are inherently more complicated and harder to understand than are simple traces, more work is needed to understand how best to compute and communicate this information. Furthermore, the new operators and analyses available in ECDAR make it natural to work with multiple pairings of system declarations and properties, but this is not yet well supported by the user interface.

References

1. Abadi, M., Alpern, B., Apt, K.R., Francez, N., Katz, S., Lamport, L., Schneider, F.B.: Preserving liveness: Comments on "Safety and liveness from a methodological point of view". Information Processing Letters 40(3), 141–142 (1991)

2. Abadi, M., Lamport, L.: The existence of refinement mappings. Theoretical Computer Science 82(2), 253–284 (1991)
3. Alur, R., Dill, D.L.: A theory of timed automata. Theoretical Computer Science 126(2), 183–235 (1994)
4. Behrmann, G., Cougnard, A., David, A., Fleury, E., Larsen, K.G., Lime, D.: UPPAAL-Tiga: Time for Playing Games! In: Damm, W., Hermanns, H. (eds.) CAV 2007. LNCS, vol. 4590, pp. 121–125. Springer, Heidelberg (2007)
5. Behrmann, G., David, A., Larsen, K.G.: A Tutorial on UPPAAL. In: Bernardo, M., Corradini, F. (eds.) SFM-RT 2004. LNCS, vol. 3185, pp. 200–236. Springer, Heidelberg (2004)
6. Bourke, T.: Modelling and Programming Embedded Controllers with Timed Automata and Synchronous Languages. PhD thesis, University of New South Wales, Sydney (2009)
7. Bourke, T., Sowmya, A.: Automatically transforming and relating Uppaal models of embedded systems. In: EMSOFT, pp. 59–68 (2008)
8. Cassez, F., David, A., Fleury, E., Larsen, K.G., Lime, D.: Efficient On-the-Fly Algorithms for the Analysis of Timed Games. In: Abadi, M., de Alfaro, L. (eds.) CONCUR 2005. LNCS, vol. 3653, pp. 66–80. Springer, Heidelberg (2005)
9. Chakrabarti, A., de Alfaro, L., Henzinger, T.A., Stoelinga, M.: Resource interfaces. In: Alur, R., Lee, I. (eds.) EMSOFT 2003. LNCS, vol. 2855, pp. 117–133. Springer, Heidelberg (2003)
10. David, A., Larsen, K.G., Legay, A., Nyman, U., Wąsowski, A.: ECDAR: An Environment for Compositional Design and Analysis of Real Time Systems. In: Bouajjani, A., Chin, W.-N. (eds.) ATVA 2010. LNCS, vol. 6252, pp. 365–370. Springer, Heidelberg (2010)
11. David, A., Larsen, K., Legay, A., Nyman, U., Wąsowski, A.: Timed I/O Automata: A Complete Specification Theory for Real-Time Systems. In: HSCC 2010, pp. 91–100. ACM (2010)
12. de Alfaro, L., da Silva, L.D., Faella, M., Legay, A., Roy, P., Sorea, M.: Sociable Interfaces. In: Gramlich, B. (ed.) FroCos 2005. LNCS (LNAI), vol. 3717, pp. 81–105. Springer, Heidelberg (2005)
13. de Alfaro, L., Henzinger, T.A.: Interface automata. In: FSE, Vienna, Austria, pp. 109–120. ACM Press (September 2001)
14. de Alfaro, L., Henzinger, T.A.: Interface-based design. In: Marktoberdorf Summer School. Kluwer Academic Publishers (2004)
15. de Alfaro, L., Henzinger, T.A., Majumdar, R.: Symbolic Algorithms for Infinite-State Games. In: Larsen, K.G., Nielsen, M. (eds.) CONCUR 2001. LNCS, vol. 2154, pp. 536–550. Springer, Heidelberg (2001)
16. de Alfaro, L., Henzinger, T.A., Stoelinga, M.: Timed Interfaces. In: Sangiovanni-Vincentelli, A.L., Sifakis, J. (eds.) EMSOFT 2002. LNCS, vol. 2491, pp. 108–122. Springer, Heidelberg (2002)
17. http://www.cs.aau.dk/~adavid/ecdar/
18. Jensen, H.E., Larsen, K.G., Skou, A.: Scaling up Uppaal: Automatic Verification of Real-Time Systems Using Compositionality and Abstraction. In: Joseph, M. (ed.) FTRTFT 2000. LNCS, vol. 1926, pp. 19–30. Springer, Heidelberg (2000)
19. Kaynar, D.K., Lynch, N.A., Segala, R., Vaandrager, F.W.: Timed I/O Automata: A mathematical framework for modeling and analyzing real-time systems. In: RTSS, pp. 166–177. IEEE Computer Society (2003)
20. Larsen, K.G.: Modal Specifications. In: Sifakis, J. (ed.) CAV 1989. LNCS, vol. 407, pp. 232–246. Springer, Heidelberg (1990)
21. Maler, O., Pnueli, A., Sifakis, J.: On the Synthesis of Discrete Controllers for Timed Systems (An Extended Abstract). In: Mayr, E.W., Puech, C. (eds.) STACS 1995. LNCS, vol. 900, pp. 229–242. Springer, Heidelberg (1995)
22. Sharp Corp. GP2D02: Compact, high sensitive distance measuring sensor (1997)
23. Stoelinga, M.: Alea Jacta est: Verification of probabilistic, real-time and parametric systems. PhD thesis, Katholieke Universiteit Nijmegen (2002)

Combining Graph Transformation and Algebraic Specification into Model Transformation

Hans-Jörg Kreowski, Sabine Kuske, and Caroline von Totth[*]

University of Bremen, Department of Computer Science,
P.O. Box 33 04 40, 28334 Bremen, Germany
{kreo,kuske,caro}@informatik.uni-bremen.de

Abstract. In this paper, we propose a new framework of model transformation that combines graph transformation with algebraic specification. While graph transformation is well-suited to describe the transformation of visual models, one can observe that models are often composite structures with visual, graphical and diagrammatic components accompanied by all kinds of data objects like strings, sets, numbers, etc. that are not adequately represented by graphs. We advocate algebraic specification to cover these parts of models and tupling to combine the graph and the data components.

1 Introduction

According to the basic idea of model-driven architecture (cf., e.g., [13]), the goal of model transformation is to transform platform-independent models (PIMs) into platform-specific models (PSMs) in a systematic way. Quite often this entails starting from visual modeling paradigms such as UML diagrams, Petri nets or business process models and ending up with JAVA or C^{++} programs. Various graph-transformation-based approaches that cover the part of visual models are discussed with some success (see, e.g., [21,2,17,20,9,29,10,4]).

Although visual modeling and the transformation of visual models are very important, one encounters many other kinds of models like grammars, automata, specifications, and various notions of systems and transformations between them in theoretical computer science. These models are usually tuples of sets, strings, numbers, truth values, and such and may have graphs and visual models as components, among others like the state graph of an automaton.

In [19], we have introduced the notion of model transformation units to combine and cover both kinds of model transformation. Models are tuples of graphs, strings, sets, numbers, truth values, or tuples again. The transformation is done component by component due to the type of each component. Graphs are transformed by graph transformation rules, while other components are transformed by operations of the respective types.

---

[*] The authors would like to acknowledge that their research is partially supported by the Collaborative Research Centre 637 (Autonomous Cooperating Logistic Processes: A Paradigm Shift and Its Limitations) funded by the German Research Foundation (DFG).

T. Mossakowski and H.-J. Kreowski (Eds.): WADT 2010, LNCS 7137, pp. 193–208, 2012.

In this paper, we generalize model transformation units in such a way that the concrete non-graph types are replaced by arbitrary algebraic specifications. In this way, model components and their transformation can be specified in a quite problem-specific way rather than by using standard types. As a running example, we discuss the reduction of the satisfiability problem for propositional formulas in conjunctive normal form with three literals per clause into the independence problem of undirected graphs starting in Section 2. In Section 3, the prerequisites of graph transformation are recalled. The generalized notion of model transformation units is introduced in Section 4 and investigated with respect to termination, functionality and correctness in Section 5. Section 6 briefly considers related work and Section 7 concludes the paper. We assume that the reader is familiar with the basic concepts of algebraic specification (cf., e.g. [1]).

2   Example: Reduction of *SAT3* to *INDEP*

A typical example of a model transformation as one can encounter in theoretical computer science is the reduction of the satisfiability problem for propositional formulas in conjunctive normal form with three literals (*SAT3*) into the independence problem for undirected graphs (*INDEP*) (cf., e.g., [14] and for many further examples [16], among others).

A propositional formula in conjunctive normal form is a conjunction of clauses, which can be (and often is) represented as a set of clauses. A *clause* is a disjunction of literals. In particular, one may consider clauses $c = c_1 \vee c_2 \vee c_3$ with three literals. A *literal* is a Boolean variable or its negation. Let X be a finite set of Boolean variables and $\overline{X} = \{\overline{x} \mid x \in X\}$ the corresponding set of negations, then the disjoint union $X + \overline{X}$ of X and $\overline{X}$ denotes the set of literals over X, $C_3(X)$ the set of clauses with three literals, and $\mathcal{P}_{fin}(C_3(X))$ the set of all finite subsets of $C_3(X)$, each representing a conjunction of clauses. Then the pair (X, F) with $F \in \mathcal{P}_{fin}(C_3(X))$ is an element of *CNF3*, the set of conjunctive normal forms with three literals per clause. (X, F) is *satisfiable* if there is an assignment $a \colon X \to \{true, false\}$ such that each clause $c \in F$ contains an $x \in X$ with $a(x) = true$ or an $\overline{x}$ with $a(x) = false$. The problem whether (X, F) is satisfiable or not for all $(X, F) \in$ *CNF3* is the well-known satisfiability problem for *CNF3*, which may be denoted by *SAT3*.

The independence problem *INDEP* is defined for pairs (G, k) where G is a finite undirected simple graph, i.e. $G = (V, E)$ with a finite set of nodes V and a set E of edges, each edge being a 2-element subset of V, and $k \in \mathbb{N}$. The problem is to decide whether there is a subset $X \subseteq V$ with k elements such that $\{x_1, x_2\} \notin E$ for all $x_1, x_2 \in X$ meaning that no two nodes in X are connected by an edge.

The reduction *SAT3-to-INDEP* can be defined by mapping each $(X, F) \in$ *CNF3* to the pair $(G(X, F) = (V(X, F), E(X, F)), \#F)$ where $\#F$ denotes the number of clauses, $V(X, F) = F \times [3]$ [1] and $E(X, F)$ contains the following edges assuming that the literals of each $c \in F$ are numbered c_1, c_2 and c_3:

---

[1] [k] for $k \in \mathbb{N}$ denotes the set $\{1, \ldots, k\}$.

1. $\{(c, i), (c, j)\}$ for $c \in F, i, j \in [3], i \neq j$,
2. $\{(c, i), (c', j)\}$ for $c, c' \in F, c \neq c', i, j \in [3]$, and $c_i = x$ and $c'_j = \bar{x}$ or $c_i = \bar{x}$ and $c'_j = x$ for some $x \in X$.

This means that there is a node for every literal in F and two literal nodes form an edge if they either stem from the same clause or contradict each other. For example, the pair $(\{x, y\}, \{x \vee \bar{x} \vee y, y \vee \bar{y} \vee x, y \vee \bar{y} \vee \bar{x}\}) \in CNF3$ is reduced to the pair $(G, 3)$ where G is the graph shown in in Fig. 1.

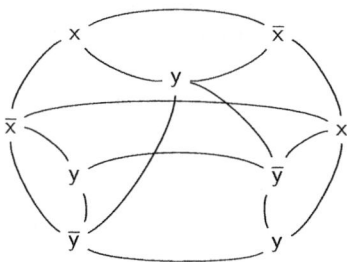

Fig. 1. Graph resulting from $(x \vee \bar{x} \vee y) \wedge (y \vee \bar{y} \vee x)(\wedge y \vee \bar{y} \vee \bar{x})$

If F is satisfiable, then there is an assignment such that each clause contains a literal that evaluates to *true*. These literals stem from different clauses and do not contradict each other. In other words, they form an independent set of nodes. All the arguments can be reversed so that the mapping turns out to be correct. Moreover, it is easy to see that it can be constructed by a polynomial algorithm. Such an algorithm may count the elements of F, copy all literals as nodes and check for each two literal nodes whether they are connected by an edge or not. The number of steps is quadratic in the number of literals. Altogether, this proves that *SAT3-to-INDEP* is a reduction between *NP*-problems. *SAT3* and *INDEP* are known to be *NP*-complete. Moreover, as reductions preserve *NP*-completeness, the *NP*-completeness of *INDEP* is implied by the *NP*-completeness of *SAT3*.

Looking at this example and many other similar examples, one can observe the following:

1. Models are tuples with components being strings, sets, symbols, numbers, graphs, and tuples again.
2. The tuples may be restricted by constraints.
3. Their transformations are constructions on the components and may require computation.
4. Input types may be different from output types, and the transformation from input models to output models may need intermediate working types.

To come up with a formal notion of model transformation that covers all these features, we propose to employ the Cartesian product and its nice properties for the tupling, graph-transformational rule bases for the graph components and algebraic specifications for all other components.

3   Graph Transformational Rule Bases

In this section, we recall the basic notions and notations of graph transformation as far as needed.

A (graph-transformational) *rule base* is a system $\mathcal{B} = (\mathcal{G}, \mathcal{R}, \Longrightarrow)$ where $\mathcal{G}$ is a class of *graphs*, $\mathcal{R}$ is a class of *rules*, and $\Longrightarrow$ is a *rule application operator* with $\underset{r}{\Longrightarrow} \subseteq \mathcal{G} \times \mathcal{G}$ for $r \in \mathcal{R}$.

This allows one to choose the favorite kind of graphs or those that fit best for the intended purpose like, for example, directed or undirected graphs, labeled or unlabeled graphs, acyclic, connected or planar graphs, hypergraphs, etc. The same applies to the kind of rules and rule application for which one encounters quite a variety of possible choices in the literature (cf, e.g., [26]).

An explicit example is the class $\mathcal{G}_{undir}(\Sigma)$ of undirected simple graphs with edge labels in Σ. Such a graph is a system $G = (V, E)$ where V is the set of *nodes*, and E the set of *edges*, each edge being a pair $e = (att(e), l(e))$ of *attachment* and *label* such that $att(e)$ is a 1- or 2-element subset of V and $l(e) \in \Sigma$. If the attachment $att(e)$ is a singleton set, then e is called a *loop*. The components of G are also denoted by V_G and E_G.

We assume that Σ contains a special symbol $*$ that is invisible in drawings of graphs so that unlabeled edges can be represented as $*$-labeled ones. In particular, we get $\mathcal{G}_{undir} \subseteq \mathcal{G}_{undir}(\Sigma)$ if one identifies a 2-element edge $e = \{v_1, v_2\}$ with the pair $(\{v_1, v_2\}, *)$ where $\mathcal{G}_{undir}$ denotes the set of unlabeled undirected graphs introduced in Section 2.

As an explicit example, we use rules of the form $r = (L \supseteq K \subseteq R)$ where L and R are graphs and K is a common subgraph, i.e., $V_K \subseteq V_L \cap V_R, E_K \subseteq E_L \cap E_R$.

To apply a rule $r = (L \supseteq K \subseteq R)$ to a graph G, one needs an injective graph morphism $g: L \to G$, i.e. an injective mapping $g: V_L \to V_G$ with $g(e) = (\{g(v_1), g(v_2)\}, a) \in E_G$ for all $e = (\{v_1, v_2\}, a) \in E_L$. Then the resulting graph H is constructed as follows:

- Remove all nodes in $g(V_L - V_K)$ and all incident edges as well as all edges $g(e)$ for $e \in E_L - E_K$.
- Add all nodes in $V_R - V_K$ disjointly and add all edges in $E_R - E_K$ where each incident $v \in V_K$ is replaced by $g(v)$.

The application of a rule is denoted by $G \underset{r}{\Longrightarrow} H$. $\mathcal{B}_{undir}(\Sigma)$ denotes the rule base that consists of $\mathcal{G}_{undir}(\Sigma)$, the rules and the rule application, as introduced. $\mathcal{B}_{undir}$ denotes the somewhat odd rule base that consists of $\mathcal{G}_{undir}$ and the empty set of rules. Since in the following this rule base is only used as output type it does not need to have transformation rules.

Consider the rules in Figures 2. The rule *addclause* adds its right-hand side disjointly to any given graph because there is always the empty graph morphism, nothing is removed and all is added. The rule *addcontradiction* adds an edge between two nodes if they carry loops with contradicting literals. The dotted line in the left-hand side is a *negative application condition* and means that the rule is not applied if the two nodes are already connected in the host graph. This prevents the rule from being applied twice to the same two nodes.

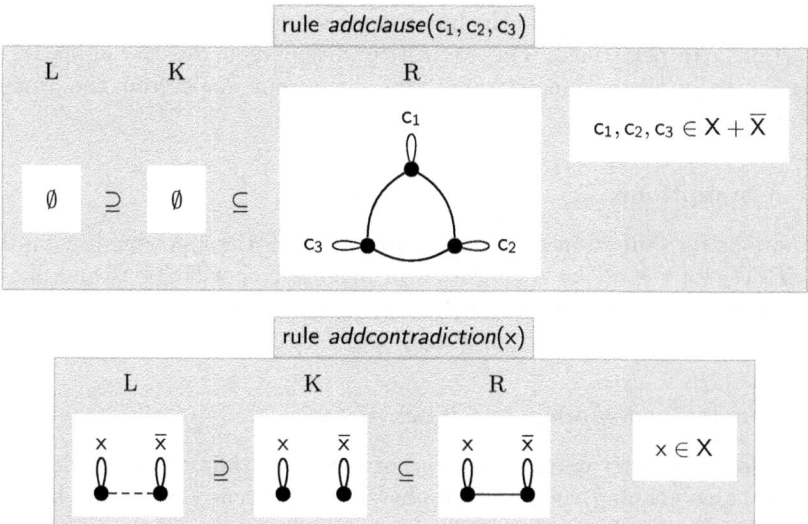

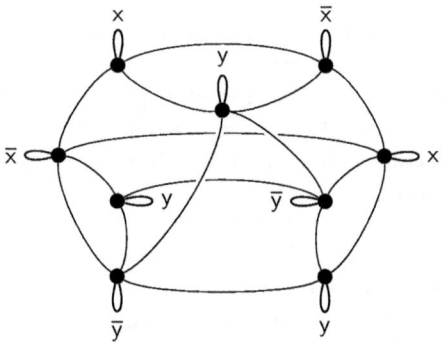

Fig. 2. The graph transformation rules *addclause* and *addcontradiction*

Starting from the empty graph, three applications of *addclause* (with the proper literals) and seven applications of *addcontradiction* yield the graph in Fig. 3.

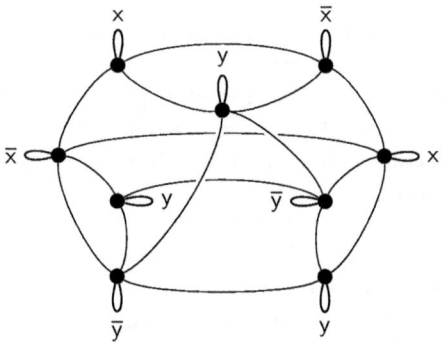

Fig. 3. A graph resulting from rule applications

4   Model Transformation

In this section, the notion of model transformation units is introduced. It is based on models being tuples comprised of graphs and elements of other data types as well as tuples again. A transformation step is given by the application of an action which is a tuple of rules, terms, and actions corresponding to the respective types of the components of models. A set of actions together with a control condition, an initialization and a terminalization forms a model transformation unit. The initialization adapts the input models to the working models on which the actions are performed. The control condition regulates the application of actions. Finally, the terminalization filters the output model from the processed working models.

4.1 Assumptions

The notions introduced in this section are based on the following assumptions: Let $SPEC_i$ for $i \in [l]$ be a collection of algebraic specifications and A_i be a $SPEC_i$-algebra for each $i \in [l]$. Let $\mathcal{B}_j = (\mathcal{G}_j, \mathcal{R}_j, \Longrightarrow)$ for $j \in [n]$ be a collection of rule bases.

4.2 Constrained Models and Their Types

Models are defined recursively as tuples of models initialized by the elements of the given algebras and classes of graphs. They are typed in a straightforward way:

(i) $m \in A_{i,s}$ is a *model of type* s for some sort s of $SPEC_i$ and $i \in [l]$,
(ii) $m \in \mathcal{G}_j$ is a *model of type* $\mathcal{B}_j$ for some rule base $\mathcal{B}_j, j \in [n]$, and
(iii) $(m_1, \ldots, m_k)$ is a *model of type* $T_1 \times \cdots \times T_k$ if m_i is a model of type T_i for all $i \in [k]$ and some $k \in \mathbb{N}$.

The set of all models of type T is denoted by $\mathfrak{M}(T)$.

Instead of the elements of the data domains in (i) and the graphs in (ii), one may consider the corresponding 1-tuples. Hence, one can assume that all models are tuples without loss of generality.

In many cases, one is not interested in all tuples of some type, rather one may like to deal with tuples that have some specific properties and interrelations. To cover this, we use constraints where we allow any syntactic entity that may restrict the class of models:

Let T be a model type and $\mathcal{X}(T)$ be a class of *constraints* with $SEM(x) \subseteq \mathfrak{M}(T)$ for $x \in \mathcal{X}(T)$. Then $\langle T \text{ with } x \rangle$ for some $x \in \mathcal{X}(T)$ is a *constrained model type* with $\mathfrak{M}(\langle T \text{ with } x \rangle) = SEM(x)$.

The definition can be used recursively meaning that the assumed type T may be itself constrained. This yields types of the form $\langle \ldots \langle T \text{ with } x_1 \rangle \ldots \text{ with } x_p \rangle$ which may be denoted by $\langle T \text{ with } x_1, \ldots, x_p \rangle$.

For further use, types may be named and provided with standard variables $(M_1, \ldots, M_k)$ for the components. Clearly, the variables get their types from the components. Constraints may be Boolean terms and graph properties using these variables. As constraints may contradict each other, the class of constrained models may be empty.

4.3 Actions Combining Rules and Operations

The most basic syntactic construct for model transformation is an action which is defined componentwise. If the component is a graph, the action provides a rule. If the component is an element of some data domain, the action provides a term of the corresponding sort, which specifies an operation on this sort. If the component is a composite model, the action component is – recursively – an action. Alternatively to these three cases, the action component may be the special symbol "$-$" that requests "no change" in this component.

Let $\langle T_1 \times \cdots \times T_k \text{ with } x_1, \ldots, x_l \rangle$ be a model type. Then an *action* $a = (a_1, \ldots, a_k)$ is a k-tuple such that for each $i \in [k]$ one of the following holds:

1. $a_i = -$, called *void*,
2. $a_i \in \mathcal{R}_j$ provided that $T_i = \mathcal{B}_j$ for some $j \in [n]$,
3. a_i is a term of type T_i provided that T_i is some sort, and
4. a_i is an action for type T_i provided T_i is a constrained product type which is neither a rule base nor a sort.

4.4 Application of Actions

The application of an action transforms models into models of the same type componentwise. Let $T_1 \times \cdots \times T_k$ be a model type with the standard variables $(M_1, \ldots, M_k)$ and let $m = (m_1, \ldots, m_k) \in \mathfrak{M}(T_1 \times \cdots \times T_k)$. Let *assign* be an assignment with $assign(M_j) = m_j$ for $j \in [k]$ and free choice for all other occurring variables. Then the action $a = (a_1, \ldots, a_k)$ may be performed on m yielding $m' = (m'_1, \ldots, m'_k) \in \mathfrak{M}(T_1 \times \cdots \times T_k)$ (denoted by $m \underset{a}{\Longrightarrow} m'$) if the following holds for $i = 1, \ldots, k$:

1. $m'_i = m_i$ if $a_i = -$,
2. $m_i \underset{a_i}{\Longrightarrow} m'_i$ if $a_i \in \mathcal{R}_j$ for some $j \in [n]$,
3. $m'_i = a_i[assign]$ if a_i is a term and $a_i[assign]$ is the term obtained from a_i by substituting each variable M in a_i by $assign(M)$, and
4. $m_i \underset{a_i}{\Longrightarrow} m'_i$ if a_i is an action.

Given a set of actions A, $\underset{A}{\Longrightarrow}$ denotes the union of all relations $\underset{a}{\Longrightarrow}$ for $a \in A$, and $\underset{A}{\overset{*}{\Longrightarrow}}$ denotes the reflexive and transitive closure of $\underset{A}{\Longrightarrow}$. It may be noticed that the application of an action is defined in such a way that it is an induced mapping into the product for each fixed choice of the free (non-standard) variables.

4.5 Model Transformation Units

A *model transformation unit* is a system $mtu = (ITD, OTD, WT, A, C)$ where

- $WT = T_1 \times \cdots \times T_k$ is the *working type*.
- A is the *set of actions* on the working type,
- C is the *control condition*,
- ITD is the *input type declaration* consisting of the *input type* $IT = \langle I_1 \times \cdots \times I_m \text{ with } x \rangle$ and an *initialization initial*: $IT \to WT$ specified by projections and constants, and
- OTD is the *output type declaration* consisting of the *output type* $OT = \langle O_1 \times \cdots \times O_n \text{ with } y \rangle$ and a *terminalization terminal*: $WT \to OT$ specified by projections.

Here, a control condition is any syntactic entity that may cut down the nondeterminism of action applications, meaning that $SEM(C) \subseteq \underset{A}{\overset{*}{\Longrightarrow}}$. $SEM(C)$ may be denoted by $\underset{A,C}{\overset{*}{\Longrightarrow}}$.

4.6 Semantics

The *model transformation* specified by *mtu* is a relation between input and output models

$$SEM(mtu) : \mathfrak{M}(IT) \to \mathfrak{M}(OT)$$

defined by the sequential composition

$$\mathfrak{M}(IT) \xrightarrow{f_{initial}} \mathfrak{M}(WT) \underset{A,C}{\overset{*}{\Longrightarrow}} \mathfrak{M}(WT) \xrightarrow{f_{terminal}} \mathfrak{M}(OT),$$

where $f_{initial}$ is the mapping induced by *initial*, i.e., by the projections and constants given by *initial* restricted to $\mathfrak{M}(IT)$, $f_{terminal}$ is the mapping induced by *terminal* restricted to $\mathfrak{M}(OT)$ and $\underset{A,C}{\overset{*}{\Longrightarrow}}$ is the iterated action application relation obeying the control condition C.

4.7 Example

Continuing our running example, Fig. 4 displays the reduction of *SAT3* to *INDEP* as a model transformation unit. It is based on the algebraic specification **clause3(data)** which specifies clauses as a disjunction of literals $c_1 \lor c_2 \lor c_3$ where a literal is a Boolean variable or its negation. The variables are given by an actualization of the formal parameter **data**. If this is chosen as a set of identifiers ID, the free **clause3**-algebra provides the set $C_3(ID)$ described in Section 2. Moreover, we use the standard algebraic specifications **set(data)** and **nat** with the algebras $\mathcal{P}_{fin}(ID)$, $\mathcal{P}_{fin}(C_3(ID))$ and $\mathbb{N}$ and the rule bases $\mathcal{B}_{undir}(\Sigma)$ and $\mathcal{B}_{undir}$ as described in Section 3.

The working type of the unit combines sets of identifiers, sets of clauses with the identifiers as literals, undirected graphs and natural numbers. In addition, the four standard variables X, F, G and k are declared. Due to the mapping in "initialize" the first two components form the input type, which is additionally constrained by the requirement that the literals in all clauses of component 2 stem from the actual set of Boolean variables of component 1. The components 3 and 4 are initialized by the standard constants empty graph and 0 respectively. At the end, they are used as the output components due to the mapping in "terminalize". However, the resulting graph is only accepted if it is unlabeled.

The unit in Fig. 4 has three actions. The action *clause* removes a clause from the second component. Its literals are given as variables that do not belong to the standard repertoire and can therefore be chosen freely. At the same time, the rule *addclause* of Fig 2 must be applied in the third component using the same literals and the counter in the fourth component is increased by 1. Due to the

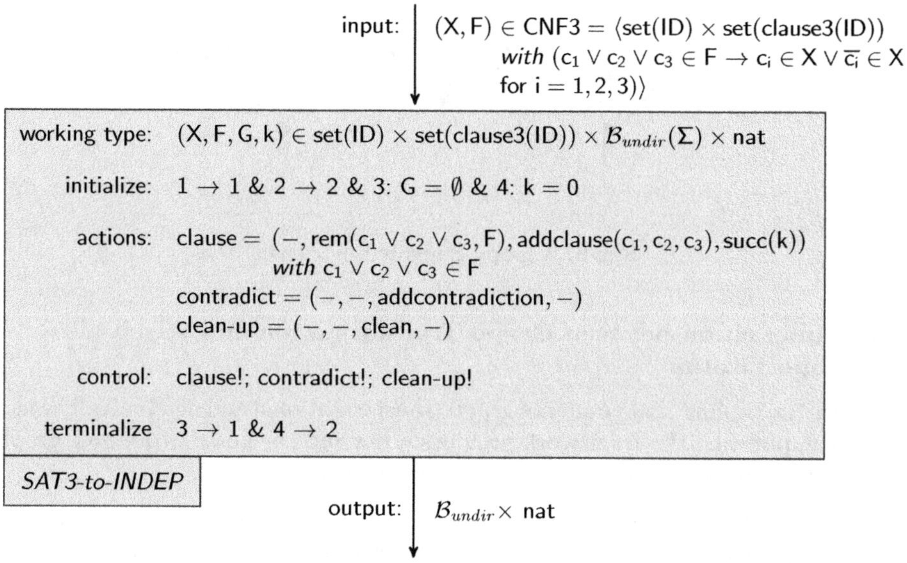

input: $(X, F) \in CNF3 = \langle set(ID) \times set(clause3(ID))$
$with\ (c_1 \vee c_2 \vee c_3 \in F \rightarrow c_i \in X \vee \overline{c_i} \in X$
$for\ i = 1, 2, 3) \rangle$

working type: $(X, F, G, k) \in set(ID) \times set(clause3(ID)) \times \mathcal{B}_{undir}(\Sigma) \times nat$

initialize: $1 \rightarrow 1\ \&\ 2 \rightarrow 2\ \&\ 3{:}\ G = \emptyset\ \&\ 4{:}\ k = 0$

actions: $clause = (-, rem(c_1 \vee c_2 \vee c_3, F), addclause(c_1, c_2, c_3), succ(k))$
$with\ c_1 \vee c_2 \vee c_3 \in F$
$contradict = (-, -, addcontradiction, -)$
$clean\text{-}up = (-, -, clean, -)$

control: $clause!;\ contradict!;\ clean\text{-}up!$

terminalize $3 \rightarrow 1\ \&\ 4 \rightarrow 2$

SAT3-to-INDEP

output: $\mathcal{B}_{undir} \times nat$

Fig. 4. The reduction *SAT3-to-INDEP* as a model transformation unit

exclamation mark after *clause* the control condition requires that this action be applied as long as possible so that no clause is left in component 2, all of them are represented as disjoint triangular subgraphs in component 3, and component 4 provides their number at the end.

The subexpression "; *contradict!*" in the control condition requires to continue the transformation by the application of *contradict* as long as possible. This means that in the graph component edges are added between each two nodes with contradicting loops that are not yet connected. Then the rest "; *clean-up!*" requires to remove all loops in the graph component to end up with an unlabeled graph. More precisely, every application of the rule *clean* removes one loop from the current graph. For reasons of space limitations, the rule is not depicted.

It should be noted that the action *clause* is constrained. Formally, this is a further control condition which requires that the Boolean term $c_1 \vee c_2 \vee c_3 \in F$ evaluates to *true* whenever *clause* is applied. In this way, the removal operation removes a clause successfully. As this part of the control condition concerns only the action *clause*, it is placed next to it.

An example of a model transformation performed by the unit *SAT3-to-INDEP* is

$$(\{x, y\}, \{x \vee \bar{x} \vee y, y \vee \bar{y} \vee x, y \vee \bar{y} \vee \bar{x}\}) \xrightarrow{f_{initial}}$$
$$(\{x, y\}, \{x \vee \bar{x} \vee y, y \vee \bar{y} \vee x, y \vee \bar{y} \vee \bar{x}\}, \emptyset, 0) \underset{clause}{\overset{*}{\Longrightarrow}} (\{x, y\}, \emptyset, G_1, 3) \underset{contradict}{\overset{*}{\Longrightarrow}}$$
$$(\{x, y\}, \emptyset, G_2, 3) \underset{clean\_up}{\overset{*}{\Longrightarrow}} (\{x, y\}, \emptyset, G_3, 3) \xrightarrow{f_{terminal}} (G_3, 3)$$

where G_1 is given in Fig. 5, G_2 is the graph of Fig. 3 and G_3 obtained from G_2 by removing each loop.

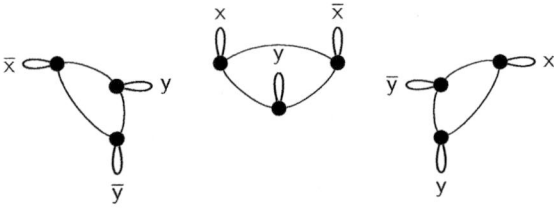

Fig. 5. A graph generated by *clause*!

4.8 Interaction between Graph Transformation and Algebraic Specification

Besides the tupling that combines graph transformational and algebraically specified components, the framework provides a few further possibilities how graph transformation and algebraic specification can be intertwined.

1. If a parameter of some algebraic specification is a single sort or a sort with an equality predicate, then the class of graphs of a rule base or their set of labels can be used as actual parameter. For instance, the set of identifiers is an actual parameter as well as a subset of the set of labels used in the running example.
2. The other way round, the graphs of some rule base may be labeled by elements of some algebra domain.
3. The constraints can use the standard variables and determine in this way how all the components of models must be interrelated.
4. Rules and terms may use the same variables which must be identically instantiated if they occur in the same action. For instance, the action *clause* removes a clause from the set of clauses and adds at the same time a triangular subgraph that is labeled by the literals of the removed clause because the variables c_1, c_2 and c_3 that occur in the term as well as in the rule must be instantiated by the same assignment.
5. Analogously to Points 3 and 4, the variables that can occur in control conditions may regulate the application of actions depending on the interrelation of the involved components.

5   Termination, Functionality and Correctness

Model transformation units are devices to specify the stepwise transformation of input into output models in a precise way. But in most cases, it is not enough to construct such a relation. In addition, one would like to guarantee certain desired properties. Looking at our running example, three main properties come to mind. The model transformation unit *SAT3-to-INDEP* is intended to define a reduction of *SAT3* to *INDEP* meaning that the transformation must be polynomial, functional and correct as discussed in Section 2. As these three properties are typical for many model transformations, we want to sketch some very first ideas how they may be proved in the context of model transformation units.

5.1   Termination

Termination can be shown in the traditional way by finding a valuation of models that decreases if actions are performed.

Let us recall some well-known facts about termination based on arbitrary binary relations.

Let C be a set of configurations, $size \colon C \to \mathbb{N}$ a size function, and $\longrightarrow \subseteq C \times C$ a binary relation, the elements of which are considered as elementary computational steps. Then a computation from c to c' is a sequence of steps $c = c_0 \longrightarrow c_1 \longrightarrow \ldots \longrightarrow c_k = c'$ for $k \in \mathbb{N}$, which is shortly denoted by $c \overset{k}{\longrightarrow} c'$. $(C, size, \longrightarrow)$ is *terminating* with the bound $b \colon \mathbb{N} \to \mathbb{N}$ if $c \overset{k}{\longrightarrow} c'$ implies $k \leq b(size(c))$ for all computations. $(C, size, \longrightarrow)$ is *polynomial* if b is a polynomial. The bound is given on the size rather than for each configuration to cover the special case of polynomiality as it is used for the famous complexity classes P and NP, for example.

To prove termination, one can often use the following observation.

Observation 1. Let $val \colon C \to \mathbb{N}$ be a valuation function with $val(c) > val(c')$ for all $c \longrightarrow c'$. Then $(C, size, \longrightarrow)$ is terminating with the bound

$$b(n) = max\{val(c) \mid size(c) \leq n\} \text{ for } n \in \mathbb{N}.$$

Termination is undecidable for graph transformation in general [24]. However, all model transformations specified by the example unit *SAT3-to-INDEP* are terminating. To illustrate how this works, consider the models (X, F, G, k) of the working type of the model transformation unit *SAT3-to-INDEP* as configurations and the applications of the action *clause* as the step relation. Then the number of clauses $\#F$ decreases in each step by 1. Therefore, all iterations of *clause*-applications are terminating in a number of steps that is linear in the number of clauses. Moreover, the bound is sharp as *clause* can be applied as long as there is some clause left.

The situation is similar for the actions *contradict* and *clean-up*. As both change only the graph component, we can focus on this. The application of the rule *addcontradiction* adds an edge between two given nodes, but only if there is no such edge. In other words, the number of pairs of nodes that are not connected by an edge decreases by 1 whenever the rule is applied. This proves that the iteration of *contradict* terminates with a bound that is quadratic in the number of nodes. With respect to the rule *clean*, the argument is even simpler because it removes a loop whenever applied (and does nothing else).

Due to the control condition, *SAT3-to-INDEP* is the sequential composition of the three polynomial computations given by the three actions. Such a composition is polynomial itself if the steps increase the size of configurations by a constant at most. This is the case with *clause*, *contradict* and *clean-up* if one chooses the number of clauses plus the number of nodes as *size*. Only the application of *addclause* increases the size by three nodes.

Termination criteria for graph transformation systems have been intensively studied (see [6,11,3,7] for some examples).

5.2 Functionality

Consider $(C, \longrightarrow, I, T)$ where $\longrightarrow$ is a step relation on C (as in 5.1) and I and T are subsets of C providing the initial and terminal configurations respectively. Then this system is *functional* if, for each $c \in I$, there is a unique $\bar{c} \in T$ with $c \xrightarrow{*} \bar{c}$ (where $\xrightarrow{*}$ denotes the reflexive and transitive closure of $\longrightarrow$). $(C, \longrightarrow)$ is *confluent* if, for each two $c \xrightarrow{*} c'$ and $c \xrightarrow{*} c''$, there are $c' \xrightarrow{*} \bar{c}$ and $c'' \xrightarrow{*} \bar{c}$ for some $\bar{c}$.

The following observation states how confluence can yield functionality.

Observation 2.   1. Let $(C, \longrightarrow, I, T)$ be a system with $\longrightarrow \subseteq C \times C$ and $I \cup T \subseteq C$ subject to the conditions:
  (a) $(C, \longrightarrow)$ is confluent,
  (b) T is reduced, i.e., $c \xrightarrow{k} c'$ and $c \in T$ implies $k = 0$, and
  (c) $\longrightarrow$ is I, T-complete, i.e., for each $c \in I$, there is a $\bar{c} \in T$ with $c \xrightarrow{*} \bar{c}$.
  Then the system is functional.
  2. $(C, \longrightarrow)$ is confluent if it has the strong local Church-Rosser property, i.e., $c \longrightarrow c'$ and $c \longrightarrow c''$ implies $c' \longrightarrow \bar{c}$ and $c'' \longrightarrow \bar{c}$ for some $\bar{c}$.

This applies very nicely to our running example considering the application of each action separately as in 5.1. Obviously, each two applications of the same rule have the strong local Church-Rosser property so that confluence is guaranteed. If one chooses the models that are reduced with respect to the actions in consideration as terminal configurations, then the reducedness condition holds by definition. And finally, the completeness condition holds in all three cases for any choice of initial configuration because we know already that the three step relations yield terminating computations. In particular, a computation as long as possible holds and ends in a reduced form.

In other words, the three computations corresponding to the three actions are functional so that their sequential composition is functional, too, yielding the functionality of *SAT3*-to-*INDEP*.

In the area of graph transformation, confluence can be shown by analyzing so-called critical pairs [23,15,25].

5.3   Correctness

Correctness is an essential issue of model transformation. It means that the semantics of input models is preserved by the transformation into output models. For instance, if the input model describes a secure bank transfer system, then one would like the result of the transformation to still be secure and handle bank transfers properly. A formal notion of correctness of model transformation units requires the semantics of the input and output models to be comparable as in the following example.

Let *mtu* be a model transformation unit with $SEM(mtu) : \mathfrak{M}(IT) \to \mathfrak{M}(OT)$. Let $SEM(IT) : \mathfrak{M}(IT) \to DOM$ and $SEM(OT) : \mathfrak{M}(OT) \to DOM$ be semantic relations of the input and output models into a common semantic domain DOM. Then mtu is *correct* if

$$SEM(IT) = SEM(OT) \circ SEM(mtu).$$

In the case of our running example, the common semantic domain is $BOOL = \{true, false\}$ and the semantics are the satisfiability $SAT3$ on one hand and the solvability of the independence problem $INDEP$ on the other hand. As pointed out in Section 2, the mapping $SAT3$-*to*-$INDEP$ is a correct reduction, i.e.

$$SAT3 = INDEP \circ SAT3\text{-to-}INDEP.$$

By induction on the number of clauses in F and on the number of contradicting literals in F and taking into account the considerations concerning termination of the model transformation unit $SAT3$-*to*-$INDEP$ in 5.1, it can be shown that the semantics of the unit coincides with the reduction mapping so that the unit is correct in the sense of the definitions above.

Besides correctness proofs for reductions between *NP*-problems, one encounters quite many transformations of some kinds of grammars and automata into other kinds of grammars and automata in the literature. In each of these these cases, the common semantic domain is the set of formal languages and the correctness proof shows that the input model specifies the same language as its transformed counterpart.

It is beyond the scope of this paper to elaborate the topic of correctness, but it will be a focus of future research. The hope is to learn from all the correctness proofs in theoretical computer science and to develop proof methods for model transformation units that apply to the model transformation in software engineering and business process modeling.

6   Related Work

In the literature, one encounters quite a variety of graph-transformational approaches to model transformation. One approach to define model transformations based on graph transformation is by triple grammars [27,17,28]. They generate corresponding source and target models simultaneously, but they can also be considered as transformations from source to target and from target to source. In [12], models are graphs equipped with a semantics given as a set of simulation rules, and a model transformation is composed of generating first an integrated model by graph transformation rules and restricting it then to the target model. In [20], an approach to model transformation is presented that uses graph transformation units [18] based on typed attributed graph transformation.

Examples of model transformation tools based on graph transformation are VIATRA2 [29], GReAT [2], ATOM³ [22], and MOMENT2-GT where the latter is a graph transformation variant of MOMENT2 [5]. VIATRA2 integrates graph transformation and abstract state machines. GReAT mainly consists of a pattern specification language, a transformation rule language and a control flow language. ATOM[3] focuses on modeling complex systems composed of various formalisms and allows to transform them into a single common formalism. MOMENT2-GT translates model transformation specifications into theories in Maude [8] such that models correspond to terms and applications of graph transformation rules to term rewriting.

All these approaches to model transformation consider a model as a graph. Therefore, each of them could be used in our approach to deal with some graph component of our models. To cover the algebraically specified components in the other approaches, a graph encoding of arbitrary data types would be necessary.

7   Conclusion

In this paper, we have proposed a formal approach to model transformation by combining graph transformation and algebraic specification generalizing our work in [19]. This is based on the observation that models are often not just diagrams or pieces of programs, but composite structures with components being strings, sets, numbers, trees, graphs, etc. Especially, the wealth of models one encounters in theoretical computer science like grammars, automata, specifications and systems of various kinds are structured in this way. Our notion of a model transformation unit takes this into account by providing models that are tuples of components of various types which are transformed componentwise according to their type. Graph components are transformed by graph transformation rules. And other components are assumed to be data objects of data types with algebraic specifications so that operations can do the transformation. The first considerations are somewhat promising and indicate that the framework may provide a solid fundament for the formal modeling of model transformation. However, future research should shed more light on the significance of this approach including the following points:

1. In Section 5, we have shown that well-known methods can help to show interesting properties of model transformation like termination, polynomiality and functionality. Clearly, one would like to have methods that are more specific and reach farther.
2. In particular, correctness is an essential property. Our hope is that model transformation units help to study and prove correctness.
3. For the practical use of this novel approach, it will be important to provide tool support especially to prove properties.

References

1. Astesiano, E., Kreowski, H.-J., Krieg-Brückner, B. (eds.): Algebraic Foundations of Systems Specification. Springer, Heidelberg (1999)
2. Balasubramanian, D., Narayanan, A., van Buskirk, C.P., Karsai, G.: The graph rewriting and transformation language: GReAT. Electronic Comunications of the EASST 1 (2006)

3. Bisztray, D., Heckel, R.: Combining Termination Criteria by Isolating Deletion. In: Ehrig, H., Rensink, A., Rozenberg, G., Schürr, A. (eds.) ICGT 2010. LNCS, vol. 6372, pp. 203–217. Springer, Heidelberg (2010)

4. Bisztray, D., Heckel, R., Ehrig, H.: Compositionality of model transformations. Electr. Notes Theor. Comput. Sci. 236, 5–19 (2009)

5. Boronat, A., Meseguer, J.: An algebraic semantics for MOF. Formal Asp. Comput. 22(3-4), 269–296 (2010)

6. Bottoni, P., Hoffmann, K., Parisi-Presicce, F., Taentzer, G.: High-level replacement units and their termination properties. J. Vis. Lang. Comput. 16(6), 485–507 (2005)

7. Bottoni, P., Parisi-Presicce, F.: A termination criterion for graph transformations with negative application conditions. Electronic Communications of the EASST 30 (2010)

8. Clavel, M., Durán, F., Eker, S., Lincoln, P., Martí-Oliet, N., Bevilacqua, V., Talcott, C.: All About Maude - A High-Performance Logical Framework, How to Specify, Program and Verify Systems in Rewriting Logic. LNCS, vol. 4350. Springer, Heidelberg (2007)

9. Ehrig, H., Ehrig, K., Ermel, C., Hermann, F., Taentzer, G.: Information Preserving Bidirectional Model Transformations. In: Dwyer, M.B., Lopes, A. (eds.) FASE 2007. LNCS, vol. 4422, pp. 72–86. Springer, Heidelberg (2007)

10. Ehrig, H., Ehrig, K., Hermann, F.: From model transformation to model integration based on the algebraic approach to triple graph grammars. Electronic Communications of the EASST 10 (2008)

11. Ehrig, H., Ehrig, K., de Lara, J., Taentzer, G., Varró, D., Varró-Gyapay, S.: Termination Criteria for Model Transformation. In: Cerioli, M. (ed.) FASE 2005. LNCS, vol. 3442, pp. 49–63. Springer, Heidelberg (2005)

12. Ehrig, H., Ermel, C.: Semantical Correctness and Completeness of Model Transformations Using Graph and Rule Transformation. In: Ehrig, H., Heckel, R., Rozenberg, G., Taentzer, G. (eds.) ICGT 2008. LNCS, vol. 5214, pp. 194–210. Springer, Heidelberg (2008)

13. Frankel, D.S.: Model Driven Architecture. Applying MDA to Enterprise Computing. Wiley, Indianapolis (2003)

14. Garey, M.R., Johnson, D.S.: Computers and Intractability: A Guide to the Theory of NP-Completeness. W.H. Freeman (1979)

15. Heckel, R., Küster, J.M., Taentzer, G.: Confluence of Typed Attributed Graph Transformation Systems. In: Corradini, A., Ehrig, H., Kreowski, H.-J., Rozenberg, G. (eds.) ICGT 2002. LNCS, vol. 2505, pp. 161–176. Springer, Heidelberg (2002)

16. Hopcroft, J.E., Motwani, R., Ullman, J.D.: Introduction to automata theory, languages, and computation, 3rd edn. Addison-Wesley Longman (2007)

17. Königs, A., Schürr, A.: Tool integration with triple graph grammars - a survey. Electr. Notes Theor. Comput. Sci. 148(1), 113–150 (2006)

18. Kreowski, H.-J., Kuske, S.: Graph transformation units with interleaving semantics. Formal Aspects of Computing 11(6), 690–723 (1999)

19. Kreowski, H.-J., Kuske, S., von Totth, C.: Stepping from graph transformation units to model transformation units. Electronic Communications of the EASST 30 (2010)

20. Küster, J.M.: Definition and validation of model transformations. Software and System Modeling 5(3), 233–259 (2006)

21. de Lara, J., Taentzer, G.: Automated Model Transformation and Its Validation Using AToM[3] and AGG. In: Blackwell, A.F., Marriott, K., Shimojima, A. (eds.) Diagrams 2004. LNCS (LNAI), vol. 2980, pp. 182–198. Springer, Heidelberg (2004)

22. de Lara, J., Vangheluwe, H., Alfonseca, M.: Meta-modelling and graph grammars for multi-paradigm modelling in AToM^3. Software and System Modeling 3(3), 194–209 (2004)
23. Plump, D.: Hypergraph rewriting: Critical pairs and undecidability of confluence. In: Sleep, M.R., Plasmeijer, R., van Eekelen, M. (eds.) Term Graph Rewriting. Theory and Practice, pp. 201–213. Wiley & Sons (1993)
24. Plump, D.: Termination of graph rewriting is undecidable. Fundamenta Informaticae 33(2), 201–209 (1998)
25. Plump, D.: Checking graph-transformation systems for confluence. Electronic Communications of the EASST 26 (2010)
26. Rozenberg, G. (ed.): Handbook of Graph Grammars and Computing by Graph Transformation. Foundations, vol. 1. World Scientific, Singapore (1997)
27. Schürr, A.: Specification of Graph Translators with Triple Graph Grammars. In: Mayr, E.W., Schmidt, G., Tinhofer, G. (eds.) WG 1994. LNCS, vol. 903, pp. 151–163. Springer, Heidelberg (1995)
28. Schürr, A., Klar, F.: 15 years of Triple Graph Grammars. In: Ehrig, H., Heckel, R., Rozenberg, G., Taentzer, G. (eds.) ICGT 2008. LNCS, vol. 5214, pp. 411–425. Springer, Heidelberg (2008)
29. Varró, D., Balogh, A.: The model transformation language of the VIATRA2 framework. Sci. Comput. Program. 68(3), 214–234 (2007)

Towards Bialgebraic Semantics for the Linear Time – Branching Time Spectrum[*]

Ana Paula Maldonado[1], Luís Monteiro[1], and Markus Roggenbach[2]

[1] CITI, Departamento de Informática, Faculdade de Ciências e Tecnologia,
Universidade Nova de Lisboa, 2829-516 Caparica, Portugal
[2] Swansea University, Wales, UK

Abstract. Process algebra, e.g. CSP, offers different semantical observations (e.g. traces, failures, divergences) on a single syntactical system description. These observations are either computed algebraically from the process syntax, or "extracted" from a single operational model. Bialgebras capture both approaches in one framework and characterize their equivalence; however, due to use of finality, lack the capability to simultaneously cater for various semantics. We suggest to relax finality to quasi-finality. This allows for several semantics, which also can be coarser than bisimulation. As a case study, we show that our approach works out in the case of the CSP failures model.

1 Introduction

Giving semantics to process algebra in the form of SOS has become standard since Plotkin's seminal paper [1]. Besides the transition system of a process, however, one is often interested in a more abstract description of a process, based on observations such as traces or failures. Naturally, this leads to the questions of how to obtain observations from a transition system. This can be done via an algebraic approach using initiality of the term algebra, or coalgebraically via finality of the intended observational model.

Bialgebras [2, 3] host these two approaches within one framework, and allow one to study conditions under which they are equal. In this paper we are interested in observations that lead to notions of process equivalence coarser than bisimulation. This is motivated, e.g., by the various semantics of the process algebra CSP [4, 5], which lie within the linear time (trace semantics) – branching time (bisimulation semantics) spectrum [6]. As final objects are unique up to isomorphisms, any given category can cater for just one semantics given by finality. If one is interested in various semantics for one process algebra – as is, e.g., the case for CSP – there are two possible approaches: either one defines a specialized category for each semantic model, or, one relaxes the requirement of finality. Here, we follow the latter approach and relax finality to quasi-finality [7, 8]. Briefly, quasi-finality has a characterization similar to that of finality, except that the defining conditions are required to hold only in the underlying

---

[*] This research was supported in part by Grid-Tools Ltd, UK.

T. Mossakowski and H.-J. Kreowski (Eds.): WADT 2010, LNCS 7137, pp. 209–225, 2012.

category (of sets and functions, for simplicity) rather than in the category of coalgebras.

Relatively to a functor B describing the type of transition structure, we introduce the notion of an observational model. Such a model consists of observations, e.g. trace languages. These observations carry a transition structure of type B. Additionally, an observational model defines an observation function from any B-coalgebra to the set of observations of the model. Furthermore, the observational model is required to be quasi-final in the category of B-coalgebras w.r.t. these observation functions.

With these notions we obtain: Given the signature Σ of a process algebra P and its transition rules for a functor B; and an observational model such that a certain equation holds for all operators in Σ, then the derived denotational semantics is equivalent to the defined operational semantics.

The starting point of our work was the realization that when studying a certain class of systems viewed as coalgebras, we may be interested in observations or behaviours other than those described by final coalgebras. To deal with these problems we suggested in [8] to relax the requisite of finality to quasi-finality, and proceeded to show that typical behaviours in the linear-time – branching-time hierarchy [6], like traces, ready-traces, failures and synchronization trees, give rise to quasi-final coalgebras. The next step was to apply these ideas to the semantics of an actual process calculus, as an extension of bialgebraic semantics [2, 3]; we have chosen CSP mainly because different semantics have been studied for it, based on different kinds of behaviours [5]; this is essentially the work reported in this paper. Other coalgebraic semantics of CSP include [9, 10, 11]; the main difference with our work is that ours pretends to be an instance of a general framework intended to capture a general notion of behaviour in terms of quasi-final coalgebras, while the work of these authors seems to be tailored for the specific language at hand; it would be interesting, however, to look deeper into Boreale and Gadducci's model to see if or how it could be adapted to our framework. Other attempts have been made at trying to capture in coalgebraic terms traces of several kinds of systems [12, 13, 14]. We should also mention van Glabbeek's work [6] on the characterization of process equivalences other than bisimilarity using modal logics, as well as the more recent work of Jacobs and Solokova [15] based on dual adjunctions. Klin [16, 17] is interested in proving compositionality of process equivalences, and relies heavily on bialgebraic semantics and SOS congruence formats. The goal of Klin appears to be similar to ours, because compositionality of process equivalences is basically the same as compositionality of operational semantics; furthermore, despite being based on different approaches, there seem to exist deep connections between the two works; but these need to be further investigated.

This paper is organized as follows: In Sections 2 and 3 we review the background of our construction, namely the concept of bialgabraic semantics in Section 2, and the notion of quasi-finality in Section 3. In Section 4 we discuss how to generalize bialgebraic semantics to the quasi-final case. Finally, in Section 5 we apply our approach to the CSP failures semantics.

2   Bialgebraic Semantics

Bialgebras [2, 3] host initial or denotational semantics and final or operational semantics within one framework, and allow one to study conditions under which they are equal. In this section we discuss the basic concepts of bialgebraic semantics.

2.1   Syntax, Signature, Σ-Algebra

A (single-sorted) signature Σ consists of a set of symbols together with a function giving their arities. Figure 1 illustrates how the grammar of (a sublanguage of) the process algebra CSP over a given set of communications A corresponds to such a signature: The signature of our small CSP fragment has operators *Stop* (for the deadlock process) of arity 0, $a \to$ (_) (for action prefix) for all $a \in A$ of arity 1, and $\Box$ (for external choice) and $|||$ (for interleaving) of arity 2. The terms over this signature correspond to the expressions of CSP derived from the grammar.

$$\text{Grammar } P ::= Stop \mid a \to P \mid P \Box P \mid P \mid\mid\mid P$$

$$\text{Signature } \Sigma = \{ Stop, a \to {\_}, {\_} \Box {\_}, {\_} \mid\mid\mid {\_} \mid a \in A \}$$

Fig. 1. One sorted signature for CSP sublanguage

A Σ-algebra is a pair $< M, \mu >$ where M is a set and μ is a function $\mu : \Sigma M \to M$. Here we overload notation and treat Σ also as the functor defined on sets X by setting $\Sigma X = \{ < \sigma, x_1, \ldots, x_n >: \sigma \in \Sigma, \text{arity } \sigma = n, x_1, \ldots, x_n \in X \}$. The operation $\sigma_M : M^n \to M$ associated with $\sigma \in \Sigma$ of arity n is given by $\sigma_M(m_1, \ldots, m_n) = \mu(< \sigma, m_1, \ldots, m_n >)$. A homomorphism from $< M, \mu >$ to another Σ-algebra $< M', \mu' >$, or Σ-homomorphism, is a function $f : M \to M'$ such that the following diagram commutes:

$$
\begin{array}{ccc}
\Sigma M & \xrightarrow{\Sigma f} & \Sigma M' \\
\mu \downarrow & & \downarrow \mu' \\
M & \xrightarrow{f} & M'.
\end{array}
$$

The set $T0$ of Σ-terms can be turned into a Σ-algebra $< T0, \alpha_0 >$ by defining $\alpha_0(< \sigma, t_1, \ldots, t_n >) = \sigma(t_1, \ldots, t_n)$ where $\sigma \in \Sigma$ has arity n and $t_1, \ldots, t_n$ are terms; this is the initial Σ-algebra: there is a unique Σ-homomorphism ι_M from $< T0, \alpha_0 >$ to any Σ-algebra $< M, \mu >$. In this context, ι_M is also known as the *initial* or *denotational* semantics of the language $T0$ defined by Σ in the semantic model M.

The free Σ-algebra generated by a set X is defined similarly; it will be denoted $< TX, \alpha_X >$, with "inclusion of generators" $\eta_X : X \to TX$. The characterizing property of the free algebra is that any function $f : X \to M$ has a unique extension to a Σ-homomorphism $\bar{f}$ from $<TX, \alpha_X>$ to $<M, \mu>$. This may be visualized in the following commutative diagram:

$$X \xrightarrow{\eta_X} TX \xleftarrow{\alpha_X} \Sigma TX$$
$$f \searrow \quad \downarrow \bar{f} \qquad\qquad \downarrow \Sigma\bar{f}$$
$$M \xleftarrow[\mu]{} \Sigma M .$$

2.2 SOS Rules, Transition System, Coalgebra

Recall that a labelled transition system (LTS) over a set A of communications or actions is a pair $<S, \rightarrow>$ where S is the set of states and $\rightarrow \subseteq S \times A \times S$ is the transition relation; as usual, $(s, a, s') \in \rightarrow$ is written $s \xrightarrow{a} s'$. Below, besides A we consider also the set $A \cup \{\tau, \checkmark\}$ of labels, where τ represents an internal transition and $\checkmark$ represents termination, but for now we stick to A.

Action prefix: $$\frac{}{a \rightarrow P \xrightarrow{a} P}$$

External choice: $$\frac{P \xrightarrow{a} P'}{P \square Q \xrightarrow{a} P'} \qquad \frac{Q \xrightarrow{a} Q'}{P \square Q \xrightarrow{a} Q'}$$

Interleaving: $$\frac{P \xrightarrow{a} P'}{P \;|||\; Q \xrightarrow{a} P' \;|||\; Q} \qquad \frac{Q \xrightarrow{a} Q'}{P \;|||\; Q \xrightarrow{a} P \;|||\; Q'}$$

Fig. 2. SOS rules for the fragment of CSP

Given a set of SOS rules the set $T0$ of terms of the language is turned into an LTS. As an example, Figure 2 shows the rules of our CSP fragment: *Stop* cannot perform any action, thus there is no rule for this process. Action prefix with a can perform the action a. For external choice, the first action decides which branch is chosen; both branches are possible. In interleaving, the involved processes run independently of each other.

Such dynamic behaviour can also be captured with the notion of a coalgebra: Given an endofunctor B on the category **Set** of sets and functions, a B-coalgebra is a pair $<S, \varphi>$ where S is a set and $\varphi : S \rightarrow BS$ is a function, called the transition structure of the coalgebra. A B-morphism from $<S, \varphi>$ to $<S', \varphi'>$ is function $f : S \rightarrow S'$ such that $\varphi' \circ f = Bf \circ \varphi$, i.e., the following diagram commutes:

$$S \xrightarrow{f} S'$$
$$\varphi \downarrow \qquad\qquad \downarrow \varphi'$$
$$BS \xrightarrow[Bf]{} BS' .$$

A transition system $<S, \rightarrow>$ will be identified with a coalgebra $<S, \varphi>$ for the functor $B = \mathcal{P}(-)^A$; the structure function $\varphi : S \rightarrow \mathcal{P}(S)^A$ is defined by

$\varphi(s)(a) = \{s' : s \xrightarrow{a} s'\}$; conversely, given such a coalgebra, the corresponding transition relation is defined by $s \xrightarrow{a} s'$ iff $s' \in \varphi(s)(a)$; the two notations will be used interchangeably. For the LTS $< T0, \longrightarrow >$ on the set $T0$ of Σ-terms we denote the structure function with ψ_0; thus, we have a structure $< T0, \alpha_0, \psi_0 >$ which is both a Σ-algebra and a B-coalgebra; such structures are called *bialgebras* and play a fundamental role in the sequel; a morphism of bialgebras is a function that is both a Σ-homomorphism and a B-morphism.

When the category of B-coalgebras has a final object $< Z, \zeta >$, whose elements are interpreted as observations or behaviours, there is a unique B-morphism $\beta_S : S \to Z$ from any B-coalgebra $< S, \varphi >$ to $< Z, \zeta >$. When S is $T0$, the B-morphism β_{T0} is the *final* or *operational* semantics of $T0$. Note in passing that for cardinality reasons the functor $\mathcal{P}(-)^A$ does not have a final coalgebra, which precludes the existence of a final semantics; to obtain a final semantics in this case the standard procedure is to replace $\mathcal{P}$ with the finite powerset functor $\mathcal{P}_{\text{fi}}$, when possible. But $\mathcal{P}(-)^A$ has (infinitely) many quasi-final coalgebras, which justifies our interest in semantics based on them.

In order to compare the denotational and the operational semantics of $T0$, we must be able to define a Σ-algebra $< Z, \theta >$ to turn Z into a denotational semantic model; in other words, we need to endow Z with a bialgebra structure $< Z, \theta, \zeta >$; the next result explains why.

Proposition 1. *If there is a bialgebra morphism $< T0, \alpha_0, \psi_0 > \to < Z, \theta, \zeta >$, it must be equal to ι_Z by initiality, and to β_{T0} by finality; in particular, the denotational and the operational semantics coincide.*

To define θ we need, in general, to define a transition structure not only in $T0$ but also in any free algebra TS, where S is the underlying set of a B-coalgebra $< S, \varphi >$. Structurally this coalgebra, say $< TS, \psi_S >$, must have the properties:

Assumption 1. *The unique Σ-homomorphism $\iota_{TS} : T0 \to TS$ and the inclusion map of generators $\eta_S : S \to TS$ are both B-coalgebra morphisms.*

Again, we have a bialgebra $< TS, \alpha_S, \psi_S >$ and the assumption on ι_{TS} implies that it is a bialgebra morphism. In the case of LTSs, the transition structure on TS is specified by the set of rules $R \cup R_S$ where R is the set of rules for the operators in Σ and R_S is just the set of all transitions in S turned into rules with no premises. For rules in "well-behaved" formats it is guaranteed that ι_{TS} and η_S are indeed B-morphisms (in [2] this was shown for the *tyft/tyxt* format [18]); in the sequel we take this for granted. A more abstract approach was proposed by Turi and Plotkin [3], but the one just outlined is enough for our purposes in this paper.

2.3   Bialgebraic Semantics

Assuming that there exists a final B-coalgebra $< Z, \zeta >$, with $\beta_S : S \to Z$ the only B-morphism from any B-coalgebra $< S, \varphi >$, and assuming that the free algebras TS can be given transition structures as described above, the set Z

of behaviours of interest is made into a Σ-algebra by defining $\theta : \Sigma Z \to Z$ as $\theta \mathrel{\widehat{=}} \beta_{TZ} \circ \alpha_Z \circ \Sigma \eta_Z :$

$$
\begin{array}{ccc}
\Sigma TZ & \xleftarrow{\;\;\Sigma \eta_Z\;\;} & \Sigma Z \\
\alpha_Z \downarrow & & \downarrow \theta \\
TZ & \xrightarrow[\;\;\beta_{TZ}\;\;]{} & Z \,.
\end{array}
\tag{1}
$$

Here, β_{TZ} is the unique coalgebra morphism into Z thanks to finality, α_Z is the structure map of the Σ-algebra $< TZ, \alpha_Z >$ as defined in subsection 2.1, and $\Sigma \eta_Z$ arises from the inclusion map of generators.

Now consider the diagram:

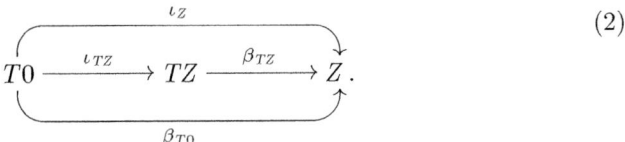

$$\tag{2}$$

If we can prove that β_{TZ} is a Σ-homomorphism, then it is a bialgebra morphism; by Assumption 1, ι_{TZ} is a bialgebra morphism, so $\beta_{TZ} \circ \iota_{TZ}$ is also a bialgebra morphism; by Proposition 1, we conclude that $\iota_Z = \beta_{T0}$: the denotational and the operational semantics coincide.

Proposition 2. *Suppose that for any B-coalgebra $< S, \varphi >$, a B-coalgebra has been defined on TS such that $\iota_{TS} : T0 \to TS$ and $\eta_S : S \to TS$ satisfy Assumption 1. Suppose furthermore that $< Z, \zeta >$ is a final B-coalgebra and $\theta : \Sigma Z \to Z$ is defined by (1). If $\beta_{TZ} : TZ \to Z$ is a Σ-homomorphism, then the denotational semantics ι_Z is equal to the operational semantics β_{T0}.*

Provided that the structural operational rules are in some "well-behaved" format, β_{TZ} is indeed a Σ-homomorphism: [2] shows this for rules in the *tyft/tyxt* format [18]; in the more abstract setting of [3], this is guaranteed by the abstract GSOS format [19].

Overall, bialgebraic semantics provides an "automatic" framework for defining equivalent operational and denotational semantics: Given a set of SOS rules R in a suitable format, and given a suitable final coalgebra $< Z, \zeta >$ of behaviours, it derives a denotational semantics equivalent to the operational semantics given via finality of $< Z, \zeta >$.

2.4   Limitations of Finality

While bialgebras provide an elegant semantical framework, from our point of view the requirement of finality is too strong: It leaves out many interesting types of behaviours and, thus, process equivalences. Typical examples include traces, ready-traces, failures. In fact most of Van Glabbeek's linear-time – branching-time hierarchy [6] for transition systems is not covered. This leaves two possible solutions:

- To restrict the category to make the coalgebra of behaviours of interest final, see [7].
- To relax the requirement of finality as in [8].

The first approach often gives rise to complex constructions and also fails to offer "natural" solutions: To begin with, there may be several categories where the coalgebra of behaviours of interest is final, which raises the problem of justifying any particular choice. For example, the transition system of trace languages, see below, is final in the category of deterministic transition systems, as well as in the full subcategory formed by itself alone. Another problem, which arises once a category is chosen, is how to associate behaviours with coalgebras that are outside that category. In the rest of the paper we follow the second approach.

3   The Concept of Quasi-finality

We begin our discussion with a non-standard formulation of finality:

Proposition 3. *An object Z in a category $\mathbf{C}$ is final if, and only if, there is a morphism $\beta_S : S \to Z$ for every object S in $\mathbf{C}$ such that*

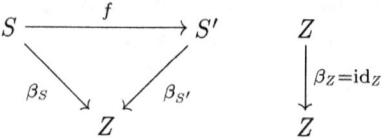

commutes for every morphism f.

Proof. If Z is final and the $\beta_S : S \to Z$ are the unique morphisms into Z, then the two conditions expressed by the previous diagram hold by uniqueness. Conversely, if those conditions hold, each β_S is the unique morphism from S to Z. Indeed, if $f : S \to Z$ is another morphism, then, by the previous diagram with $S' = Z$, we obtain: $\beta_S = \beta_Z \circ f = \mathrm{id}_Z \circ f = f$.

Taking this characterization of finality as a starting point, we define the concept of quasi-finality, as introduced and discussed in [8]. Let $\mathbf{C}$ be a concrete category with forgetful functor $U : \mathbf{C} \to \mathbf{Set}$.

Definition 1. *An object Z in $\mathbf{C}$ is quasi-final if there is a function $\beta_S : US \to UZ$ for every object S in $\mathbf{C}$ such that*

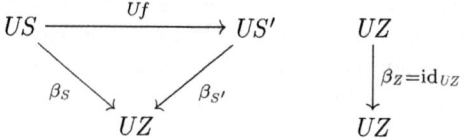

commutes for every morphism f. (Note that f is a morphism in $\mathbf{C}$, the β_S are functions in $\mathbf{Set}$.)

Typical examples of behaviours in the linear time – branching time spectrum [6] have been shown [7, 8] to give rise to quasi-final coalgebras (for LTSs without internal transitions). These include traces, ready-traces and failures, as well as synchronization trees. We review now briefly the case of traces; later we consider also failures, in the more general setting where internal transitions and termination are also taken into account.

Example 1 (Trace languages). If $s, t \in S$ are states of an LTS $< S, \rightarrow >$ labelled with A and $x = < a_0, a_1, \ldots, a_{n-1} > \in A^*$, we put $s \overset{x}{\mapsto} t$ if there exists a sequence of states $s = s_0, s_1, \ldots, s_n = t$, $n \geq 0$, such that $s_i \overset{a_i}{\rightarrow} s_{i+1}$ for all $0 \leq i < n$; the sequence x is a trace of s and we let $Tr_S(s)$ be the set of all traces of s. The set $Tr_S(s)$ is nonempty and prefix closed, and is called a trace language over A. The set T of all trace languages is made into a LTS with transitions $L \overset{a}{\rightarrow} L'$ whenever $\langle a \rangle \in L$ and $L' = \{x \mid \langle a \rangle \frown x \in L\}$. The LTS $< T, \rightarrow >$ is quasi-final with behaviour Tr. Indeed, for any morphism $f : S \rightarrow S'$ of LTSs, every $s \in S$ and its image $f(s)$ have the same traces, that is, $Tr_S(s) = Tr_{S'}(f(s))$; thus, $Tr_S = Tr_{S'} \circ f$. On the other hand, the set of traces of a prefix-closed language L as a state of $< T, \rightarrow >$ is L itself, that is, $Tr_T(L) = L$; thus, $Tr_T = \mathrm{id}_T$.

Note that unlike final objects, quasi-final objects need not be unique up to isomorphism—there may be even infinitely many on the same underlying set. For example, if $1 = \{*\}$ is a singleton, any object Z in **C** such that $UZ = 1$ is quasi-final. More concretely still, in the category of LTSs over A, any LTS with a single state and any number of transitions from the state to itself is quasi-final. Thus, if A is an infinite set this already gives infinitely many quasi-final LTSs. This does not mean that quasi-final objects do not satisfy any uniqueness properties; in fact it is easy to find two categories in which quasi-final objects turn out to be final: one is a super-category of **C** with the same objects and additional arrows, the other is a full sub-category of **C**.

By definition of quasi-finality, a morphism $f : S \rightarrow S'$ in **C** "preserves behaviours" in the sense that $\beta_{S'} \circ Uf = \beta_S$; we now turn this property into a concept. For simplicity, from now on we identify a morphism f in **C** with the function Uf in **Set**, as is customary when dealing with forgetful functors. For any objects S, S' in **C**, a function $f : US \rightarrow US'$ is called a β-map if $\beta_{S'} \circ f = \beta_S$. Morphisms f in **C**, with the identification $Uf = f$, are β-maps. Any $\beta_S : S \rightarrow Z$ is also a β-map since $\beta_Z \circ \beta_S = \mathrm{id}_Z \circ \beta_S = \beta_S$. Furthermore, β_S is the only β-map from S to Z: another β-map $f : S \rightarrow Z$ satisfies $\beta_S = \beta_Z \circ f = f$. Thus, Z is final in the category $\mathbf{C}/\beta$ that has the same objects as **C** and β-maps as morphisms. Now let **B** be the full sub-category of **C** determined by all objects S such that β_S is actually a morphism $S \rightarrow Z$ and not just a function $US \rightarrow UZ$; first note that Z is in **B** since $\beta_Z = \mathrm{id}_Z$ is a morphism; as we have seen, any morphism $f : S \rightarrow Z$, being a β-map, is equal to β_S; thus, β_S is the only morphism from S to Z; this shows that Z is final in **B**. We summarize these observations in the following proposition.

Proposition 4. *Let* **C** *be a concrete category over* **Set** *with forgetful functor U and let Z be a quasi-final object of* **C** *with behaviour β.*

1. Z is final in the category $\mathbf{C}/\beta$ which has the same objects as $\mathbf{C}$ and β-maps ("behaviour preserving" functions) as morphisms.
2. Z is final in the full subcategory $\mathbf{B}$ of $\mathbf{C}$ determined by all objects S such that $\beta_S : US \to UZ$ is a morphism in $\mathbf{C}$.

We next show how bialgebraic semantics can be extended to the case where the behaviours (observations) of interest give rise to a quasi-final coalgebra rather than a final coalgebra.

4 Towards Quasi-final Semantics

We apply notions related to quasi-finality to the case where $\mathbf{C}$ is the category $\mathbf{Coalg}_B$ of coalgebras of a functor B, and the forgetful functor maps a coalgebra to its underlying set and a B-morphism to itself as a function. Given again the signature Σ and the set $T0$ of Σ-terms, the operational semantics is defined as before as $\beta_{T0} : T0 \to Z$, except that now $< Z, \zeta >$ is a quasi-final coalgebra with respect to a behaviour β. The algebraic structure $\theta : \Sigma Z \to Z$ is again defined by (1), which allows to define the denotational semantics by the unique Σ-homomorphism $\iota_Z : T0 \to Z$. Let us call a *bialgebra β-map* any function between underlying sets of bialgebras that is a Σ-homomorphism and a β-map. We have the following generalization of Proposition 1:

Proposition 5. *Let $< Z, \zeta >$ be a quasi-final B-coalgebra with behaviour β. If there is a bialgebra β-map $< T0, \alpha_0, \psi_0 > \to < Z, \theta, \zeta >$, it must be equal to ι_Z by initiality, and to β_{T0} by quasi-finality; in particular, the denotational and the operational semantics coincide.*

Once again, the equality $\beta_{T0} = \iota_Z$ holds whenever $\beta_{TZ} : TZ \to Z$ is a Σ-homomorphism. Indeed, if that is the case, referring to (2), $\beta_{TZ} \circ \iota_{TZ}$ is a Σ-homomorphism and a β-map, because ι_{TZ} is a B-morphism by Assumption 1, hence a β-map, and β_{TZ} is also a β-map, hence so is their composition. The equality $\beta_{T0} = \iota_Z$ now follows from the previous proposition.

Proposition 6. *Proposition 2 still holds if $< Z, \zeta >$ is a quasi-final B-coalgebra with behaviour β.*

For LTSs, that β_{TZ} is a Σ-homomorphism is no longer a consequence of the fact that the transition rules are in some well-behaved format, as is the case for finality—additional conditions are needed. In [20] we give sufficient conditions for β_{TZ} to be a Σ-homomorphism, embodied in the notion of "adequate" quasi-final coalgebra. These conditions seem to be general enough to ensure that quasi-final coalgebras studied before, like traces and failures of LTSs without τ's, are adequate. When τ-transitions are used, however, the conditions may fail, as is the case with CSP; in those cases, currently we have to verify for each set of SOS rules that β_{TZ} is a Σ-homomorphism. Concretely, we have to show that the following diagram commutes:

$$\begin{array}{ccc} \Sigma TZ & \xrightarrow{\Sigma \beta_{TZ}} & \Sigma Z \\ \alpha_Z \downarrow & & \downarrow \theta \\ TZ & \xrightarrow{\beta_{TZ}} & Z \,. \end{array}$$

Let us illustrate what is required by considering, without loss of generality, a binary operator $\otimes$ written in infix notation:

$$\begin{array}{ccc} <\otimes, t, u> & \xmapsto{\Sigma \beta_{TZ}} & <\otimes, \beta_{TZ}(t), \beta_{TZ}(u)> \\ \alpha_Z \Big\uparrow & & \Big\downarrow \theta \\ & & \beta_{TZ}(\beta_{TZ}(t) \otimes \beta_{TZ}(u)) \\ & & = \\ t \otimes u & \xmapsto{\beta_{TZ}} & \beta_{TZ}(t \otimes u) \,. \end{array}$$

Thus, we must have

$$\beta_{TZ}(t \otimes u) = \beta_{TZ}(\beta_{TZ}(t) \otimes \beta_{TZ}(u)) \tag{3}$$

for all $t, u \in TZ$.

4.1   A Transformation of the SOS Rules

In order to present conditions under which Assumption 1 holds, we use a transformation of the SOS rules at hand that was proposed by Turi and Plotkin [3]; we illustrate the approach with our running example. We can reformulate the rules in Figure 2 to define directly the transition map $S \to \mathcal{P}(S)^A$, or rather its graph, whose pairs we write in the form $P \mapsto f$ for readability. The new rules are depicted in Figure 3, where an informal lambda notation was used for function definition.

Null:
$$\frac{}{Stop \mapsto \lambda\, a.\emptyset}$$

Action prefix:
$$\frac{}{a \to P \mapsto \lambda\, b. \begin{cases} \{P\} & \text{if } b = a \\ \emptyset & \text{if } b \neq a \end{cases}}$$

External choice:
$$\frac{P \mapsto f \qquad Q \mapsto g}{P \,\square\, Q \mapsto \lambda\, a.f(a) \cup g(a)}$$

Interleaving:
$$\frac{P \mapsto f \qquad Q \mapsto g}{P \,|||\, Q \mapsto \lambda\, a.\{P' \,|||\, Q : P' \in f(a)\} \cup \{P \,|||\, Q' : Q' \in g(a)\}}$$

Fig. 3. Inductive definition of the transition map

Each rule has as many premises as the arity of the main operator of the rule,[1] and in fact can be seen as specifying a function of that arity on $TS \times BTS$ with values again in $TS \times BTS$. Actually, it is enough to consider the values in BTS; thus, treating together all operators in Σ, the new set of rules defines a function $\varrho_S : \Sigma(TS \times BTS) \to BTS$. For example,

$$\varrho_S(|||, P \mapsto f, Q \mapsto g) = \lambda\ a.\{P' \ ||| \ Q : P' \in f(a)\} \cup \{P \ ||| \ Q' : Q' \in g(a)\}.$$

It is easy to see that the ϱ_S define a natural transformation $\varrho : \Sigma(T \times BT) \to BT$, since the definition of ϱ_S does not depend on any particular feature of S. More generally, any set of SOS rules in the GSOS format [19] gives rise to a natural transformation $\varrho : \Sigma(T \times BT) \to BT$ [3]. In [3], Theorem 5.1, it is shown that ϱ allows to define by structural recursion a B-coalgebra $< TS, \psi_S >$ from $< S, \varphi >$, the function ψ_S being the only one that makes the following diagram commute:

$$
\begin{array}{ccccc}
S & \xrightarrow{\ \eta_S\ } & TS & \xleftarrow{\quad \alpha_S \quad} & \Sigma TS \\
\varphi \downarrow & & \downarrow \psi_S & & \downarrow \Sigma<\mathrm{id}_{TS},\psi_S> \\
BS & \xrightarrow[\ B\eta_S\ ]{} & BTS & \xleftarrow[\ \varrho_S\ ]{} & \Sigma(TS \times BTS).
\end{array}
\tag{4}
$$

The definition already shows that η_S is a B-morphism, as required. Furthermore, the mapping of $< S, \varphi >$ to $< TS, \psi_S >$ extends to a functor by mapping any B-morphism f from $< S, \varphi >$ to $< S', \varphi' >$ to Tf, which can be shown to be a morphism from $< TS, \psi_S >$ to $< TS', \psi_{S'} >$. This applies in particular to $\iota_{TS} : T0 \to TS$, which results from the unique B-morphism from the initial B-coalgebra on 0 to $< S, \varphi >$, so that ι_{TS} is a B-morphism. These remarks show that the existence of the natural transformation $\varrho : \Sigma(T \times BT) \to BT$ implies that Assumption 1 holds.

4.2   A Proof Approach

To prove (3) we use the fact that the transition structure of coalgebras preserves behaviours: if $< S, \varphi >$ is a B-coalgebra, then $< BS, B\varphi >$ is also a B-coalgebra and $\varphi : S \to BS$ is a B-morphism; thus, in particular, φ is a β-map, so $\beta_S = \beta_{BS} \circ \varphi$. In the case at hand we have $\beta_{TZ} = \beta_{BTZ} \circ \psi_Z$, with $\psi_Z : TZ \to BTZ$ defined as in (4). With this result, (3) may be rewritten as

$$\beta_{BTZ}(\psi_Z(t \otimes u)) = \beta_{BTZ}(\psi_Z(\beta_{TZ}(t) \otimes \beta_{TZ}(u))). \tag{5}$$

The advantage of (5) over (3) is that the behaviour of expressions has been replaced by the behaviour of their "next steps" under ψ_Z, whose structure is determined by the operational rules of the language and allows a form of inductive reasoning. We shall use this equation only implicitly, however: we continue to use (3) but compute $\beta_{TZ}(t)$ with $\beta_{BTZ}(\psi_Z(t))$.

---

[1] Except for action prefix, to which we can add a premise $P \mapsto f$, which is not actually used.

Let us illustrate our approach with the trace semantics and the interleaving operator (for the other operators in the fragment of CSP we have been considering the verification is immediate):

Example 2. Let Z be the set T of trace languages and let β be the family Tr of trace functions.[2] Then, the appropriate instance of (3) is

$$Tr(P \;|||\; Q) = Tr(Tr(P) \;|||\; Tr(Q)),$$

where we abbreviated Tr_{TT} to Tr for readability; we prove by induction on n that both sides have the same traces of length less than or equal to n. This is immediate for $n = 0$, since both sides contain ε, the only trace of length zero. Now let us assume the result for n. Since the transition structures of coalgebras preserve behaviours, we have, in general , $Tr(P) = Tr(\lambda a.\{P' \mid P \xrightarrow{a} P'\}) = \{\varepsilon\} \cup \bigcup\{a \cdot Tr(P') \mid P \xrightarrow{a} P'\}$, where $a \cdot Tr(P') = \{ax : x \in Tr(P')\}$. We can write

$$Tr(P \;|||\; Q) = \{\varepsilon\} \cup \bigcup_{a \in A} a \cdot (\bigcup_{P \xrightarrow{a} P'} Tr(P' \;|||\; Q) \cup \bigcup_{Q \xrightarrow{a} Q'} Tr(P \;|||\; Q')),$$

$Tr(Tr(P) \;|||\; Tr(Q)) =$
$$\{\varepsilon\} \cup \bigcup_{a \in A} a \cdot (\bigcup_{Tr(P) \xrightarrow{a} L} Tr(L \;|||\; Tr(Q)) \cup \bigcup_{Tr(Q) \xrightarrow{a} M} Tr(Tr(P) \;|||\; M)).$$

(The equality of the two right-end sides of these equations is nothing but the instance of (5) for this case.) Note that $P \xrightarrow{a}$ iff $Tr(P) \xrightarrow{a}$. When this is the case, there is a unique L such that $Tr(P) \xrightarrow{a} L$, namely $L = \bigcup_{P \xrightarrow{a} P'} Tr(P')$, so the union $\bigcup_{Tr(P) \xrightarrow{a} L} Tr(L \;|||\; Tr(Q))$ reduces to $Tr((\bigcup_{P \xrightarrow{a} P'} Tr(P')) \;|||\; Tr(Q))$. It is not difficult to see that the last expression is equal to $\bigcup_{P \xrightarrow{a} P'} Tr(Tr(P') \;|||\; Tr(Q))$, so we are ready for the inductive step. By the induction hypothesis, $Tr(P' \;|||\; Q)$ and $Tr(Tr(P') \;|||\; Tr(Q))$ have the same traces of length at most n, so the same happens to $\bigcup_{P \xrightarrow{a} P'} Tr(P' \;|||\; Q)$ and $\bigcup_{P \xrightarrow{a} P'} Tr(Tr(P') \;|||\; Tr(Q)) = \bigcup_{Tr(P) \xrightarrow{a} L} Tr(L \;|||\; Tr(Q))$. A symmetric reasoning applies to the two other unions in the previous equations. We conclude that $Tr(P \;|||\; Q)$ and $Tr(Tr(P) \;|||\; Tr(Q))$ have the same traces of length at most $n + 1$.

We summarize our approach with a definition and a corollary:

Definition 2. *An* observational model *is a pair* $< O, obs >$, *where* O *is a quasi-final object in* $\mathbf{Coalg}_B$ *with respect to a family* obs *of functions* obs_S *from any* B-*coalgebra* S *to* O.

Corollary 1. *Given a set of SOS rules and an observational model* $< O, obs >$ *such that equation (3) holds for all operators in* Σ *(with the necessary adaptations for operators of different arities), then the derived denotational semantics is equivalent to the defined operational semantics.*

---

[2] For conciseness we denote here the null trace by ε and concatenation of traces x and y by xy.

5   Application to CSP

In this section we apply Corollary 1 to the failure semantics of CSP. First we define an observational model, and then we prove that (3) holds for all operators of the language, with the necessary adaptations. The SOS rules of CSP considered here are a variant of the rules in [5], in that we use a different version of the transition rules with tick for parallel operators, more adequate for our purposes but still defining the same failures and divergences. As an example, Figure 4 shows the rules for the synchronous parallel operator, where Ω is a unique state which can only be reached after successful termination. The complete set of rules and the proofs of the results below can be found in [21]. The rules are in the GSOS format, which as noted in subsection 4.1 is enough to guarantee that Assumption 1 holds.

$$\frac{P \xrightarrow{\tau} P'}{P \parallel Q \xrightarrow{\tau} P' \parallel Q} \qquad\qquad \frac{Q \xrightarrow{\tau} Q'}{P \parallel Q \xrightarrow{\tau} P \parallel Q'}$$

$$\frac{P \xrightarrow{a} P' \quad Q \xrightarrow{a} Q'}{P \parallel Q \xrightarrow{a} P' \parallel Q'} \qquad \frac{P \xrightarrow{\checkmark} P' \quad Q \xrightarrow{\checkmark} Q'}{P \parallel Q \xrightarrow{\checkmark} \Omega}$$

Fig. 4. SOS rules for the synchronous parallel operator ($a \in A$)

Let A be as before an alphabet of actions or communications. Let τ denote an internal action, and let $\checkmark$ stand for successful termination. The set $A^{\checkmark}$ will denote the extended alphabet $A \cup \{\checkmark\}$, $A^{\tau\checkmark} = A \cup \{\checkmark, \tau\}$ and $A^{*\checkmark} = A^* \cup \{x\checkmark : x \in A^*\}$. Let $\mathsf{S} = <S, \to>$ be a labelled transition system, where S is the set of states and $\to \subseteq S \times A^{\tau\checkmark} \times S$. We say that S is a transition system in A with internal actions and termination if for every pair of states such that $s \xrightarrow{\checkmark} t$, t is a state without transitions. For any $a \in A$, define $\xRightarrow{a}$ as the relational composition $(\xrightarrow{\tau})^* \xrightarrow{a} (\xrightarrow{\tau})^*$ and $\xRightarrow{\checkmark}$ as $(\xrightarrow{\tau})^* \xrightarrow{\checkmark}$, where $(\xrightarrow{\tau})^*$ is the reflexive-transitive closure of $\xrightarrow{\tau}$; for arbitrary strings over A, define $\xRightarrow{\langle\rangle}$ as $(\xrightarrow{\tau})^*$ and $\xRightarrow{\widehat{xy}}$ as $\xRightarrow{x}\xRightarrow{y}$; finally, define $\xRightarrow{\tau}$ as $\xrightarrow{\tau} (\xrightarrow{\tau})^*$. In the sequel, for conciseness, we write ε for $\langle\rangle$ and abbreviate $x \frown y$ to xy.

Given a transition system $<S, \to>$ over A with internal actions and termination, the relations $\xRightarrow{a}$ ($a \in A^{\tau\checkmark}$) define a transition system $\mathsf{S}^* = <S, \Rightarrow>$. The operation $*$ is idempotent, i.e. $(\mathsf{S}^*)^* = \mathsf{S}^*$. As a consequence, all notions defined in S using $\Rightarrow$ coincide with the same notions in S^* using $\Rightarrow$ again. For example, any $s \in S$ has the same traces in S and in S^*; the same can be said of initials, refusals and failures, to be introduced below.

It is easy to see that a morphism $f : \mathsf{S}_1 \to \mathsf{S}_2$ of transition systems, as a function $f : S_1 \to S_2$, is also a morphism $f : \mathsf{S}_1^* \to \mathsf{S}_2^*$ (but the converse is not true).

Definition 3. *A function* $\mathsf{S}_1 \to \mathsf{S}_2$ *is a* weak morphism *if it is a morphism* $\mathsf{S}_1{}^* \to \mathsf{S}_2{}^*$.

In particular, 1_S is a weak morphism $S \to S^*$. We noticed that every $s \in S$ has the same traces in S and in S^*; more generally, if $f : S_1 \to S_2$ is a weak morphism, then $Tr(s) = Tr(f(s))$ for every $s \in S_1$.

Given a transition system $S \; =< S, \to >$ over A with internal actions and termination and a state $s \in S$, we study various observations related to s:

- The set of *initials* of s after $x \in A^{*\checkmark}$ is $I(s) = \{a \in A^{\checkmark} : \exists\, t, s \overset{x}{\Rightarrow} t\}$.
- The set of *failures* of s is

$$Fl(s) = \{(x, X) \in A^* \times \mathcal{P}(A^{\checkmark}) : \exists\, t, s \overset{x}{\Rightarrow} t \text{ and } t \; ref\; X\}$$
$$\cup \{(x\checkmark, X) \in A^{*\checkmark} \times \mathcal{P}(A^{\checkmark}) : \exists\, t, s \overset{x\checkmark}{\Rightarrow} t\},$$

  were $s \; ref\; X \subseteq A^{\checkmark}$ if and only if either s is a stable state (one without τ or $\checkmark$ transitions) and $I(s) \cap X = \emptyset$ or $s \overset{\checkmark}{\to}$ and $X \subseteq A$.
- A *refusal* of s is a set $X \subseteq A^{\checkmark}$ such that (ε, X) is a failure of s; the set of refusals of s is written $Rf(s)$.

Weak morphisms preserve these observations:

Lemma 1. *If $f : S_1 \to S_2$ is a weak morphism and $s \in S_1$, then $Fl(s) = Fl(f(s))$. In particular, $Rf(s) = Rf(f(s))$.*

A *failure-set* over A is any set $P \subseteq A^{*\checkmark} \times \mathcal{P}(A^{\checkmark})$ such that the following closure conditions hold:

F1 $(\varepsilon, \emptyset) \in P$.
F2 $(xy, \emptyset) \in P \Rightarrow (x, \emptyset) \in P$.
F3 $(x, X) \in P \wedge Y \subseteq X \Rightarrow (x, Y) \in P$.
F4 $(x, X) \in P \wedge \forall\, a \in Y, (xa, \emptyset) \notin P \Rightarrow (x, X \cup Y) \in P$.
F5 $(x\checkmark, \emptyset) \in P \Rightarrow \forall X \subseteq A, (x, X) \in P$.
F6 $(x\checkmark, \emptyset) \in P \Rightarrow \forall X \subseteq A^{\checkmark}, (x\checkmark, X) \in P$.

Let $\mathcal{F}$ be the set of all failure sets. We turn $\mathcal{F}$ into a transition system (in coalgebraic notation) $F =< \mathcal{F}, \zeta_{Fl} >$ by defining the transitions associated with ζ_{Fl} as follows. Let P and Q be failure-sets and $a \in A^{\checkmark}$; we write

$$P \overset{\tau}{\to} Q \quad \text{iff} \quad Q \subseteq P \wedge Q \neq P;$$
$$P \overset{a}{\to} Q \quad \text{iff} \quad \forall (x, X) \in Q, \quad (ax, X) \in P.$$

Remark 1. It is easy to show that $P \overset{\checkmark}{\to} Q$ iff $Q = \{(\varepsilon, X) : X \subseteq A^{\checkmark}\}$ and $(\checkmark, X) \in P$ for all $X \subseteq A^{\checkmark}$; and $\{(\varepsilon, X) : X \subseteq A^{\checkmark}\} \overset{a}{\not\to}$ for every $a \in A^{\tau\checkmark}$.

The transition system of failures F is well behaved with regard to various observations:

Lemma 2. *Let P be a failure-set. In $F =< \mathcal{F}, \zeta_{Fl} >$ the following equalities hold:*

1. $Tr(P) = \{x \in A^{*\checkmark} : (x, \emptyset) \in P\}$.
2. $I(P) = \{a \in A^{\checkmark} : (a, \emptyset) \in P\}$.

3. $Rf(P) = \{X \subseteq A^\checkmark : (\varepsilon, X) \in P\}$.
4. $Fl(P) = P$.

With Lemma 1 and the last statement of Lemma 2 we obtain:

Proposition 7. $<\mathsf{F} = <\mathcal{F}, \zeta_{Fl}>, Fl>$ is an observational model.

Finally, we note that (3) holds, with the necessary adaptations, for all operators of CSP [5]: action prefix, prefix choice, external choice, internal choice, general parallel, alphabetized parallel, synchronous parallel, interleaving, sequential, hiding, renaming and conditional.

Proposition 8. For every $t, u \in TZ$,
1. $Fl(a \to t) = Fl(a \to Fl(t))$
2. $Fl(?x : K \to t) = Fl(?x : K \to Fl(t))$
3. $Fl(t \,\square\, u) = Fl(Fl(t) \,\square\, Fl(u))$
4. $Fl(t \,\sqcap\, u) = Fl(Fl(t) \,\sqcap\, Fl(u))$
5. $Fl(t \,[\![\,K\,]\!]\, u) = Fl(Fl(t) \,[\![\,K\,]\!]\, Fl(u))$
6. $Fl(t \,[\![\,K_1 \,|\, K_2\,]\!]\, u) = Fl(Fl(t) \,[\![\,K_1 \,|\, K_2\,]\!]\, Fl(u))$
7. $Fl(t \parallel u) = Fl(Fl(t) \parallel Fl(u))$
8. $Fl(t \,|||\, u) = Fl(Fl(t) \,|||\, Fl(u))$
9. $Fl(t;\, u) = Fl(Fl(t);\, Fl(u))$
10. $Fl(t \setminus K) = Fl(Fl(t) \setminus K)$
11. $Fl(t[\![R]\!]) = Fl(Fl(t)[\![R]\!])$
12. $Fl(\text{if } \varphi \text{ then } t \text{ else } u) = Fl(\text{if } \varphi \text{ then } Fl(t) \text{ else } Fl(u))$

Recursive definitions are easily accommodated with the bialgebraic framework.

Proposition 9. Suppose a set $PVar \subseteq \Sigma$ of operators with arity zero acting as "process variables" has been given, together with an "equation" $x = t_x$ for every $x \in PVar$. Let the SOS rules of the $x \in PVar$ be the transitions $x \xrightarrow{\tau} t_x$ with no premises. Then in the Σ-algebra defined on $\mathcal{F}$ by (1), the constant assigned to $x \in PVar$ is $Fl_{T\mathcal{F}}(x)$, which in turn is equal to $Fl_{T\mathcal{F}}(t_x)$ since the transition structures of coalgebras preserve behaviours. Furthermore, the version of (3) for process variables (and in fact for any operator with arity zero) holds vacuously.

The previous two propositions and Corollary 1 immediately yield:

Proposition 10. The operational and the denotational (failure) semantics of CSP with respect to the observational model $<\mathsf{F}, Fl>$ are equivalent.

In [5] it is stated that failures equivalence of CSP processes obtained from the SOS rules via the function Fl is equivalent to the one [5] defines in terms of denotational rules such as $Fl(P \sqcap Q) = Fl(P) \cup Fl(Q)$. Using this result, we can conclude that the initial semantics we define is equivalent to the denotational semantics given in [5].

6   Conclusion and Future Work

In this paper, we have successfully generalized the concept of bialgebraic semantics from using final coalgebras to the quasi-final case. This generalization opens the possibility to study notions of process equivalence coarser than bisimulation in a bialgebraic framework. To this end, we suggested a proof method and demonstrated it in terms of the CSP failures semantics.

Future work will include the development of SOS rule formats that guarantee equation (3); [20] already gives sufficient conditions, however, only on transition systems without the silent action τ.

Yet another topic is the systematic study of further CSP semantics: CSP denotations usually combine various observations, e.g., failures and divergences. Such observations are coupled via healthiness conditions, e.g., that every trace that leads to a divergence must be a possible system run. It is an interesting question to study how the bialgebraic approach can support such constructions.

Finally, it will be interesting to see how our framework can deal with a language such as CSP-CASL [22], which integrates both data and process specication.

Acknowledgements. The authors would like to thank the anonymous reviewers for their constructive comments and Erwin R. Catesbeiana (Jr) for his exemplary behaviour over the full range of the Linear Time – Branching Time Spectrum.

References

[1] Plotkin, G.D.: A structural approach to operational semantics. Technical Report DAIMI FN-19, University of Aarhus (1981)

[2] Rutten, J., Turi, D.: Initial Algebra and Final Coalgebra Semantics for Concurrency. In: de Bakker, J.W., de Roever, W.-P., Rozenberg, G. (eds.) REX 1993. LNCS, vol. 803, pp. 530–582. Springer, Heidelberg (1994)

[3] Turi, D., Plotkin, G.: Towards a mathematical operational semantics. In: Proc. 12th LICS Conference, pp. 280–291. IEEE, Computer Society Press (1997)

[4] Hoare, C.A.R.: Communicating Sequential Processes. Series in Computer Science. Prentice-Hall International (1985)

[5] Roscoe, A.: The Theory and Practice of Concurrency. Prentice-Hall, Englewood Cliffs (1998)

[6] van Glabbeek, R.: The linear time–branching time spectrum I: the semantics of concrete, sequential processes. In: Bergstra, J., Ponse, A., Smolka, S. (eds.) Handbook of Process Algebra, pp. 3–99. Elsevier (2001)

[7] Monteiro, L.: A Coalgebraic Characterization of Behaviours in the Linear Time – Branching Time Spectrum. In: Corradini, A., Montanari, U. (eds.) WADT 2008. LNCS, vol. 5486, pp. 251–265. Springer, Heidelberg (2009)

[8] Freire, E., Monteiro, L.: Defining Behaviours by Quasi-finality. In: Oliveira, M.V.M., Woodcock, J. (eds.) SBMF 2009. LNCS, vol. 5902, pp. 290–305. Springer, Heidelberg (2009)

[9] Aczel, P.: Final Universes of Processes. In: Brookes, S., Main, M., Melton, A., Mislove, M., Schmidt, D. (eds.) MFPS 1993. LNCS, vol. 802, pp. 1–28. Springer, Heidelberg (1994)

[10] Wolter, U.: CSP, partial automata, and coalgebras. Theoretical Computer Science 280, 3–34 (2002)

[11] Boreale, M., Gadducci, F.: Processes as formal power series: A coinductive approach to denotational semantics. Theoretical Computer Science 360, 440–458 (2006)

[12] Power, J., Turi, D.: A coalgebraic foundation for linear time semantics. In: Hofmann, M., Rosolini, G., Pavlovic, D. (eds.) Conference on Category Theory and Computer Science, CTCS 1999. Electronic Notes in Theoretical Computer Science, vol. 29, pp. 259–274. Elsevier (1999)

[13] Jacobs, B.: Trace semantics for coalgebras. In: Adamek, J., Milius, S. (eds.) Coalgebraic Methods in Computer Science. Electronic Notes in Theoretical Computer Science, vol. 106, pp. 167–184. Elsevier (2004)

[14] Hasuo, I., Jacobs, B., Sokolova, A.: Generic trace semantics via coinduction. Logical Methods in Computer Science 3(4:11), 1–36 (2007)

[15] Jacobs, B., Sokolova, A.: Exemplaric expressivity of modal logic. Journal of Logica and Computation 5(20), 1041–1068 (2010)

[16] Klin, B.: Bialgebraic methods and modal logic in structural operational semantics. Inf. Comput. 207(2), 237–257 (2009)

[17] Klin, B.: Structural operational semantics and modal logic, revisited. In: Jacobs, B., Niqui, M., Rutten, J., Silva, A. (eds.) Coalgebraic Methods in Computer Science. Electronic Notes in Theoretical Computer Science, vol. 264, pp. 155–175. Elsevier (2010)

[18] Groote, J.F., Vaandrager, F.W.: Structural operational semantics and bisimulations as a congruence. Information and Computation 100(2), 202–260 (1992)

[19] Bloom, B., Istrail, S., Meyer, A.R.: Bisimulation can't be traced. J. ACM 42(1), 232–268 (1995)

[20] Monteiro, L., Maldonado, A.P.: Towards bialgebraic semantics based on quasi-final coalgebras. Technical Report FCT/UNL-DI 1-2011, CITI and DI, Faculdade de Ciências e Tecnologia, UNL (2011), http://ctp.di.fct.unl.pt/~lm/publications/MM-TR-1-2011.pdf

[21] Maldonado, A.P., Monteiro, L., Roggenbach, M.: Towards bialgebraic semantics for CSP. Technical Report FCT/UNL-DI 3-2010, CITI and DI, Faculdade de Ciências e Tecnologia, UNL (2010), http://ctp.di.fct.unl.pt/~apm/publications/MMR-TR-3-2010.pdf

[22] Roggenbach, M.: CSP-CASL—A new integration of process algebra and algebraic specification. Theoretical Computer Science 354, 42–71 (2006)

Algebraic Signatures Enriched
by Dependency Structure

Grzegorz Marczyński

Institute of Informatics, University of Warsaw
gmarc@mimuw.edu.pl

Abstract. Classical single-sorted algebraic signatures are defined as sets
of operation symbols together with arities. In their many-sorted variant
they also list sort symbols and use sort-sequences as operation types. An
operation type not only indicates sorts of parameters, but also constitutes dependency between an operation and a set of sorts. In the paper
we define algebraic signatures with dependency relation on their sym-
bols. In modal logics theory, structures like $\langle W, R \rangle$, where W is a set and
$R \subseteq W \times W$ is a transitive relation, are called transitive Kripke frames
[Seg70]. Part of our result is a definition of a construction of non-empty
products in the category of transitive Kripke frames and p-morphisms.
In general not all such products exist, but when the class of relations is
restricted to bounded strict orders, the category lacks only the final object to be finitely (co)complete. Finally we define a category **AlgSigDep**
of signatures with dependencies and we prove that it also has all finite
(co)limits, with the exception of the final object.

1 Introduction

Classical single-sorted algebraic signatures are defined as sets of operation symbols together with arities. In their many-sorted variant they also list sort symbols
and use sort-sequences as operation types. One should notice that an operation
type not only indicates sorts of parameters, but also constitutes dependency between an operation and a set of sorts. Informally, one can say that in order to
define the operation, all sort carriers from its type must be present in a model.

In architectural approach to system specification [ST97], a signature represents a software module interface. The whole system (or, to be precise, its model)
is obtained as a series of applications of so-called generic modules [BST99]
also known as constructor implementations [ST88]. Modules are put together
and constitute a whole only if all required parameter-modules are instantiated.
Clearly this reveals a dependency relation between modules and, as a consequence, between operation symbols they define.

Our work on architectural models led us to a need of dependency structures put
directly on operation symbols right in signatures. In most approaches to stepwise
construction of software systems there exist dependencies of one entities on another. Take components, services, SML functors, C libraries or Java packets, ev-
erywhere, in order to link and run the application, one needs first to provide the

T. Mossakowski and H.-J. Kreowski (Eds.): WADT 2010, LNCS 7137, pp. 226–250, 2012.

implementation of the required (parameter) symbols. The dependency structure allows one not only to track the provenience of symbols, but also to estimate the impact of the their use, removal or change within the system in question.

The idea is to extend the classical many-sorted signatures by explicitly defining the dependency of sorts and operations. In the paper we define a category of algebraic signatures with dependency structures being bounded strict orders and p-morphisms (short for pseudo-epimorphisms). Unfortunately the category lacks the final object. However, it has all other limits and all finite colimits.

Results presented in the paper are part of the ongoing work on covariant semantics of generic software modules and their architectures.

The paper is organized as follows. First in Sect. 2 we present the motivation to our work. In Sect. 3 we define the categories of dependency relations. In Sections 3.3 and 3.4 we analyze the existence of (co)limits in these categories and in Sect. 4 we define the category of signatures with dependency structure and prove its properties. Finally, Sect. 5 contains conclusion and future work. Proofs of most lemmas and theorems are in the Appendix A.

2   Motivation

Classical algebraic many-sorted signatures naturally capture dependence of operation symbols on their parameters' sorts. In architectural specifications [BST99], signature of a generic module is an injective signature morphism $\sigma\colon \Sigma_{Form} \to \Sigma_{Res}$, where Σ_{Form} is a formal parameters signature and Σ_{Res} is a result signature. This renders a dependency between all symbols from the result signature and those from the parameters signature.

The above-described dependency is weak in the sense that it is not required that actual implementation of result symbols uses the parameter symbols intrinsically. It rather conveys the negative information, leaving some symbols definitely independent of others. One may think of it as of a *potential dependency*.

The generic module application along a fitting morphism $\varphi\colon \Sigma_{Form} \to \Sigma_{Act}$ on the signature-level is simply the pushout of φ and σ. Consider the following simple example.

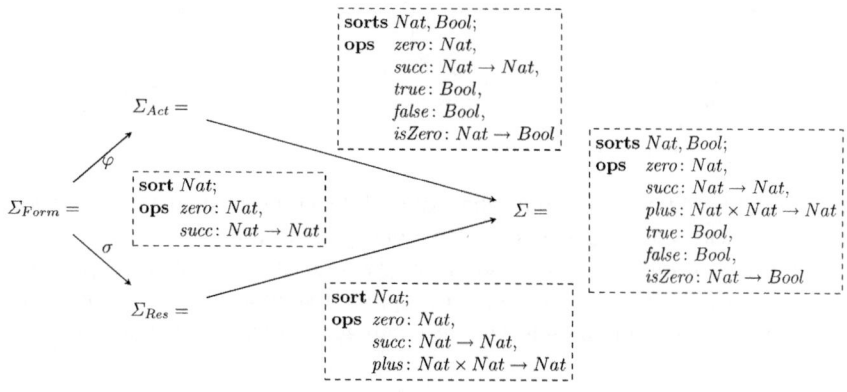

The construction signature σ defines an operation $plus\colon Nat \times Nat \to Nat$, provided it is given a sort Nat with operations $zero\colon Nat$ and $succ\colon Nat \to Nat$. The actual parameters signature Σ_{Act} enriches the formal parameters signature Σ_{Form} by several symbols like $Bool$, $true\colon Bool$ etc. The pushout signature contains all symbols together. However, the information about a potential dependency of the operation $plus$ on $zero$ and $succ$ is lost from the signature. One needs to keep the whole pushout diagram to track down these dependencies.

Here comes the idea to *enrich the signatures by dependency structure* to explicitly show how symbols may depend on other symbols. Dashed lines on the diagram below expose a transitive dependency relation.

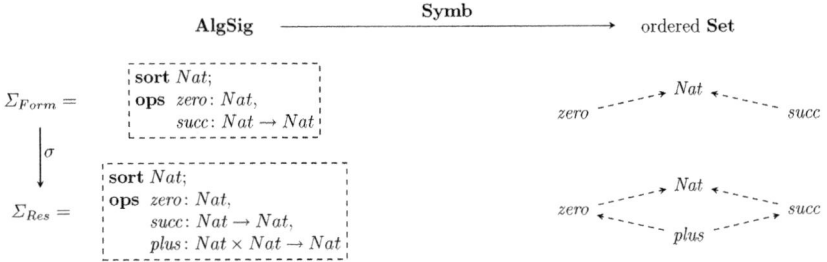

In the pushout, the dependencies should be preserved, as on the following diagram. It is visible that the operation $isZero$ doesn't depend on $plus$. Neither the latter depends on the former.

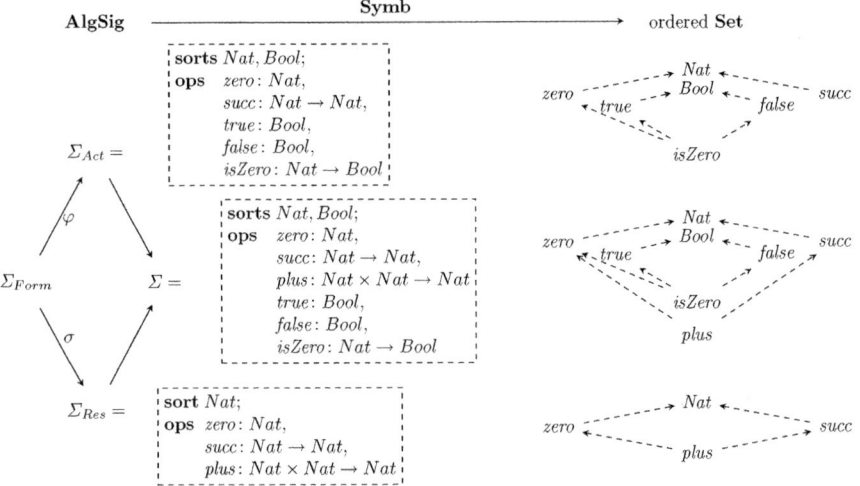

In our paper we try to find out what kind of dependency relation shall we use and how to enrich the signatures by dependency structure.

The simple example given above already says something about the dependency relation – it needs to be transitive. We require that morphisms do not change (in)dependencies of symbols and that the category of enriched signatures have all finite pushouts and pullbacks. While pushouts are needed to compute sums of signatures, pullbacks are used to get their intersections.

3 Dependency Relation

We investigate properties of several categories of sets ordered by various transitive relations with morphisms that not only preserve the dependencies, but also weakly reflect their structure.

3.1 Category Rset↓ and Its Subcategories

The dependency relations have to be transitive. We begin from the most general setting and formalise the category of R-sets and p-morphisms.

Definition 1 (R-sets). *An R-set is a pair $\langle A, R_A \rangle$ where $R_A \subseteq A^2$ is a transitive relation on a set A. In what follows we sometimes write A_R instead of $\langle A, R_A \rangle$. We may use the infix notation for R_A and for $a_1, a_2 \in A_R$ we may also write $a_1 \, R \, a_2$ instead of $a_1 \, R_A \, a_2$, when decorations are clear from the context.*

Definition 2 (Category Rset↓ of R-sets and P-morphisms). *Rset↓ has R-sets as objects and pseudo-epimorphisms, or p-morphisms, as morphisms. A p-morphism is a function that preserves the relation R and weakly reflects R-set down-closures, i.e. a morphism $f : \langle A, R_A \rangle \to \langle B, R_B \rangle$ is a function $f : A \to B$ such that:*

1. *(monotonicity) for all $a_1, a_2 \in A$, $a_1 \, R_A \, a_2$ implies $f(a_1) \, R_B \, f(a_2)$.*
2. *(weakly reflected R-down-closures) for all $a_2 \in A, b_1 \in B$, $b_1 \, R_B \, f(a_2)$ implies that there exists $a_1 \in A$, that $a_1 \, R_A \, a_2$ and $f(a_1) = b_1$.*

Identities and composition are defined as expected.

In modal logics, R-sets are called transitive Kripke frames [Seg70] and p-morphisms are sometimes called bounded morphisms. It is well known that the category of Kripke frames and p-morphisms is the category $\mathbf{Set_P}$ of coalgebras of the powerset functor (cf. [GS01]). It makes the category **Rset↓** a full subcategory of $\mathbf{Set_P}$.

Definition 3 (Sub R-set). *Given an R-set $A_R = \langle A, R_A \rangle$ and $a \in A$, its closed down sub R-set induced by an element a is defined as $A_R{}^a{\downarrow} = A'_R$, where $A'_R = \langle A', R_{A'} \rangle$ with $A' = \{a' \in A \mid a' \, R_A \, a\} \cup \{a\}$ and $R_{A'} = R_A|_{A'}$.*

It is important to notice that $a \in A_R{}^a{\downarrow}$, for any set A, $a \in A$ and a relation R.

The table below summarises the results[1] that we prove in the rest of Sect. 3. We investigate the existence of (co)limits[2] in the category **Rset↓** and three of its

---

[1] "No?" in the table means that the absence of the property is a plausible conjecture.

[2] The final object is a product of an empty class. In the table the final object and nonempty products are presented in separate columns, because in some categories apparently their existence does not coincide.

full subcategories – **Preord**↓ (reflexive), **Soset**↓ (asymmetric), **Soset$_b$**↓ (asymmetric and bounded). Clearly **Soset$_b$**↓ is the category with the most suitable properties to work with.

category	relation	eq.	final obj.	nonempty product	coeq.	coprod.
Rset↓	transitive	yes	no?	no?	yes	yes
Preord↓	preorder	yes	yes	no?	yes	yes
Soset↓	strict order	yes	no	no?	no	yes
Soset$_b$↓	bounded strict order	yes	no	yes	yes	yes (finite)

Objects of **Soset$_b$**↓ are bounded strict orders, i.e. strict orders $A_<$ with descending chains limited by a natural number $\widetilde{A_<}$.

We don't consider partial orders as a candidate for dependency relation, because the antisymmetric "closure", existing in coequalisers of posets, may require "gluing" of incompatible operation symbols, therefore, coequalisers won't exist in the category of algebraic signatures enriched by poset dependency structure.

3.2   R-multisets and Dependency Bisimulation

This section defines R-multisets that become handy when it comes to definition of products in the category **Rset**↓ and its subcategories (cf. Sect. 3.3). The reader may skip this section at the first reading.

The idea is to take an R-set and define a multiset of its elements without adding new dependencies; however, some original dependencies may be dropped.

Definition 4 (Labeled R-set). *A labeled R-set is a triple* $\langle A_R, P_R, \mu \rangle$ *where* A_R *and* P_R *are R-sets and* $\mu \colon A_R \to P_R$ *is a monotonic labeling function.*

Definition 5 (Labeled R-set Isomorphism). *Two labeled R-sets* $\langle A_R, P_R, \mu \rangle$, $\langle A'_R, P_R, \mu' \rangle$ *are isomorphic[3] iff there exists a bijection* $\tau \colon A \to A'$ *such that for all* $a \in A$, $\mu(a) = \mu'(\tau(a))$ *and for all* $a, a' \in A$ $a\,R_A\,a'$ *iff* $\tau(a)\,R_{A'}\,\tau(a')$.

Definition 6 (R-mset – R-multiset). *An R-multiset, or R-mset,* $[A_R, P_R, \mu]$ *is the isomorphism class[4] of a labeled R-set* $\langle A_R, P_R, \mu \rangle$.

Definition 7 (R-submultiset). *Given an R-mset* $[A_R, P_R, \mu]$ *and* $a \in A$, *its R-submultiset* $[A_R, P_R, \mu]^a{\downarrow}$ *induced by* a *is defined as an R-mset:*

$$[A_R, P_R, \mu]^a{\downarrow} = [A_R{}^a{\downarrow}, P_R, \mu_{(A_R{}^a{\downarrow})}]$$

---

[3] We use the word "isomorphism" in categorical sense here. For a given R-set P_R one can define a category of labeled R-sets **LRset$_{P_R}$** where morphism are functions with compatibility requirements, as in Def. 5.

[4] For technical convenience we will sometimes define concepts and constructions on R-msets by introducing them on representatives – leaving to the reader the details of generalization to their isomorphism classes.

Lemma 8. *Given an R-mset $[A_R, P_R, \mu]$ and $a, a' \in A$, such that $a'Ra$, it holds that*

$$[A_R, P_R, \mu]^{a'}{\downarrow} = ([A_R, P_R, \mu]^a{\downarrow})^{a'}{\downarrow}$$

For a multiset we can easily calculate the set of its distinct elements. The similar question can be asked with regard to R-multisets, but here the matter is a bit more complex, because the labeling is not required to be a p-morphism, therefore two elements of the same label may have different dependency structure. In the following definition we use the bisimulation-like relation to find the *kernel relation* of the given R-mset. In state transition systems the bisimilar states share the same behavior; here the situation in analogous – the kernel of an R-multiset relates these elements that share the same dependency structure.

Definition 9 (Kernel Relation of R-mset). *Given an R-mset $[A_R, P_R, \mu]$ its* kernel relation $K([A_R, P_R, \mu]) \subseteq A^2$ *is the greatest dependency bisimulation relation on A. A dependency bisimulation is a relation $\sim \subseteq A^2$, such that for $a_1, a_2 \in A$*

$$if\ a_1 \sim a_2\ then\ \mu(a_1) = \mu(a_2)\ and$$
$$for\ all\ a_1' \in A\ such\ that\ a_1'\,R\,a_1$$
$$there\ exists\ a_2' \in A,\ a_2'\,R\,a_2,\ a_1' \sim a_2'$$
$$and\ for\ all\ a_2' \in A\ such\ that\ a_2'\,R\,a_2$$
$$there\ exists\ a_1' \in A,\ a_1'\,R\,a_1,\ a_1' \sim a_2'$$

The family of dependency bisimulations is non-empty (it contains id_A) and it is closed under unions, hence the kernel relation exists for every R-mset $[A_R, P_R, \mu]$. Moreover, the kernel relation, as the greatest dependency bisimulation, is an equivalence relation.

Definition 10 (R-class of Compatible Dependencies). *Given a family of monotonic functions $(w_i \colon P_R \to W_R^i)_{i \in I}$ where P_R and W_R^i for $i \in I$ are* **Rset**$\downarrow$-*objects, an R-class[5] of compatible dependencies for $(w_i)_{i \in I}$ is a pair $\langle Q_R, l \rangle$ of an R-class $Q_R = \langle Q, R_Q \rangle$ and a monotonic function $l \colon Q \to P$ defined as:*

$$Q = \bigcup_{p \in P} \{\langle [X_R, P_R, \mu], p \rangle \mid [X_R, P_R, \mu]\ is\ an\ R\text{-}mset\colon such\ that\colon$$

- $K([X_R, P_R, \mu]) = id_X$,
- *for any $i \in I$, $\mu; w_i$ are* **Rset**$\downarrow$*-morphisms,*
- *there exists $x \in X$, such that $\mu(x) = p$*
  and for all $x' \in X$, if $x' \neq x$ then $x'\,R_X\,x\}$

with a relation

---

[5] Like to R-set, but without carriers limited to sets, because in general, for a given nonempty set of labels P, the class of all P-labeled R-msets may be a proper class.

$$R_Q = \{\langle\langle[X'_R, P_R, \mu'], p'\rangle, \langle[X_R, P_R, \mu], p\rangle\rangle \in Q^2 \mid$$
$$\text{there exists } x' \in X, \mu(x') = p', p' \, R \, p$$
$$\text{and } [X'_R, P_R, \mu'] = [X_R, P_R, \mu]^{x'}\!\downarrow\}$$

and a monotonic function $l: Q_R \to P_R$ simply given by $l(\langle[X_R, P_R, \mu], p\rangle) = p$.

Lemma 11 (Universal R-set of Compatible Dependencies). *If an R-class of compatible dependencies for $(w_i)_{i \in I}$ is a set, then it is a universal R-set of compatible dependencies for $(w_i)_{i \in I}$, i.e. for all $i \in I$, $(l; w_i): Q_R \to W^i_R$ is an* **Rset↓***-morphism and for any pair $\langle T_R, \theta: T \to P\rangle$ with the same property, there exists the unique* **Rset↓***-morphism $u: T_R \to Q_R$, such that $u; l = \theta$.*

Proof: The proof is quite technical (see App. A). The main point is to observe that when there is any object with projections to W^i_R, it may be seen as an R-multiset. We take its kernel (cf. Def. 9) and find out that every element of the kernel (with its dependency structure) is also present in Q. This allows us to define the unique **Rset↓**-morphism from the given object to $Q_<$.   □

Lemma 12. *Given a family of monotonic functions $(w_i: P_R \to W^i_R)_{i \in I}$ where P_R and W^i_R for all $i \in I$ are* **Rset↓***-objects with relations R_P and R_{W^i} for all $i \in I$ being reflexive / irreflexive / asymmetric / strict bounded, then the relation R_Q from Def. 10 is also so.*

The proof is straightforward. For details see App. A.

3.3   Limits in Rset↓

The completeness of categories of coalgebras is problematic (cf. [GS01]) – equalisers always exist, but this is not true in case of products. The similar situation happens also in categories of multialgebras (cf. [WW08]). To our knowledge, however, there are no results concerning the completeness of categories of transitive Kripke frames and its reflexive, asymmetric and bounded subcategories. We are particularly interested in the existence of pullbacks, which we are going to define through equalisers and products of nonempty families of objects. We begin from the fact that comes as no surprise.

Theorem 13. Rset↓, Preord↓, Soset↓ *and* **Soset_b↓** *have all equalisers.*

Let us now look at final objects.

Theorem 14. Preord↓ *has a singleton ordered by identity as its final object.*

Theorem 15. Soset↓ *and* **Soset_b↓** *do not have a final object.*

Conjecture 16. Rset↓ *does not have a final object.*

Before we propose a definition of products of R-sets, let us first present an example in **Preord↓**. Let $A_\leq = \langle\{a_1, a_2\}, a_1 \leq a_1, a_2 \leq a_2, a_2 \leq a_1\rangle$ and $B_\leq = \langle\{b_1, b_2\}, b_1 \leq b_1, b_2 \leq b_2, b_2 \leq b_1\rangle$ be two **Preord↓**-objects. The product of $A_\leq$ and $B_\leq$ in **Preord↓** is an infinite R-set $Q_\leq$ that may be depicted as below.

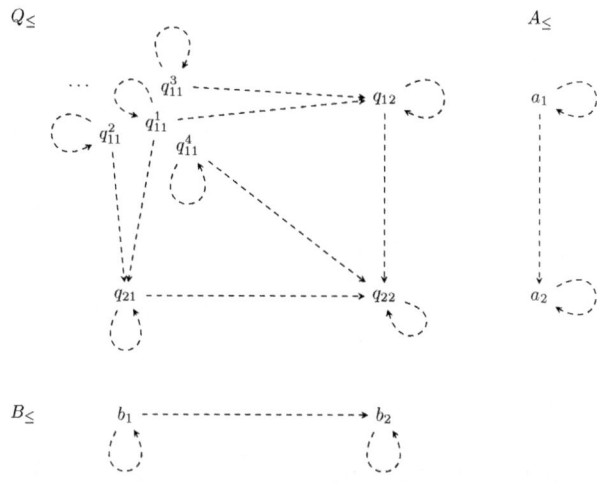

As expected there are $q_{22} = \langle a_2, b_2\rangle$, $q_{12} = \langle a_1, b_2\rangle$, $q_{21} = \langle a_2, b_1\rangle$ in $Q_\leq$. They are in relation with themselves and also $q_{22} \leq_Q q_{21}$ and $q_{22} \leq_Q q_{12}$. There is also $q_{11}^1 = \langle a_1, b_1\rangle$ that is in relation with all above mentioned elements. However, $Q_\leq$ contains as well infinitely many distinct elements also projected to a_1 and b_1, marked above as $q_{11}^2, q_{11}^3, q_{11}^4, \ldots$ that differ from each other only on their dependency structures. In fact $Q_\leq$ can be seen as an R-multiset $[Q_\leq, (A \times B)_\leq, \langle\pi_A, \pi_B\rangle]$ (cf. Def. 6), such that $K[Q_\leq, (A \times B)_\leq, \langle\pi_A, \pi_B\rangle] = id_Q$ with π_A and π_B being projections. To illustrate the need for all these elements that are projected to a_1 and b_1, let us have a **Preord↓**-object $Q'_\leq$ defined as $Q_\leq$ without q_{11}^2, that is an element with the dependencies defined exactly as $q_{11}^2 \leq_Q q_{11}^1$, $q_{21} \leq_Q q_{11}^2$ and $q_{22} \leq_Q q_{11}^2$. If q_{11}^2 was not needed in $Q_\leq$, then $Q'_\leq$ would be the product of $A_\leq$ and $B_\leq$ and there would exist the universal morphism $h: Q_\leq \to Q'_\leq$ compatible with projections. It is easy to prove that such morphism does not exist, because there is no element of $Q'_\leq$ that q_{11}^2 may be mapped to through a p-morphism (cf. Def. 2). P-morphisms are monotone and weakly reflect the dependency structure, thus, we can not map q_{11}^2 neither to q_{11}^4, because $q_{21} \not\leq q_{11}^4$, nor to q_{11}^1, because $q_{12} \leq q_{11}^1$ and this dependency would be not reflected. The same reasoning may be applied to all other $Q_\leq$-elements that are projected to a_1 and b_1.

The following definition proposes the construction of the product, hence the name – *the product candidate*. The problem is that in some cases the structure it describes is a proper class. In Lemma 18 we prove that whenever $A \prod B$ happens to be a set, then $\langle A \prod B, R_{A \prod B}\rangle$ with projections ρ_A and ρ_B is a product of the given **Rset↓**-objects.

Definition 17 (Product Candidate). *Given two* **Rset**↓*-objects* $\langle A, R_A \rangle$ *and* $\langle B, R_B \rangle$, *we define a* product candidate *as a pair of a class and a relation* $\langle A \prod B, R_{A \prod B} \rangle$ *together with two functions* $\rho_A \colon A \prod B \to A$ *and* $\rho_B \colon A \prod B \to B$. *Let* $\langle P, R_P \rangle$ *be a pair of the set* $P = A \times B$, *the product of A and B in* **Set** *together with projections* $\pi_A \colon P \to A$ *and* $\pi_B \colon P \to B$, *and a relation* $R_P \subseteq P^2$:

$$R_P = \{\langle p_1, p_2 \rangle \in P^2 \mid \pi_A(p_1) \, R_A \, \pi_A(p_2) \text{ and } \pi_B(p_1) \, R_B \, \pi_B(p_2)\}$$

Now, let the class $\langle (A \prod B)_R, l \rangle$ *be an R-class of compatible dependencies for* π_A *and* π_B *(cf. Lemma 11) and the product-candidate projection functions be defined as* $\rho_A = l; \pi_A$ *and* $\rho_B = l; \pi_B$.

The class $A \prod B$ contains every element of P taken as many times as there are distinct (wrt. the kernel relation) R-msets of elements lower than it wrt. R_P (cf. Def. 10). These R-msets are subject to the requirement that their labelings composed with Cartesian product projections are p-morphisms. This makes them weakly reflect the R-down-closures of A_R and B_R.

The following lemma guarantees that once we show that the product candidate is a set, then it is indeed the product.

Lemma 18. *Consider two* **Rset**↓*-objects* A_R *and* B_R, *if* $A \prod B$ *is a set, then the product candidate* $\langle A_R \prod B_R, R_{A \prod B} \rangle$ *is their product in* **Rset**↓.

This is a consequence of Lemma 11. For details see App. A.

Theorem 19. *The category* **Soset**$_b$↓ *has all binary products. Moreover, given two* **Soset**$_b$↓*-objects,* $A_<$, $B_<$, *their product is isomorphic to the product candidate from Def. 17.*

We prove that, given any $A_<, B_< \in$ **Soset**$_b$↓, the product candidate $A \prod B$ is a set. Then, by Lemma 18, we argue that $(A \prod B)_<$ is indeed the product of the given objects. The proof goes by induction. We bound the number of possible distinct structures labeled by every $p \in P_<$. This is possible, since dependency relations in question are bounded. For details see App. A. The products of n **Soset**$_b$↓-objects, for $n > 2$, are defined following the same idea.

We cannot find a similar proof for **Rset**↓, **Preord**↓ and **Soset**↓.

Conjecture 20. Rset↓, Preord↓ *and* Soset↓ *do not have all binary products.*

3.4 Colimits in Rset↓

The categories of coalgebras are shown to be cocomplete (cf. [GS01]). Basically the colimits in these categories are as in **Set**. The same applies to **Rset**↓ and **Preord**↓. Howevwer, the category **Soset**$_b$↓ has only finite coproducts and **Soset**↓ does not have all coequalisers.

Theorem 21. Rset↓ *and its subcategory* Preord↓ *have all coequalisers.*

In case of **Rset**↓ and **Preord**↓ we just prove that the transitivity and reflexivity is preserved in the coequaliser construction from **Set**. However, the irreflexivity is not preserved, hence the following theorem.

Theorem 22. Soset↓ *does not have all coequalisers.*

The counterexample involves two functions $id : \mathbb{N} \to \mathbb{N}$ and $succ \colon \mathbb{N} \to \mathbb{N}$ and the inverse ordering of natural numbers. See App. A for details.

It is enough to limit strict orders to bounded orders to avoid counterexamples like the one presented above.

Theorem 23. Soset$_b$↓ *has all coqualiers.*

Theorem 24. Rset↓, Preord↓, Soset↓ *have all coproducts.* **Soset$_b$↓** *has finite coproducts.*

4   Algebraic Signatures with Dependent Symbols

In this section we formalize what we hand-waved in Sect. 2. We define the category of algebraic signatures enriched by dependency structure. Dependency structures are formalized in the category of bounded strict orders **Soset$_b$↓**. The chosen dependency relation not only suits the technical needs, but also reflects the practical intuition how should such relation be defined. Let us begin by recalling the standard definition.

Definition 25 (Algebraic Signatures). *We define a category* **AlgSig** *in the standard way – with objects being algebraic signatures defined as pairs of the form $\Sigma = \langle S, \Omega_{S^+} \rangle$ where $S \in$ **Set** is a set of sorts, S^+ is a set of nonempty finite S-sequences and $\Omega_{S^+} = \langle \Omega_e \rangle_{e \in S^+}$ is an S^+-sorted set of operation names. Morphisms of* **AlgSig** *are pairs of the form $\langle \sigma_S, \sigma_{\Omega_S} \rangle \colon \Sigma \to \Sigma'$ where $\sigma_S \colon S \to S'$ and $\sigma_{\Omega_S} = \langle \sigma_{\Omega_e} \colon \Omega_e \to \Omega'_{\sigma_S^+(e)} \rangle_{e \in S^+}$. Identities and composition are defined as one expects.*

Before we allow a general dependency structure on symbols, we define a functor that recognizes the basic dependency of operation symbols on sort symbols, as discussed in Sect. 2.

Definition 26 (SigSymb Functor). *Let* **SigSymb: AlgSig $\to$ Soset$_b$↓** *be the functor that transforms algebraic signatures to bounded strict orders of signatures' symbols. Given an algebraic signature $\Sigma = \langle S, \Omega_{S^+} \rangle$ we define*

$$\mathbf{SigSymb}(\Sigma) = \langle S \uplus (\biguplus_{e \in S^+} \{o : e | o \in \Omega_e\}), <_{\mathbf{SigSymb}(\Sigma)} \rangle$$

having operation symbols naturally dependent on sorts of their result and from their arities; i.e. for all $e \in S^+, e = \langle s_0 \ldots s_n \rangle, o \in \Omega_e$, we have

$$(s_k) <_{\mathbf{SigSymb}(\Sigma)} (o : e)$$

for all $0 \leq k \leq n$. Given a morphism $\sigma = \langle \sigma_S, \sigma_{\Omega_S} \rangle \colon \Sigma \to \Sigma'$, we define a **Soset$_b$↓**-*morphism* **SigSymb$(\sigma)$: SigSymb$(\Sigma)$ $\to$ SigSymb(Σ')** *as*

$$\mathbf{SigSymb}(\sigma) = \sigma_S \uplus (\biguplus_{e \in S^+} \sigma'_{\Omega_e})$$

where $\sigma'_{\Omega_e}(o : e) = \sigma_{\Omega_e}(o) : \sigma_S^+(e)$ for $o \in \Omega_e$.

Definition 27 (SetSymb Functor). *Let a functor giving a set of signature's symbols,* **SetSymb: AlgSig → Set,** *be defined as* **SetSymb = SigSymb; U** *where* **U: Soset$_b$↓ → Set** *is the obvious forgetful functor.*

At the moment we have everything needed to define the structures from the paper's title.

Definition 28 (Algebraic Signatures with Dependent Symbols). *Objects of a category* **AlgSigDep** *of algebraic signatures with dependent symbols are pairs*

$$\Sigma_< = \langle \Sigma, <_\Sigma \rangle$$

where $\Sigma \in$ **AlgSig** *is an algebraic signature and* $<_\Sigma \subseteq$ **SetSymb**$(\Sigma)^2$ *is such dependency relation that* $<_{\text{SigSymb}(\Sigma)} \subseteq <_\Sigma$ *(cf. Def. 26) and*

$$\langle \text{SetSymb}(\Sigma), <_\Sigma \rangle \in \text{Soset}_b\!\downarrow$$

Morphisms between $\Sigma_<, \Sigma'_< \in$ **AlgSigDep** *are such algebraic signature morphisms* $\sigma\colon \Sigma \to \Sigma'$, *for which a function* **SetSymb**(σ) *is a* **Soset$_b$↓**-*morphism (cf. Def. 2).*

The above definition of dependency relation, that extends the basic dependency of operation names on sorts from their arities, implies that all minimal elements of the ordering are sort symbols. The example of a signature with dependent symbols is given at the end of Sect. 4.1.

The definition of limits in **AlgSigDep** will require some work in all four categories mentioned so far, i.e. **AlgSigDep, AlgSig, Set** and **Soset$_b$↓**. The commuting diagram below presents functors between these categories.

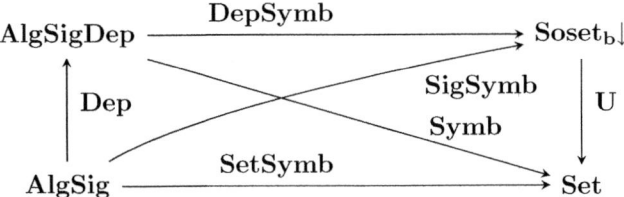

The functors **SigSymb, SetSymb** and **U**, are already given above. The functor **Dep** adds the basic dependency between operation symbols and sort symbols from their arities. The functor **DepSymb** is simply a projection and the functor **Symb** is defined as **Symb = DepSymb; U**.

4.1   Reconstructing Signatures with Dependent Symbols

In this section we define a powerful notion of the signature with dependent symbols "reconstruction", given a function into a set of signature's symbols. Since we are going to use the "reconstruction" in the definition of limits in

AlgSigDep, it has to be universally maximal, i.e. there must exist the mediating morphism from all possible reconstructions to the "reconstruction's" result.

We don't require the function to be injective, thus the number of "reconstructed" symbols may be greater than these of the original signature. It makes the "reconstruction" problem much more complicated that one would expect.

We begin by defining of the "pre-reconstruction" functor **PRec**, that is only universal in **AlgSig**. Later, the real "reconstruction" is given by the functor **Rec**, that is defined as a repetitive application of **PRec** and construction of a universal R-set of compatible dependencies from Lemma 11.

Definition 29. *Given a* $\mathbf{Soset_b}\downarrow$*-object* $A_<$*, the functor*

$$\mathbf{PRec}_{A_<} : (\mathbf{U}(A_<) \downarrow \mathbf{Symb}) \to (\mathbf{DepSymb} \downarrow A_<)$$

where both $(\mathbf{U}(A_<) \downarrow \mathbf{Symb})$ *and* $(\mathbf{DepSymb} \downarrow A <)$ *are comma categories, is defined as follows.*

For an $(\mathbf{U}(A_<) \downarrow \mathbf{Symb})$*-object,* $f : \mathbf{U}(A_<) \to \mathbf{Symb}(\Sigma_<)$*, where* $\Sigma_< = \langle\langle S, \Omega_{S+}\rangle, <_\Sigma\rangle$*, let* $\mathbf{PRec}_{A_<}(f) : \mathbf{DepSymb}(\Sigma^f) \to A_<$ *be defined by* $\Sigma^f_< = \langle\langle S^f, \Omega^f_{S^f+}\rangle, <_{\Sigma^f}\rangle$ *with* $S^f = \{s \in A' \mid f(s) \in S\}$ *and, for any* $e \in S^{f+}$*,* $\Omega^f_e = \{o \in A' \mid f(o) \in \Omega_{f(e)}\}$*, where* $A' = \{a \in A \mid (<_{\mathbf{SigSymb}(\Sigma)}\|_{\mathbf{SigSymb}(\Sigma)^{f(a)}\downarrow} \subseteq f(<_A |_{A_< a\downarrow}) \subseteq <_\Sigma |_{f(A_< a\downarrow)}\}$*, as*

$$\mathbf{PRec}_{A_<}(f)(s) = s \text{ for } s \in S^f \text{ and } \mathbf{PRec}_{A_<}(f)(o : e) = o, \text{ for } o \in \Omega^f_e, e \in S^{f+}$$

with the relation given as $<_{\Sigma^f} = <_A |_{\mathbf{PRec}_{A_<}(f)}.$

For an $(\mathbf{U}(A_<) \downarrow \mathbf{Symb})$*-morphism,* $\sigma : \Sigma^1_< \to \Sigma^2_<$*, i.e. such that* $f_2 = f_1; \mathbf{Symb}(\sigma)$*, where* $f_i : \mathbf{U}(A_<) \to \mathbf{Symb}(\Sigma^i_<)$*, for* $i \in \{1, 2\}$*, are two* $(\mathbf{U}(A_<) \downarrow \mathbf{Symb})$*-objects,* $\mathbf{PRec}_{A_<}(\sigma) = \langle\sigma'_S, \sigma'_{\Omega_{Sf_1}}\rangle : \mathbf{PRec}_{A_<}(f_1) \to \mathbf{PRec}_A <(f_2)$ *is defined by* $\sigma'_S = id_A|_{S^{f_1}}$ *and for every* $e \in S^{f_1+}$*,* $\sigma'_{\Omega_e} = id_A|_{\Omega^f_e}.$

Not all symbols from a given set stays in the "reconstructed" signature – only these from A', the subset of A, that contains all symbols fulfilling the compatibility requirements. The first requirement, $(<_{\mathbf{SigSymb}(\Sigma)}\|_{\mathbf{SigSymb}(\Sigma)^{f(a)}\downarrow} \subseteq f(<_A |_{A_< a\downarrow})$, makes the relation $<_A$ extend the basic dependency of Σ-operation symbols on sorts from their arities. It assures that all symbols that are needed to place the given symbol as a part of a signature are present in A'. The second, $f(<_A |_{A_< a\downarrow}) \subseteq <_\Sigma |_{f(A_< a\downarrow)}$, requires the function f be monotonic on the dependency structure for the given symbol a. In the example at the end of Sect. 4.1 we show the application of **PRec**.

Lemma 30. *For* $f : \mathbf{U}(A_<) \to \mathbf{Symb}(\Sigma_<)$ *there exists the* **AlgSig**-*morphism* $\xi_f : \Sigma^f \to \Sigma$*, where* Σ^f *comes from Def. 29 above, such that* $\mathbf{SetSymb}(\xi_f) = \mathbf{U}(\mathbf{PRec}_{A_<}(f)); f$ *and that is universal in* **AlgSig**, *i.e. for a* $\mathbf{Soset_b}\downarrow$*-morphism* $g : \mathbf{DepSymb}(\Sigma'_<) \to A_<$ *and an* **AlgSig**-*morphism* $\gamma : \Sigma' \to \Sigma$ *such that* $\mathbf{U}(g); f = \mathbf{SetSymb}(\gamma)$ *there exists exactly one* **AlgSig**-*morphism* $\sigma^f_g : \Sigma' \to \Sigma^f$ *such that* $\mathbf{SetSymb}(\sigma^f_g); \mathbf{U}(\mathbf{PRec}_{A_<}(f)) = \mathbf{U}(g)$ *in* **Set** *and* $\sigma^f_g; \xi_f = \gamma$ *in* **AlgSig**.

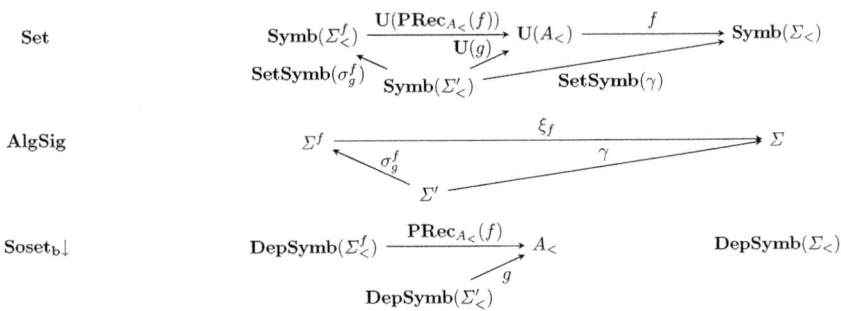

The above lemma gives an universality result, but only for **AlgSig**. What we need is a functor that has the same property as above, but with σ_g^f being an **AlgSigDep**-morphism. The true "reconstruction", universally maximal in **AlgSigDep**, is defined below.

Definition 31 (Signature Reconstruction). *Given a **Soset$_b\downarrow$**-object $A_<$, the functor*

$$\mathbf{Rec}_{A_<} : (\mathbf{U}(A_<) \downarrow \mathbf{Symb}) \to (\mathbf{DepSymb} \downarrow A_<)$$

*"reconstructing" the signature with dependent symbols, is defined as follows. For an $(\mathbf{U}(A_<) \downarrow \mathbf{Symb})$-object $f : \mathbf{U}(A_<) \to \mathbf{Symb}(\Sigma_<)$, let there be a **Soset$_b\downarrow$**-morphism $f' = \mathbf{PRec}_{A_<}(f) : \mathbf{DepSymb}(\Sigma_<^f) \to A_<$:*

- *if f' is injective then we define*

$$\mathbf{Rec}_{A_<}(f) = f'$$

- *otherwise let $l : Q_< \to \mathbf{DepSymb}(\Sigma_<^f)$ be given by the universal R-set of compatible dependencies for f' (cf. Lemma 11), we define*

$$\mathbf{Rec}_{A_<}(f) = \mathbf{Rec}_{Q_<}(l); l; f'$$

We postpone the definition of $\mathbf{Rec}_{A_<}$ on morphisms to the Lemma 34 below.

Lemma 32. $\mathbf{Rec}_{A_<}$ *is well defined on objects.*

Proof: The proof follows the observation that for any **Soset$_b\downarrow$**-morphism $h : X_< \to X'_<$ there exists a natural number $n_h = min(\{\widetilde{X_<^x\downarrow} \mid x, x' \in X, x \neq x', h(x) = h(x')\} \cup \{\widetilde{X_<}\})$, where $\widetilde{X_<}$ is the length of the maximal chain in $X_<$. It holds that $n_h \leq \widetilde{X_<} \leq \widetilde{X'_<}$. In the definition of $\mathbf{Rec}_{A_<}$, if f' is not injective, let $l' = \mathbf{PRec}_{Q_<}(l) : \mathbf{DepSymb}(\Sigma_<^l) \to Q_<$; it holds that $n_{f'} < n_{l'} \leq \widetilde{A_<}$, therefore, altogether there must be no more than $\widetilde{A_<}$ recursive steps. The reason for $n_{f'} < n_{l'}$ lies in the definition of $\mathbf{PRec}_{A_<}$ where the only source of non-injectivity in the result morphism is the non-injectivity on sorts in the parameter function and in the Def. 28 that requires all operation symbols to depend on sorts from their arities. $\square$

Lemma 33. $\mathbf{Rec}_{A_<}$ *enjoys the property described in Lemma 30 for* $\mathbf{PRec}_{A_<}$, *but in case of* $\mathbf{Rec}_{A_<}$, σ_g^f *is an* $\mathbf{AlgSigDep}$-*morphism, i.e.* $\sigma_g^f: \Sigma_<' \to \Sigma_<^f$ *is such that* $\mathbf{DepSymb}(\sigma_g^f); \mathbf{Rec}_{A_<}(f) = g$ *in* $\mathbf{Soset_b}\!\downarrow$ *and* $\sigma_g^f; \xi_f = \gamma$ *in* $\mathbf{AlgSig}$.

Lemma 34. $\mathbf{Rec}_{A_<}$ *is a functor.*

Proof: For a $(\mathbf{U}(A_<) \downarrow \mathbf{Symb})$-morphism $\sigma: \Sigma_<^1 \to \Sigma_<^2$ such that $f_2 = f_1; \mathbf{Symb}(\sigma)$ where $f_i: \mathbf{U}(A_<) \to \mathbf{Symb}(\Sigma_<^i)$, for $i \in \{1,2\}$, are two $(\mathbf{U}(A_<) \downarrow \mathbf{Symb})$-objects, $\mathbf{Rec}_{A_<}(\sigma) = \sigma_{\mathbf{Rec}_{A_<}(f_1)}^{f_2}: \Sigma_<^1 \to \Sigma_<^2$, an $\mathbf{AlgSigDep}$-morphism from Lemma 33. $\qquad\square$

Hopefully the following simple example will help the reader to grasp the idea behind the "reconstruction" defined above. Let there be a signature with dependent symbols $\Sigma_<$, defined as a signature $\Sigma = (\mathbf{sorts}\ s, t, \mathbf{ops}\ a : s,\ b : t)$ and a relation on its symbols $<_\Sigma = \{\langle s, a : s\rangle, \langle t, b : t\rangle, \langle a : s, b : t\rangle, \langle s, b : t\rangle\}$. Let $A_< = \langle A, <_A\rangle$ with $A = \{x, x', y, z\}$ and $<_A = \{\langle y, x\rangle, \langle z, x\rangle\}$ be an $\mathbf{Soset_b}\!\downarrow$-object and a function $f: \mathbf{U}(A) \to \mathbf{Symb}(\Sigma_<)$ be defined as $f(x) = a : s$, $f(x') = b : t$, $f(y) = f(z) = s$. The "reconstructed" signature with dependent symbols is $\Sigma_<^l = \langle \Sigma^l, <_{\Sigma^l}\rangle$ where $\Sigma^l = (\mathbf{sorts}\ y,\ z\ \mathbf{ops}\ x_1 : y,\ x_2 : y,\ x_1 : z,\ x_2 : z)$ and $<_{\Sigma^l} = \{\langle y, x_1 : y\rangle, \langle z, x_1 : y\rangle, \langle y, x_2 : y\rangle, \langle y, x_1 : z\rangle, \langle z, x_1 : z\rangle, \langle z, x_2 : z\rangle\}$.

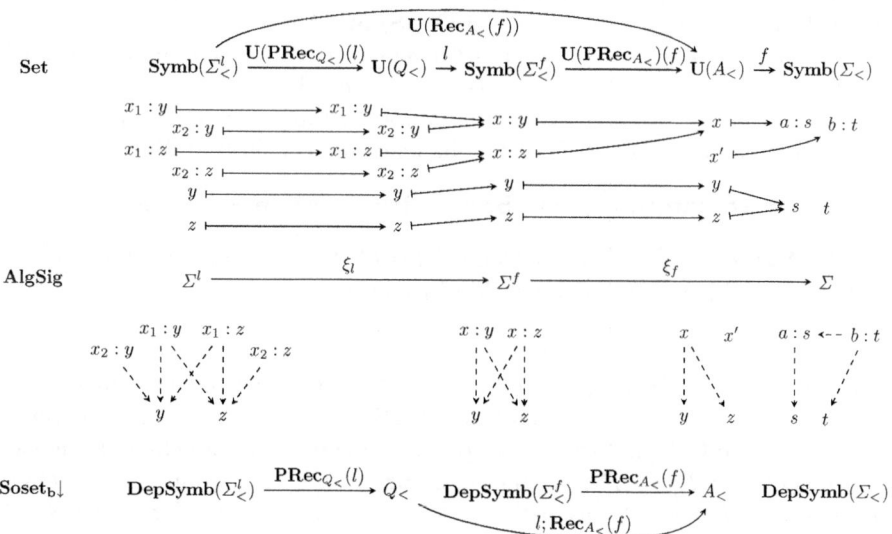

4.2   Limits and Colimits in AlgSigDep

The rich technical background from the previous sections is to be used here to prove the existence of (co)limits in **AlgSigDep**.

Before we begin we would like to remind the reader that there is a standard result that the category **AlgSig** is both complete and cocomplete.

We start our review of limits from the final object.

Theorem 35. *The category* **AlgSigDep** *does not have the final object.*

This is the straight consequence of the lack of the final object in $\mathbf{Soset_b}\!\downarrow$. In what follows we are going to show that all other limits and all finite colimits are present in **AlgSigDep**.

Theorem 36. AlgSigDep *has all equalisers.*

We use the Theorem 13 and the "reconstructing" functor from Def. 31. For the proof see App. A.

The definition of products in **AlgSigDep** follows the definition of products in $\mathbf{Soset_b}\!\downarrow$. The result is then "reconstructed" as a signature with dependency structure.

Theorem 37. AlgSigDep *has all products of non-empty families of objects.*

Proof: Let $(\Sigma^i_<)_{i\in I}$, for $I \neq \emptyset$, be a family of **AlgSigDep**-objects and let Σ together with a family of projection morphisms $(\pi^\Sigma_i : \Sigma \to \Sigma^i)_{i\in I}$ be the product of $(\Sigma^i)_{i\in I}$ in **AlgSig**. Now, let $P = \mathbf{SetSymb}(\Sigma)$, $(\pi_i = \mathbf{SetSymb}(\pi^\Sigma_i))_{i\in I}$ and $<_P= \{\langle p_1, p_2\rangle \in P \times P \mid \pi_i(p_1) <_{\Sigma_i} \pi_i(p_2)$ for all $i \in I\}$. It is easy to check that for every $i \in I$, $\pi_i : P_< \to \mathbf{DepSymb}(\Sigma^i_<)$ is a monotonic function, it doesn't have to be a p-morphism, though. Through an universal R-set[6] of compatible dependencies for $(\pi_i)_{i\in I}$ (cf. Lemma 11), $\langle Q_<, l: Q \to P\rangle$, we get the family of $\mathbf{Soset_b}\!\downarrow$-morphisms $(\rho_i = l; \pi_i : Q_< \to \mathbf{DepSymb}(\Sigma^i_<))_{i\in I}$.

By "reconstructing" the signature with dependent symbols from l (cf. Def. 31) we obtain the $\mathbf{Soset_b}\!\downarrow$-morphism $\mathbf{Rec}_{Q_<}(l): \mathbf{DepSymb}(\Sigma^l_<) \to Q_<$ and an **AlgSig**-morphism $\xi_l : \Sigma^l \to \Sigma$ such that $\mathbf{SetSymb}(\xi_l) = \mathbf{U}(\mathbf{Rec}_{Q_<}(l)); l$ in **Set** (cf. Lemma 33). For $i \in I$ we get the following equalities in **Set**

$$\mathbf{SetSymb}(\xi_l; \pi^\Sigma_i) = \mathbf{U}(\mathbf{Rec}_{Q_<}(l)); l; \pi_i = \mathbf{U}(\mathbf{Rec}_{Q_<}(l); \rho_i)$$

thus $\mathbf{SetSymb}(\xi_l; \pi^\Sigma_i)$ is a well defined $\mathbf{Soset_b}\!\downarrow$-morphism; therefore, $\rho^{\Sigma_i} = \xi_l; \pi^\Sigma_i : \Sigma^l_< \to \Sigma_<$ is a well defined **AlgSigDep**-morphism.

The product of $(\Sigma^i_<)_{i\in I}$ is an **AlgSigDep**-object $\Sigma^l_<$ together with the family of projections $(\rho^{\Sigma_i} : \Sigma^l_< \to \Sigma^i_<)_{i\in I}$.

To prove it, let $\Sigma^T_<$ be an algebraic signature with dependent symbols and let $(\theta_i : \Sigma^T_< \to \Sigma^i_<)_{i\in I}$ be a family of **AlgSigDep**-morphisms. Let us name $T_< = \mathbf{DepSymb}(\Sigma^T_<)$. Like in the proof of Lemma 18 we define the monotone function $\theta: T_< \to P_<$ as $\theta(t) = \langle \mathbf{DepSymb}(\theta_i)(t)\rangle_{i\in I}$ and the $\mathbf{Soset_b}\!\downarrow$-morphism $u: T_< \to Q_<$ as $u(t) = \langle [T_<, P_<, \theta]_{/K}^{[t]_K}\!\downarrow, \theta(t)\rangle$ for any $t \in T$. Mimicking the proof of Lemma 18, we learn that for $i \in I$, $\theta; \pi^\Sigma_i = \theta_i$, that $u(t) \in Q$, the morphism u is indeed a $\mathbf{Soset_b}\!\downarrow$-morphism and that it is unique such that $(u; \rho_i = \mathbf{DepSymb}(\theta_A))_{i\in I}$ is a family of $\mathbf{Soset_b}\!\downarrow$-morphisms. By Lemma 33, there exists the unique **AlgSigDep**-morphism, $\sigma^l_u : \Sigma^T_< \to \Sigma^l_<$ such that $\mathbf{DepSymb}(\sigma^l_u); \mathbf{Rec}_{Q_<}(l) = u$ in $\mathbf{Soset_b}\!\downarrow$ and $\sigma^l_u; \xi_l = \theta$ in **AlgSig**, therefore, for $i \in I$, $\sigma^l_u; \rho^{\Sigma_i} = \theta_i$ in **AlgSigDep**.   □

---

[6] Cf. Theorem 19 for the proof that an R-class of compatible dependencies in $\mathbf{Soset_b}\!\downarrow$ is a set

The coequalisers and finite coproducts are inherited from **AlgSig**.

Theorem 38. *The category* **AlgSigDep** *has all coequalisers and finite coproducts.*

At this point we proved enough to say that the category **AlgSigDep** is a natural extension of **AlgSig**. It simply adds dependencies between signatures' symbols. The only cost we pay, while shifting from one category to another, is the loss of the final object and restriction to finite colimits. The other vital properties are preserved.

5   Conclusion

In the paper we proposed the category of algebraic signatures with dependent symbols **AlgSigDep** and proved that it has all pushouts and pullbacks. The proposal was proceeded by the long analysis of several possible dependency orderings and proofs of the existence of the (co)limits in the respective categories. On the way we defined a product candidate in the category R-sets and p-morphisms, also known as the category of transitive Kripke frames. Finally we decided that the category of bounded strict orders and p-morphisms, that lacks only the final object to be finitely (co)complete, is the most suitable setting to describe the dependency relation.

The decision also reflects the intuition one may have regarding the matter described in Sect. 2. In the example given there we said that the pushout should preserve the dependencies in a way presented on the given diagram. Our proposal does exactly it, actually, the diagram given there is in **AlgSigDep**.

Through our paper we didn't say anything about models of signatures with dependent symbols. We believe that they may or may not reflect the dependency structure from the signature. This is not that place to discuss all possibilities in detail.

Future work concerns the use of algebraic signatures enriched by dependency structure in covariant definition of signatures of generic modules in the architectural specifications framework. Models of such signatures will take advantage of the dependency information given in their signatures.

In the paper we used the standard category **AlgSig** and added dependency structure to algebraic signatures. However, the construction is more general than that and should work with most signature categories in the similar way.

Acknowledgment. The author would like to thank Andrzej Tarlecki for his invaluable help and support and Bartek Klin for thoughtful suggestions.

References

[BST99]   Bidoit, M., Sannella, D., Tarlecki, A.: Architectural Specifications in CASL. In: Haeberer, A.M. (ed.) AMAST 1998. LNCS, vol. 1548, pp. 341–357. Springer, Heidelberg (1998)

[GS01]   Gumm, H.P., Schröder, T.: Products of coalgebras. Algebra Universalis 46, 163–185 (2001)

[Seg70]   Segerberg, K.: Modal logics with linear alternative relations. Theoria 36, 301–322 (1970)

[ST88]   Sannella, D., Tarlecki, A.: Toward formal development of programs from algebraic specifications: Implementations revisited. Acta Informatica 25(3), 233–281 (1988)

[ST97]   Sannella, D., Tarlecki, A.: Essential Concepts of Algebraic Specification and Program Development. Formal Asp. Comput. 9(3), 229–269 (1997)

[WW08]   Walicki, M., Wolter, U.: Universal multialgebra. In: New Topics in Theoretical Computer Science, pp. 27–93. Nova Science Publishers (2008)

A   Proofs of Lemmas and Theorems

Here we give proofs of lemmas and theorems for whom there was no place in the main sections. Before we get to the proofs let us introduce a bit more of technicalities.

Definition 39 (Quotient of R-mset wrt. $\equiv$). *Given an R-mset $[A_R, P_R, \mu]$ and an equivalence relation $\equiv \subseteq A^2$, a quotient R-mset is defined as an R-mset:*

$$[A_R, P_R, \mu]_{/\equiv} = [A'_R, P_R, \mu']$$

where $A' = A_{/\equiv}$, $R_{A'}$ is defined by

$$[a_1]_\equiv R_{A'} [a_2]_\equiv \text{ iff for any } a'_1 \in [a_1]_\equiv \text{ and } a'_2 \in [a_2]_\equiv, \ a'_1 R_A a'_2$$

and the labeling

$$\mu'([a]_\equiv) = \mu(a)$$

where $[a]_\equiv$ is the equivalence class of a wrt. $\equiv$.

Definition 40 (Kernel of an R-mset). *Given an R-mset $[A_R, P_R, \mu]$, its kernel is defined as*

$$[A_R, P_R, \mu]_{/K} = [A_R, P_R, \mu]_{/K([A_R,P_R,\mu])}$$

Elements of the kernel are equivalence classes, for $a \in A$, $[a]_K = [a]_{K([A_R,P_R,\mu])}$.

Lemma 41. *Given an R-mset $[A_R, P_R, \mu]$*

$$K([A_R, P_R, \mu]_{/K}) = id_{A_{/K([A_R,P_R,\mu])}}$$

Proof: Let a relation $\sim' \subseteq (A_{/K([A_R,P_R,\mu])})^2$ be a dependency bisimulation on $[A_R, P_R, \mu]_{/K}$, then also $\sim \subseteq A^2$, defined as $a_1 \sim a_2$ iff $[a_1]_K \sim' [a_2]_K$, for $a_1, a_2 \in A$, is a dependency bisimulation on $[A_R, P_R, \mu]$. Therefore, $\sim \subseteq K([A_R, P_R, \mu])$ and for any $a_1, a_2 \in A$, if $[a_1]_K \sim' [a_2]_K$ then $a_1 \sim a_2$ and $\langle a_1, a_2 \rangle \in K([A_R, P_R, \mu])$, thus $[a_1]_K = [a_2]_K$ and $\sim' = id_{A_{/K([A_R,P_R,\mu])}}$.   □

Lemma 42. *Given an R-mset $[A_R, P_R, \mu]$ and $a \in A$,*

$$K([A_R, P_R, \mu]^a\!\downarrow) = K([A_R, P_R, \mu]) \cap ([A_R, P_R, \mu]^a\!\downarrow)^2$$

Proof: The proof is straightforward. Let us just notice that the kernel relation is based solely on the structure down from the given element wrt. R in the R-mset.   □

Corollary 43. *Given an R-mset $[A_R, P_R, \mu]$, and $a \in A$, it holds that*

$$([A_R, P_R, \mu]^a\!\downarrow)_{/K} = ([A_R, P_R, \mu]_{/K})^{[a]_K}\!\downarrow$$

Lemma 44. *Given an* **Rset↓**-*morphism $f\colon A_R \to B_R$, and an element $a \in A$, the reduct $f|_{A_R{}^a\downarrow}\colon A_R{}^a\!\downarrow \to B_R$ is also an* **Rset↓**-*morphism.*

Proof: The reduct $f|_{A_R{}^a\downarrow}$ is monotone, because f is so. Since R is transitive, for any $a' \in A_R{}^a\!\downarrow$, all $a''R_Aa'$ are in $A_R{}^a\!\downarrow$ as well and so $f|_{A_R{}^a\downarrow}$ meets the requirement (2) of Def. 2, as f does.   □

Lemma 45. *Let there be two R-msets $[A_R, P_R, \mu]$ and $[A'_R, P_R, \mu']$ and an* **Rset↓**-*morphism $f\colon A_R \to A'_R$, such that $\mu = f; \mu'$. If $\sim \subseteq A^2$ is a dependency bisimulation on $[A_R, P_R, \mu]$, then $f(\sim) \subseteq f(A)^2$ is a dependency bisimulation on $[f(A)_R, P_R, \mu'|_{f(A)}]$, where $R_{f(A)} = R_{A'}|_{f(A)}$.*

Proof: Let $\sim' = f(\sim)$. Then, given $a_1, a_2 \in A$, $a_1 \sim a_2$, we have $f(a_1) \sim' f(a_2)$. By $\mu = f; \mu'$, we have $\mu'(f(a_1)) = \mu'(f(a_2))$ and for any $p' \in P$, $p' R \mu'(f(a_1))$ and $b'_1 \in A'$, if $\mu'(b'_1) = p'$ and $b'_1 R f(a_1)$, by requirement (2) of Def. 2, there exists $a'_1 \in A$, $a'_1 R a_1$, $f(a'_1) = b_1$ and, since $\sim$ is a dependency bisimulation, there exists $a'_2 R a_2$, $a'_1 \sim a'_2$, thus $f(a'_2) R f(a_2)$ and $f(a'_2) \sim' f(a'_1) = b'_1$, hence $b'_1 \in f(A)$. By symmetry, this proves that $f(\sim)$ is a dependency bisimulation on $[f(A)_R, P_R, \mu']$.   □

Corollary 46. *Consider two R-msets $[A_R, P_R, \mu]$ and $[A'_R, P_R, \mu']$ and a surjective* **Rset↓**-*morphism $f\colon A_R \to A'_R$, such that $\mu = f; \mu'$. If $\sim \subseteq A^2$ is a dependency bisimulation on $[A_R, P_R, \mu]$, then $f(\sim) \subseteq A'^2$ is a dependency bisimulation on $[A'_R, P_R, \mu']$.*

Lemma 47. *For R-msets $[A_R, P_R, \mu]$ and $[A'_R, P_R, \mu']$ and an* **Rset↓**-*morphism $f\colon A_R \to A'_R$, such that $\mu = f; \mu'$, it holds that*

$$K(f) \subseteq K([A_R, P_R, \mu])$$

where $K(f) \subseteq A^2$ is the kernel relation for the function f. Moreover, if it happens that $K([A'_R, P_R, \mu]) = id_{A'}$, then

$$K(f) = K([A_R, P_R, \mu])$$

Proof: A $\mathbf{Rset}\downarrow$-morphism f weakly reflects R-set down-closures. Therefore, if $a_1, a_2 \in A$ are such that $f(a_1) = f(a_2)$, then $\mu(a_1) = \mu(a_2)$ and for any $p' R \mu(a_1)$ and any $a_3 R a_1$ such that $\mu(a_3) = p'$, since $f(a_3) R f(a_1) = f(a_2)$ there exists $a_4 R a_2$, $f(a_4) = f(a_3)$. By symmetry, this implies that $K(f)$ is a dependency bisimulation on $[A_R, P_R, \mu]$ and since $K([A_R, P_R, \mu])$ is the greatest such relation, $K(f) \subseteq K([A_R, P_R, \mu])$. When additionally $K([A'_R, P_R, \mu]) = id_{A'}$, let $\sim \subseteq A^2$ be a dependency bisimulation on $[A_R, P_R, \mu]$ and $a_1, a_2 \in A$. If $a_1 \sim a_2$ then, by definition of f, $f(a_1) f(\sim) f(a_2)$ and, since $f(\sim)$ is a dependency bisimulation on $[f(A)_R, P_R, \mu|_{f(A)}]$ (cf. Lemma 45), $f(\sim) \subseteq K([f(A)_R, P_R, \mu|_{f(A)}]) \subseteq K([A'_R, P_R, \mu']) = id_{A'}$, thus $f(a_1) = f(a_2)$, therefore $K([A_R, P_R, \mu]) \subseteq K(f)$. Together with the previous result this gives $K(f) = K([A_R, P_R, \mu])$. □

Lemma 48. *Every R-mset component of $\langle [X_R, P_R, \mu], p \rangle \in Q_R$ from Def. 10 has a distinguished top-element x, i.e. there exists exactly one $x \in X$, such that $\mu(x) = p$ and for all $x' \neq x \in X$, $x' R x$.*

Proof: By contradiction, let x_1 be another such element, i.e. $x_1 \neq x$ and for all $x' \neq x_1 \in X$, $x' R x_1$. As a consequence $x R x_1$ and $x_1 R x$, thus, since R is transitive, $\langle x, x_1 \rangle \in K([X_R, P_R, \mu]) = id_X$. Contradiction. □

Lemma 49. *The class Q from Def. 10 is* self-adequate, *meaning that for all $\langle [X_R, P_R, \mu], p \rangle \in Q$ and for all $x \in X$*

$$\langle [X_R, P_R, \mu]^x \downarrow, \mu(x) \rangle \in Q$$

Proof of Lemma 11: By definition, R_Q is transitive, which makes Q_R indeed an $\mathbf{Rset}\downarrow$-object. For $i \in I$, let $\rho_i = l; w_i$. We show that ρ_i is an $\mathbf{Rset}\downarrow$ morphism. It obviously preserves the relation. Let $q_1, q_2 \in Q$ be $q_1 = \langle [X_R^1, P_R, \mu_1], p_1 \rangle$ and $q_2 = \langle [X_R^2, P_R, \mu_2], p_2 \rangle$, such that $q_1 R_Q q_2$, then, by definition, $p_1 R_P p_2$, thus $(\rho_i(q_1)) R_{W^i} (\rho_i(q_2))$. It also meets the requirement (2) of Def. 2, which makes it an $\mathbf{Rset}\downarrow$-morphisms. Namely, given $q = \langle [X_R, P_R, \mu], p \rangle \in Q$ and $a' \in W_i$, $a' R_{W^i} \rho_i(q)$, by definition of Q there exists $x \in X$, $\mu(x) = p$, $\mu; w_i(x) = a$ and, since $\mu; w_i$ is an $\mathbf{Rset}\downarrow$-morphism, there exists $x' R x$, $\mu; w_i(x') = a'$. By self-adequacy of Q (cf. Lemma 49), there is $q' = \langle [X_R, P_R, \mu]^x \downarrow, \mu(x') \rangle$, and of course $q' R q$ and $\rho_i(q') = w_i(\mu(x')) = a'$.

Now we show that for each object $T_R \in \mathbf{Rset}\downarrow$ and a monotone function $\theta: T_R \to P_R$, such that morphisms from the family $(\theta_i = \theta; w_i: T_R \to W_R^i)_{i \in I}$ are in $\mathbf{Rset}\downarrow$, there exists a unique $\mathbf{Rset}\downarrow$-morphism $u: T_R \to Q_R$ such that $u; l = t$.

We notice that $[T_R, P_R, \theta]$ is an R-mset. Let

$$[T'_R, P_R, \theta'] = [T_R, P_R, \theta]_{/K}$$

It is easy to prove that $\theta'; w_i$ for $i \in I$ is an $\mathbf{Rset}\!\downarrow$-morphism. Let the morphism $u: T_R \to Q_R$ be defined as

$$u(t) = \langle [T_R, P_R, \theta]_{/K}{}^{[t]_K}\!\downarrow, \theta(t) \rangle$$

for any $t \in T$, where $[t]_K$ is an equivalence class of t wrt. $K([T_R, P_R, \theta])$ (cf. Def. 9 and Def. 40). Before we proceed with the proof let us simplify the notation by naming

$$[T_R^t, P_R, \theta_t] = [T_R, P_R, \theta]_{/K}{}^{[t]_K}\!\downarrow$$

for $t \in T$.

Let us show that for every $t \in T$, $u(t) \in Q$. The R-mset $[T_R^t, P_R, \theta_t]$ indeed meets all requirements from the definition of Q. By Def. 7, $\theta_t = \theta'|_{T^t}$ and, since $\theta'; w_i$ for $i \in I$ is an $\mathbf{Rset}\!\downarrow$-morphism, by Lemma 44 $\theta_t; w_i$ is also an $\mathbf{Rset}\!\downarrow$-morphism. Moreover, the element $[t]_K$ is such that for all $x \in T^t$, $(x) R([t]_K)$ and by Lemma 41 and Corollary 43, $K([T_R^t, P_R, \theta_t]) = id_{[T_R^t, P_R, \theta_t]}$. Now, let us check that u is monotone. Let $t', t \in T$ and $t' R_T t$. To prove that $u(t') R_Q u(t)$ we need to show that there exists $x' \in T^t$, $\theta_t(x') = \theta(t')$ and that $[T_R^{t'}, P_R, \theta_{t'}] = [T_R^t, P_R, \theta_t]^{x'}\!\downarrow$. Let us take $x' = [t']_K$. Obviously $\theta_t([t']_K) = \theta(t')$. The second requirement, $[T_R^{t'}, P_R, \theta_{t'}] = [T_R^t, P_R, \theta_t]^{[t']_K}\!\downarrow$ also holds because $([t']_K) R_{T^t}([t]_K)$ and by Lemma 8. It is trivial to show that $u; l = \theta$. To finish the proof we need to show that a function u is a $\mathbf{Rset}\!\downarrow$-morphism and that the choice of u is unique.

We prove that the function u is an $\mathbf{Rset}\!\downarrow$-morphism. We already have shown that it is monotone. The requirement (2) of Def. 2 says that for any $t \in T$ and $q' = \langle [X_R', P_R, \mu'], p' \rangle \in Q$ such that $(q') R_Q(u(t))$, there must exist $t' \in T$, $t' R t$ and $u(t') = q'$. Since $(q') R_Q(u(t))$, there exists $x' \in T^t$ that $\theta_t(x') = p'$ and, by definition of u, there exists $t' \in T$ that $[t']_K = x'$ and $t' R t$. Of course $\theta(t') = p'$. Moreover, by definition of R_Q, $[X_R', P_R, \mu'] = [T_R^t, P_R, \theta_t]^{[t']_K}\!\downarrow$ and by Lemma 8 we have $[T_R^t, P_R, \theta_t]^{[t']_K}\!\downarrow = [T_R^{t'}, P_R, \theta_{t'}]$. This proves that $q' = u(t')$.

To show the uniqueness of u such that $u; l = \theta$ let us have some $\mathbf{Rset}\!\downarrow$-morphism $u': T_R \to Q_R$ that $u; l = \theta$. For any $t \in T$, $u'(t) = \langle [X_R, P_R, \mu], \theta(t) \rangle$, for some R-set X_R and a monotone function $\mu: X_R \to P_R$. Let us define a surjective $\mathbf{Rset}\!\downarrow$ morphism $u_t': T_R^t\!\downarrow \to X_R$ as follows. For any $t' \in T_R^t\!\downarrow$ we have $(u'(t')) R_Q(u(t))$, thus $u'(t') = \langle [X, P_R, \mu]^{x'}\!\downarrow, \theta(t') \rangle$ for exactly one $x' \in X$ (cf. Lemma 48), let

$$u_t'(t') = x'$$

The morphism u' meets both requirements of Def. 2 – it is monotone and, since Q is self adequate (cf. Lemma 49), for each $x'' \in X$, if $x'' R_X x'$ then there exists $q'' \in Q$ such that $(q'') R_Q(u'(t'))$ and, since u' is a $\mathbf{Rset}\!\downarrow$-morphism, there must exist $t'' \in T$ such that $t'' R t'$ and $u'(t'') = q'' = \langle [X_R, P_R, \mu]^{x''}\!\downarrow, \theta(t'') \rangle$, thus $u_t'(t'') = x''$. This also proves that u_t' is surjective. Hence, there is a bijection between X and $(T_R^t\!\downarrow)_{/K(u_t')}$, where $K(u_t')$ is the kernel of the function u_t'. By Lemma 47, since u_t' is surjective and $K([X_R, P_R, \mu]) = id_X$,

$$K(u_t') = K([T_R^t\!\downarrow, P_R, \theta_{T_R^t\!\downarrow}]) = K([T_R, P_R, \theta]^t\!\downarrow)$$

This means that $[X, P_R, \mu] = ([T_R, P_R, \theta]^t\downarrow)_{/K}$. By Corollary 43, $[X, P_R, \mu] = ([T_R, P_R, \theta]_{/K})^{[t]_K}\downarrow$, therefore, $u' = u$.

This shows the uniqueness of u and completes the proof that $\langle Q, R_Q \rangle$, together with projections ρ_A and ρ_B, is the product of A_R and B_R in **Rset↓**.   □

Proof of Lemma 12: (reflexive) Let R_P and R_{W^i}, for all $i \in I$, be reflexive. Let $q = \langle [X_R, P_R, \mu], p \rangle \in Q$. By Lemma 48 there exists a top-element $x \in X$ and we notice that $[X_R, P_R, \mu] = [X_R, P_R, \mu]^x\downarrow$. By reflexivity of R_P we have $p\,R\,p$. All together this makes $q\,R\,q$ (cf. Def.10).

(Irreflexive) Let R_P and R_{W^i}, for all $i \in I$, be irreflexive. By contradiction. Let $q = \langle [X_R, P_R, \mu], p \rangle \in Q$ and $q\,R\,q$. Thus, $p\,R\,p$ which contradicts the irreflexivity of R_P.

(Asymmetric) Let R_P and R_{W^i}, for all $i \in I$, be asymmetric. Let $q = \langle [X_R, P_R, \mu], p \rangle, q' = \langle [X'_R, P_R, \mu'], p' \rangle \in Q$, and let $q\,R\,q'$. This means that $p\,R\,p'$ and, by asymmetry, it doesn't hold that $p'\,R\,p$. Therefore, by definition of R_Q (cf. Lemma 48), it doesn't hold that $q'\,R\,q$.

(Strict bounded) Let there be a family of functions $(w_i: P_R \rightarrow W^i_R)_{i \in I}$ with R_P and R_{W^i}, for all $i \in I$, be strict (i.e. transitive, asymmetric, thus irreflexive) bounded relations. From Lemma 11 we know that, for any $i \in I$; w_i is an **Rset↓**-morphism. Since, by above points, Q_R is asymmetric, it is a **Soset↓**-morphism. It is easy to prove that any **Soset↓**-morphism reflects the length of the finite chains, i.e. for a **Soset↓**-morphism $f: A_< \rightarrow B_<$ and $a \in A$ such that $\widetilde{B_{<}{}^{f(a)}\downarrow}$ is finite, $\widetilde{A_{<}{}^a\downarrow} = \widetilde{B_{<}{}^{f(a)}\downarrow}$. Therefore, since all chains in W^i_R are bounded, we bound the size of chains in Q_R by $\widetilde{Q_R} = max\{\widetilde{W^i_R}\}$.   □

Proof of Theorem 13: Given two morphisms $f, g: A_R \rightarrow B_R$ in **Rset↓**, their equaliser is an inclusion $e: C_R \rightarrow A_R$, where

$$C = \{a \in A \mid \text{ for all } a' \in A_R{}^a\downarrow, \ f(a') = g(a')\}$$

See Def. 3 for the definition of closed down sub R-set $A_R{}^a\downarrow$ induced by a. The relation is defined as $R_C = R_A|_C$. Trivially, e is an **Rset↓**-morphism. Let us check that it is universal. Let $h: D_R \rightarrow A_R$ be such **Rset↓**-morphism that $h; f = h; g$. We need to find the unique $u: D_R \rightarrow C_R$, such that $u; e = h$. Since e is an inclusion and because $h(D) \subseteq C$, putting $u(d) = h(d)$ for $d \in D$ yields the only such morphism. The inclusion $h(D) \subseteq C$ is the consequence of the fact that h, as an **Rset↓**-morphism, weakly reflects R-set down-closures. To complete the proof, it is enough to notice that if A_R and B_R are in **Preord↓**, **Soset↓** or **Soset$_b$↓**, respectively, then so is C_R.   □

Proof of Theorem 14: The relations are reflexive, therefore there exist unique morphisms from any of their objects to the singleton ordered by identity.   □

Proof of Theorem 15: Since the relations in objects of **Soset↓** (and **Soset$_b$↓**) are irreflexive, their morphisms must not glue together any elements being in relation. If the final object existed, there would be an injective map from any ordinal (represented as an R-set with natural strict bounded "ordering") into it. Hence, such a final object can not be a proper set. □

Proof of Lemma 18: Let us assume that $A \prod B$ is a set. By Lemma 11 we get that $(A \prod B)$ is an R-set and the projections, ρ_A and ρ_B, are **Rset↓**-morphisms.

Now we show that for each object $T_R \in$ **Rset↓** and two morphisms $\theta_A \colon T_R \to A_R$ and $\theta_B \colon T_R \to B_R$ in **Rset↓**, there exists a unique **Rset↓**-morphism $u \colon T_R \to Q_R$ such that $u; \rho_A = \theta_A$ and $u; \rho_B = \theta_B$.

Let an **Rset↓**-object T_R and morphisms θ_A and θ_B be as described above. We define a function $\theta \colon T \to P$ as

$$\theta(t) = \langle \theta_A(t), \theta_B(t) \rangle$$

The monotonicity of θ follows the monotonicity of θ_A and θ_B and the definition of R_P. By definition, it holds that $\theta; \pi_A = \theta_A$ and $\theta; \pi_B = \theta_B$.

Having above, by Lemma 11 we get a unique **Rset↓**-morphism $u \colon T_R \to Q_R$ such that $u; l = \theta$, therefore $u; \rho_A = u; l; \pi_A = \theta; \pi_A = \theta_A$ and similarly we get $u; \rho_B = \theta_B$.

□

Proof of Theorem 19: By Lemma 12 we know that $<_{A \prod B}$ is a bounded strict order, thus in fact $A \prod B$, if proved to be a set, is the **Soset$_b$↓**-object. To show that it is indeed a product we need to show that $A \prod B$ is a set to be able to use Lemma 18. To do so, it is enough to bound the number of $A \prod B$ elements that share the given label $p \in P$ (using notation from Def. 17). Given $q = \langle [X_<, P_<, \mu], p \rangle \in A \prod B$ and a label $p' \in P_<^p\!\!\downarrow$, let us bound the cardinality of $\mu^{-1}(p')$ by cases:

- (base case) if $P_<^p\!\!\downarrow = \{p\}$, then $p' = p$ and $|\mu^{-1}(p')| <= 1$, i.e. $l_{p'} = 1$, because there may be only one element labeled by p' distinct wrt. the kernel relation (cf. Def. 9);
- (induction step) otherwise, let $\mathcal{L} = \sum_{p'' < p'} l_{p''}$, where $l_{p''}$ is the bound of $\mu^{-1}(p'')$ for $p'' < p'$, then $|\mu^{-1}(p')| <= 2^{\mathcal{L}}$, i.e. $l_{p'} = 2^{\mathcal{L}}$; it is impossible to have more elements labeled by p' distinct wrt. the kernel relation, than all combinations of elements lower wrt. $<_X$.

The cardinal number $l_{p'}$ is well defined for every $p' \in P_<^p\!\!\downarrow$, because $<_X$ is a bounded strict order. Finally, by definition of $<_{A \prod B}$, we conclude that the cardinality of the set of elements that share the label p is bounded by l_p, therefore, $A \prod B$ is a set. By Lemma 18 a **Soset$_b$↓**-object $\langle A \prod B, <_{A \prod B} \rangle$ together with morphisms ρ_A and ρ_B, as defined in Def. 17, is a product of $A_<, B_<$ in **Soset$_b$↓**.

□

Lemma 50. *Given two **Rset↓**-morphisms $f, g \colon A_R \to B_R$ and a relation $\sim \subseteq B^2$ defined as $b_1 \sim b_2$ iff there exists $a \in A$ such that $b_1 = f(a)$ and $b_2 = g(a)$*

and its reflexive, symmetric and transitive closure $\equiv\ =\ Trans(Sym(Ref(\sim)))$, *it holds that: for any* $b_1, b_2 \in B$, *if* $b_1 R_B b_2$ *then for any* $b_2' \equiv b_2$ *there exists* $b_1' \equiv b_1$ *such that* $b_1' R_B b_2'$.

Proof: For any $b_2 \equiv b_2'$ there exists a path from b_2 to b_2' in the undirected graph $Graph(f) \cup Graph(g)$. Let $a_2 \in A$ be such that $f(a_2) = b_2$ and let $b_1 R_B b_2$. Since f is **Rset↓**-morphism, by requirement (2) of Def. 2, there exists $a_1 \in A$ such that $f(a_1) = b_1$ and $a_1 R_A a_2$. Since g is also an **Rset↓**-morphism, by requirement (1) of the same definition, it is monotone, i.e., $g(a_1) R_A g(a_2)$ and $g(a_1) \equiv b_1$ and $g(a_2) \equiv b_2$. The above procedure executed along the path between b_2 and b_2' (the same that served the transitive closure in definition of $\equiv$) results in existence of the required $b_1' \equiv b_1$ such that $b_1' R_B b_2'$. □

Lemma 51. *Given two* **Rset↓**-*morphisms* $f, g: A_R \to B_R$, *their coequaliser in* **Set**, $e: B \to C$, *is an* **Rset↓**-*morphism* $e: B_R \to C_R$ *where the relation* R_C *is defined simply as*

$$R_C = e(R_B)$$

Proof: Let $f, g: A_R \to B_R$ be two **Rset↓**-morphisms and $e: B \to C$ be their coequaliser in **Set**. Let us notice that $C = B_{/\equiv}$, where $\equiv$ is an equivalence defined as in Lemma 50. We need to show that R_C is transitive and that e meets both conditions of Def. 2. In fact R_C is transitive. Let $c_1 R_C c_2 R_C c_3$. Function e as a coequaliser in **Set** is surjective, thus there exist $b_1, b_2, b_2', b_3 \in B$ such that $e(b_1) = c_1$, $e(b_2) = e(b_2') = c_2$, $e(b_3) = c_3$ and $b_1 r_B b_2$ and $b_2' R_B b_3$. By Lemma 50, since $b_2 \equiv b_2'$, there exists $b_1' \equiv b_1$ such that $b_1' R_B b_2'$. Relation R_B is transitive thus $b_1' R_B b_3$. Of course, since e is the coequaliser of f and g, we get $e(b_1') = e(b_1) = c_1$. Therefore $c_1 R_C c_3$. Function e trivially meets the first condition of Def. 2, because $R_C = e(R_B)$. To prove the second condition let $b_2 \in B$ and $c_1 \in C$ such that $c_1 R_C e(b_2)$. From the surjectivity of e and the definition of R_C we get the existence of $b_2', b_1' \in B$ such that $e(b_1') = c_1$, $e(b_2') = e(b_2)$ and $b_1' R_B b_2'$. Therefore, $b_2' \equiv b_2$ and by Lemma 50, there exists $b_1 \in B$ such that $b_1 \equiv b_1'$ and $b_1 R_B b_2$, and accordingly $e(b_1) = e(b_1') = c_1$. □

Proof of Theorem 21: The coequaliser of two morphisms $f, g: A_R \to B_R$ in **Rset↓** is $e: B_R \to C_R$, where e is the coequaliser of f and g in **Set** and a relation $R_C = e(R_B)$. The universal properties of e in **Rset↓** are inherited from **Set**. Namely, given a morphism $h: B_R \to D_R$ in **Rset↓** such that $f; h = g; h$, there exists a unique function $k: C \to D$ such that $e; k = h$. It is monotone. Given $c_1, c_2 \in C$, $c_1 R_C c_2$ since e in surjective (as a coequaliser in **Set**) and monotone there exist $b_1, b_2 \in B$ such that $b_1 R_B b_2$ and $e(b_1) = c_1$ and $e(b_2) = c_2$. We have $h(b_1) R_D h(b_2)$, because h is monotone, and finally

$$k(c_1) = k(e(b_1)) = h(b_1) R_D h(b_2) = k(e(b_1)) = k(c_2)$$

Function k also meets the requirement (2) of Def. 2. Let $c_1 \in C$ and $d_2 \in D$ be such that $d_2 R_D k(c_1)$. Since h is a p-morphism, for any $b_1 \in B$ such that

$h(b_1) = k(c_1)$ there exits $b_2 \in B$, $b_2 R_B b_1$ and $h(b_2) = d_2$. Let us choose the b_1 such that $e(b_1) = c_1$. It exists because e is surjective. Of course $h(b_1) = k(e(b_1)) = k(c_1)$, so there is b_2 with properties given above. Function e is monotone, thus, $e(b_2) R_C e(b_1)$. Moreover, $k(e(b_2)) = h(b_2) = d_2$. The two, above shown, properties of function k make it an **Rset↓**-morphism.

Due to Lemma 51, e is an **Rset↓**-morphism and $e(R_B)$ is a transitive relation. Note that reflexivity of R_B guarantees reflexivity of $e(R_B)$. □

Proof of Theorem 22: Let us show a counterexample. A strictly ordered set $\langle \mathbb{N}, prev \rangle$, where $prev \subseteq \mathbb{N} \times \mathbb{N}$ is defined as $prev = \{\langle n+1, n \rangle \mid n \in \mathbb{N}\}$, is a **Soset↓**-object. Let $f, g \colon \langle \mathbb{N}, prev \rangle \rightarrow \langle \mathbb{N}, prev \rangle$ be two **Soset↓**-morphisms be defined as $f = id_\mathbb{N}$ and $g = succ$, where $succ$ is a successor function. Their coequaliser in **Rset↓** is $e \colon \langle \mathbb{N}, prev \rangle \rightarrow \langle C, R_C \rangle$, where $C = \{*\}$ and $R_C = \{\langle *, * \rangle\}$, i.e. a singleton ordered by identity. However, $\langle C, R_C \rangle$ fails to be a **Soset↓**-object, because R_C is reflexive. Moreover, no other function may in the same time coequalise functions f and g and stay monotonic. Therefore, they have no coequaliser in **Soset↓**. □

Proof of Theorem 23: The coequaliser of two morphisms $f, g \colon A_R \rightarrow B_R$ in **Soset$_b$↓** is $e \colon B_R \rightarrow C_R$, where e is the coequaliser of f and g in **Set** and a relation $R_C = e(R_B)$. If all descending chains in R_B are finite, for any $b_1, b_2 \in B$, if $b_1 < b_2$ then $b_1 \not\equiv b_2$, where $\equiv$ is the equivalence relation as in Lemma 50. To prove this fact, let us assume that there are $b_1, b_2 \in B$ such that $b_1 < b_2$, the proof goes by induction on the length of the descending chain lower to b_1 wrt. $<$. In the base case let for all $b \in B$, $b \not< b_1$. By contradiction let $b_1 \equiv b_2$. Due to Lemma 50, since $b_1 \equiv b_2$ and $b_1 < b_2$, there must exist $b_1' < b_1$. Contradiction. In the induction step let us assume that for all $b < b_1$, $b \not\equiv b_1$. Again by contradiction, let $b_1 \equiv b_2$. Using once more Lemma 50, since $b_1 \equiv b_2$ and $b_1 < b_2$, we get the existence of $b_1' < b_1$, that $b_1' \equiv b_1$. Contradiction. Therefore, irreflexivity of bounded R_B guarantees irreflexivity of $e(R_B)$ in **Soset$_b$↓**. □

Proof of Theorem 24: The empty set ordered by the empty relation is an initial object in all above-listed categories. A binary coproduct of $\langle A, R_A \rangle$ and $\langle B, R_B \rangle$ is $\langle A \uplus B, R_A \uplus R_B \rangle$. Other finite coproducts are defined in the same way. The infinite coproducts do not exist in **Soset$_b$↓**, because the resulting structure may be not bounded. □

Lemma 52. *If $f \colon \mathbf{U}(A_<) \rightarrow \mathbf{Symb}(\Sigma_<)$ is a* **Soset$_b$↓***-morphism, i.e. $f \colon A_< \rightarrow$* **DepSymb**$(\Sigma_<)$, *then $\xi_f \colon$* **DepSymb**$(\Sigma_<^f) \rightarrow$ **DepSymb**$(\Sigma_<)$ *(from Lemma 30) is a well defined* **AlgSigDep**-*morphism.*

Proof of Lemma 30: Let $\Sigma^f = \langle S^f, \Omega_{S^f+}^f \rangle$, the **AlgSig**-morphism $\xi_f = \langle \xi_S^f, \Omega_{S^f}^f \rangle \colon \Sigma^f \rightarrow \Sigma$ is defined as $\xi_S^f = f|_{S^f}$ and for every $e \in S^{f+}$, $\xi_{\Omega_e}^f = f|_{\Omega_e^f}$. For a **Soset$_b$↓**-morphism $g \colon$ **DepSymb**$(\Sigma_<') \rightarrow A_<$ and **AlgSig**-morphism $\gamma \colon \Sigma' \rightarrow \Sigma$ such that $\mathbf{U}(g); f = \mathbf{SetSymb}(\gamma)$ there exists an **AlgSig**-morphism

$\sigma_g^f = \langle \sigma_s, \sigma_{\Omega'_{S'}} \rangle \colon \Sigma' \to \Sigma^f$ where $\sigma_s \colon S' \to S^f$ and $\sigma_{\Omega'_{S'}} \colon \Omega_{S'+} \to \Omega_{S^f+}$ are given as $\sigma_s(s) = g(s)$, for $s \in S'$, $\sigma_{\Omega_e}(o) = g(o : e)$, for $e \in S'^+$ and $o \in \Omega'_e$. The uniqueness of σ_g^f comes by construction.   □

Proof of Lemma 33: If f is injective it is easy to check that by definition (cf. the proof of Lemma 30) σ_g^f is an **AlgSigDep** morphism. Otherwise, we use the universality of the construction of $Q_<$ and $l \colon Q_< \to$ **DepSymb**$(\Sigma_<^f)$ from Lemma 11.   □

Proof of Theorem 35: The category **Soset$_b\!\downarrow$** is embeddable into **AlgSigDep** as a full subcategory (of sort-symbols-only signatures); thus, by Theorem 15, the final object does not exist.   □

Proof of Theorem 36: Given two **AlgSigDep**-morphisms $f, g \colon \Sigma_<^A \to \Sigma_<^B$, let $e \colon C_< \to$ **DepSymb**$(\Sigma_<^A)$ be the equalizer of

$$\mathbf{DepSymb}(f), \mathbf{DepSymb}(g) \colon \mathbf{DepSymb}(\Sigma_<^A) \to \mathbf{DepSymb}(\Sigma_<^B)$$

in **Soset$_b\!\downarrow$**. The equaliser of f and g in **AlgSigDep** is

$$\xi_e \colon \Sigma_<^e \to \Sigma_<^A$$

from Lemma 52. See Def. 31 for a definition of signature with dependent symbols morphism "reconstruction".   □

Proof of Theorem 38: A coequalizer of two morphisms $f, g \colon \langle \Sigma_A, <_A \rangle \to \langle \Sigma_B, <_B \rangle$ in **AlgSigDep** is $e \colon \langle \Sigma_B, <_B \rangle \to \langle \Sigma_C, <_C \rangle$, where e is the coequalizer of f and g in **AlgSig** and the strict order $<_C =$ **Symb**$(e)(<_B)$ (cf. Theorem 13). The initial object in **AlgSigDep** is the empty signature with the empty relation. Binary coproducts in **AlgSigDep** are binary coproducts in **AlgSig** ordered by the union of the component orders. Other finite coproducts are defined in the same way.   □

Compositional Modelling and Reasoning in an Institution for Processes and Data

Liam O'Reilly[1], Till Mossakowski[2], and Markus Roggenbach[1]

[1] Swansea University, Wales, UK
[2] DFKI GmbH Bremen, Bremen, Germany

Abstract. The language CSP-CASL combines specifications of data and processes. We give an institution based semantics to CSP-CASL that allows us to re-use the institution independent structuring mechanisms of CASL. Furthermore, we extend CSP-CASL with a notion of refinement that reconciles the differing philosophies behind the refinement notions for CSP and CASL. We develop a compositional proof calculus for refinement along the CASL structuring mechanisms, and demonstrate that compositional proof techniques along parallel process composition from the context of CSP lifts to structured CSP-CASL specifications.

1 Introduction

Distributed computer applications like flight booking systems, web services, and electronic payment systems such as the EP2 standard [1] involve the parallel processing of data. Consequently, these systems exhibit concurrent aspects (e.g., deadlock freedom) as well as data aspects (e.g., functional correctness). Often, these aspects depend on each other. In [20], we present the language CSP-CASL, which is tailored to the specification of distributed systems. CSP-CASL integrates the process algebra CSP [10, 21] with the algebraic specification language CASL [15].

In [8] we apply CSP-CASL to the EP2 standard and demonstrate that CSP-CASL can deal with problems of industrial strength. Interestingly enough, CSP alone is not expressive enough to model the EP2 standard: The abstract system descriptions included in this standard require loose semantics of data. However, the exercise in [8] also demonstrates the need to enrich CSP-CASL by means for specification in the large: While the CASL structuring mechanisms are available for data to be plugged into a CSP-CASL specification, this has yet no counterpart on the process side.

Based on an institution for CSP [14], here we extend this language by loose processes and give it an institution-based semantics. The institutional setting [9] allows for specifications with loosely specified data and process parts. Moreover, the institution independent structuring mechanisms of CASL can be applied in the process algebraic setting in a methodologically meaningful way.

Furthermore, we study refinement in the context of CSP-CASL. Refinement in CASL is usually reduced to simple model class inclusion, given the power of the

T. Mossakowski and H.-J. Kreowski (Eds.): WADT 2010, LNCS 7137, pp. 251–269, 2012.

CASL structuring mechanisms that can be used to massage the involved specifications if needed. We show that a similar approach can be used for capturing CSP's traditional notion of refinement also in the setting of loose semantics. Moreover, we show that reasoning about refinements can be done in a modular way, using the CASL structuring mechanisms.

CSP-CASL also inherits from the process algebraic side: For CSP, [18] presents a compositional approach for deadlock analysis on networks of processes. We lift this technique to CSP-CASL, and show by means of an extended example, how to use it in combination with the structuring constructs inherited from CASL.

To the best of our knowledge, we were the first to suggest loose process specifications in [14]. Here, we combine this idea with loose data specifications. Accordingly, our notion of refinement for loose data and processes is new as well. Other approaches of combining data and processes, e.g., CSP-M [23], CSP-Z [7], and CSP-OZ [24], use tight semantics of both data and processes and provide only limited structuring. The WRIGHT architectural description language [2] allows reasoning on typed processes for a sublanguage of CSP; semantically, it is restricted to a single CSP model. Moreover, WRIGHT does not cover data refinement. Temporal logics offer a declarative approach to the specification of reactive behaviour. Here, [25] studies structuring of reactive systems using CASL architectural specifications over an institution of transition systems and CTL^* formulae. This again differs from our work, as we consider structured specification with loose semantics (classes of models), whereas architectural specifications focus on the structuring of individual models. In other reactive CASL extensions, e.g., MODALCASL [12] or CASL-LTL [19], the concept of refinement and its interaction with structuring has not been studied yet.

Our paper is organised as follows: In Section 2 we motivate our notion of "loose processes". Then we develop, to some extent, institutions for CSP-CASL: one institution for each of the main CSP models, namely the CSP traces model, the CSP failures/divergences model, and the CSP stable failures model. Section 4 defines CSP-CASL refinement and gives compositional proof rules along the CASL structuring mechanisms. Then we discuss how to lift a compositional deadlock analysis rule from the CSP context to CSP-CASL. We conclude the paper with an extended example.

We assume that the reader is familiar with CSP ([10, 21] provide introductions) and with CASL ([4] is a gentle introduction). Moreover, we use the notion of institutions [9] as a formalisation of the notion of logical system. The reader unfamiliar with institutions should be able to understand most parts of this paper when replacing the word "institution" by "logical system".

2   Loose Process Semantics

CSP-CASL [20] is a novel specification language which combines *processes* written in CSP [10, 21] with the specification of *data types* in CASL [15]. The general idea is to describe reactive systems in the form of processes based on CSP operators, where the communications of these processes are the values of data types, which are loosely specified in CASL. All standard CSP operators are available, such as multiple prefix, the various parallel operators, operators for non-deterministic choice, and communication over channels. Concerning CASL features, the full language is available to specify data types, namely many-sorted first order logic with sort-generation constraints, partiality, and subsorting.

CSP-CASL supports the three main CSP semantics: The traces model $\mathcal{T}$, in which one can verify safety properties; the failures/divergences model $\mathcal{N}$, which allows one to study the phenomenon of livelock, i.e., the possibility that the system can indefinitely engage in internal actions only; and the stable failures model $\mathcal{F}$, which is best suited for deadlock analysis. The traces model $\mathcal{T}$ records only the possible traces of a process; the failures/divergences model records two different behaviours: The failures – i.e., action sets which a process can refuse after executing a trace – and the divergences – i.e., traces that lead to a livelock; the stable failures model, finally, records two behaviours: The system traces exactly like the traces model, and the failures for "stable" states, i.e., states which can't perform an internal action. The main means of verification in CSP is to prove that one process, say P, refines to another one, say Q, in signs $P \sqsubseteq Q$. Each CSP model gives rise to one notion of refinement. Here, the following relations have been established: $\sqsubseteq_{\mathcal{N}} \subsetneq \sqsubseteq_{\mathcal{T}}$, $\sqsubseteq_{\mathcal{F}} \subsetneq \sqsubseteq_{\mathcal{T}}$, $\sqsubseteq_{\mathcal{N}} \not\subseteq \sqsubseteq_{\mathcal{F}}$, and $\sqsubseteq_{\mathcal{F}} \not\subseteq \sqsubseteq_{\mathcal{N}}$, see [21].

In this paper, we extend the setting of CSP-CASL as defined in [20] by adding loose semantics for processes, following the ideas of [14]. Loose process semantics offers advantages in terms of methodology, furthermore, it is required for generic specifications and instantiation.

For the methodological aspect, consider the specification ARCH_CUSTOMER of the customer of an electronic shop, see Figure 1 – in the context of our example, to be discussed in more detail in Section 6 – we call this the "architectural level". The data part written in CASL provides a type system, namely that $LoginReq$ ("Login Request") and $Logout$ are subsorts of D_C ("Data Customer"), which comprises of all data the customer can deal with. The customer communicates to the outside world over a channel C_C ("Channel Customer"), which allows for messages of type D_C. The suffix $\_def$ on sort names excludes the "error" element of the sort, i.e., we are specifying the system under the assumption that only valid messages are exchanged.[1] In the process part, the customer's behaviour is described in terms of several processes, devoted to different activities. The purpose of the architectural level is to describe how to combine these activities in order to describe the customer. The detailed description of such an activity, e.g., $Customer\_GoodLogin$, however, is postponed to a later design step. We only state that there is a process $Customer\_GoodLogin$, whose behaviour is underspecified, i.e., in semantical terms it is "loose".

With such loose specifications available for the customer, warehouse, payment system, and the coordinator, we can model the whole shop as their parallel composition over various channels, see the process part of the specification ARCH_SHOP. Here, the specifications ARCH_CUSTOMER, ARCH_WAREHOUSE,

---

[1] For simplicity we refrain from error handling.

ARCH_PAYMENTSYSTEM and ARCH_COORDINATOR serve as parameters in a generic construction. They provide the names and properties of data and processes involved. But what instances do we want to allow? Obviously, any refinement of these parameters shall be possible. To this end, we define an operator $RefCl$ – to be discussed in detail in Section 4 – which closes the model class of a CSP-CASL specification under refinement.

Loose or underspecified processes differ from non-deterministic processes in CSP. The process $P = a \rightarrow Stop \sqcap b \rightarrow Stop$ is non deterministic. For this equation, there is only *one* denotation possible for P which makes the equation true. In the traces model $\mathcal{T}$, e.g., this is the interpretation $I(P) = \{\langle\rangle, \langle a\rangle, \langle b\rangle\}$. Specifying a loose process Q by saying Q shall be any process that refines to P, written $Q \sqsubseteq P$, however, leads to infinitely many different possible denotations of Q. In the traces model $\mathcal{T}$, e.g., we have the interpretations $J(Q) = I(P)$ and $K(Q) = \{\langle\rangle, \langle a\rangle, \langle a, a\rangle, \langle b\rangle\}$. Note that this example also demonstrates that the set of interpretations of loose processes is not necessarily refinement closed: $\{\langle\rangle\}$ is not a possible interpretation for Q, however it is an element of every refinement closed set in $\mathcal{T}$.

> **spec** ARCH_CUSTOMER =
>   **data sorts** *LoginReq, Logout* < *D_C*
>   **channel** *C_C* : *D_C*
>   **process** *Customer* : *C_C* ; *Customer_GoodLogin* : *C_C* ;
>   *Customer_BadLogin* : *C_C* ; *Customer_AddItem* : *C_C* ;
>   *Customer_Body* : *C_C* ; *Customer_Quit* : *C_C* ; ...
>   *Customer* = *C_C* ! *x* :: *LoginReq_def* →
>   (*Customer_GoodLogin* ; *Customer_Body* □
>   *Customer_BadLogin* ; *Customer*)
>   *Customer_Quit* = *C_C* ! *x* :: *Logout_def* → *SKIP*
>   *Customer_Body* = *Customer_AddItem* ⊓ ... ⊓ *Customer_Quit*
> **end**

> **spec** ARCH_SHOP [*RefCl*(ARCH_CUSTOMER)] [*RefCl*(ARCH_WAREHOUSE)]
>   [*RefCl*(ARCH_PAYMENTSYSTEM)] [*RefCl*(ARCH_COORDINATOR)] =
>   **process** *System* : *C_C, C_W, C_PS* ;
>   *System* = *Coordinator* ⟦ *C_C, C_W, C_PS* ‖ *C_C, C_W, C_PS* ⟧
>   (*Customer* ⟦ *C_C* ‖ *C_W, C_PS* ⟧
>   (*Warehouse* ⟦ *C_W* ‖ *C_PS* ⟧ *PaymentSystem*))
> **end**

Fig. 1. Selections of CSP-CASL specifications of our online shop example

3 CSP-CASL Institutions for Different CSP Models

In order to give a precise semantics to (possibly structured) CSP-CASL specifications, we formalise CSP-CASL as an institution [9]; to be more precise: three institutions – one for each of the main CSP models, namely: the CSP traces model, the CSP failures/divergences model, and the CSP stable failures model.

These institutions share the notions of signatures and sentences. Their respective model categories and satisfaction relations are defined following a common scheme. We only sketch the institutions, for full details see [16]. The institutions for CSP-CASL are naturally based on institutions for CASL [15] and for CSP [14], using the ideas for the CSP-CASL semantics [20] for the combination.

3.1 Signatures

A CSP-CASL signature Σ_{CC} is a pair $\Sigma_{CC} = (\Sigma_{Data}, \Sigma_{Proc})$ where:

- $\Sigma_{Data} = (S, TF, PF, P, \leq)$ is a subsorted first-order signature consisting of a set of sort symbols S, a set of total functions symbols TF, a set of partial function symbols PF, a set of predicate symbols P, and a reflexive and transitive subsort relation $\leq \subseteq S \times S$ – see [15] for details – where the set of sorts S is finite and the subsort relation has local top elements, i.e., if $u, u' \geq s$ then there exists $t \in S$ with $t \geq u, u'$, see [20].
- $\Sigma_{Proc} = (N_{w,comms})_{w \in S^*, comms \in S_\downarrow}$ is a family of finite sets of process names. Such a process name n is typed in the sort symbols S of the data signature part:
  - a string $w = \langle s_1, \ldots, s_k \rangle$, $s_i \in S$ for $1 \leq i \leq k$, $k \geq 0$, which is n's parameter type. A process name without parameters has the empty sequence $\langle \rangle$ as its parameter type.
  - a set $comms \subseteq S$ which collects all types of events in which the process n can possibly engage in. We require the set $comms$ to be downward closed under the subsort relation, i.e., $comms \in S_\downarrow = \{X \subseteq S \mid X = \downarrow X\}$, where $\downarrow X = \{y \in S \mid \exists x \in X : y \leq x\}$ for $X \subseteq S$.

Given CSP-CASL signatures $\Sigma_{CC} = (\Sigma_{Data}, \Sigma_{Proc})$, $\Sigma'_{CC} = (\Sigma'_{Data}, \Sigma'_{Proc})$, with S as the sort set of Σ_{Data} and S' as the sort set of Σ'_{Data}, a CSP-CASL signature morphism is a pair $\theta = (\sigma, \nu) : \Sigma_{CC} \to \Sigma'_{CC}$ where:

- $\sigma : \Sigma_{Data} \to \Sigma'_{Data}$ is a CASL signature morphism for which the following additionally hold:
  refl $\sigma^S(s_1) \leq_{S'} \sigma^S(s_2)$ implies $s_1 \leq_S s_2$ for all $s_1, s_2 \in S$
  (reflection of the subsort relation), and
  weak non-extension $\sigma^S(s_1) \leq_{S'} u'$ and $\sigma^S(s_2) \leq_{S'} u'$ implies that there exists a sort $t \in S$ with $s_1 \leq_S t$, $s_2 \leq_S t$ and $\sigma^S(t) \leq_{S'} u'$.[2]
- $\nu = (\nu_{w,comms})_{w \in S^*, comms \in S_\downarrow}$ is a family of functions such that $\nu_{w,comms} : N_{w,comms} \to \bigcup_{comms' \in (\downarrow(\sigma(comms)))_\downarrow} N'_{\sigma(w),comms'}$ is a mapping of process names. Another way to express this is that a process name $n \in N_{w,comms}$ is mapped to $\nu_{w,comms}(n) = n'$, where $n' \in N'_{\sigma(w),comms'}$ and $\forall y \in comms'$: $\exists x \in comms : y \leq_{S'} \sigma(x)$ ("the target is dominated by the source"). We also write $\nu(n : w, comms) = n' : \sigma(w), comms'$.

---

[2] [20] works with the condition 'non-extension'. One can show however that the results of [20] also hold with the more liberal notion that we use here. The difference from the original version is the more liberal choice of sort t (originally, we have required t to be a pre-image of u'). Further note that for $s_1 = s_2$, the condition trivially holds.

The conditions on process translations ensure that both the parameter types as well as the communication set, are translated with the signature morphism σ of the data part. While preserving the parameter structure, the communication set is allowed to "shrink". This "non-expansion" of the communication sets – see also [14] – guarantees that the reduct functor is defined and that the satisfaction condition holds. In [16] we demonstrate that our notion of a signature morphisms is quite liberal. All morphisms arising in our shop example fulfil these conditions.

3.2   Sentences

Sentences are either data or process sentences. A Σ_{CC}-data sentence is a CASL sentence over $(S, TF, PF, P, \leq)$. A Σ_{CC}-process sentence is a process definition

$$n(x_1, \ldots, x_k) = pt$$

where $n \in N_{\langle s_1, \ldots, s_k \rangle, comms}$, x_i are global variables of type s_i, $1 \leq i \leq k$, and pt is a process term such that $sorts(pt) \subseteq comms$, i.e., the process term pt communicates only in events which are allowed for n. For further details see [16].

3.3   The Alphabet Construction

The alphabet construction takes a data (i.e., CASL) model and uses its elements as alphabet letters, which then form the alphabet for CSP. CSP-CASL's alphabet construction takes the subsort structure into account in order to determine whether two events are equal or not. More precisely, given a CASL model M, its corresponding alphabet

$$Alph(M) = (\biguplus_{s \in S} M_s \cup \{\bot_s\})/{\sim_M}$$

is constructed by taking the disjoint union of all its carrier sets extended by a bottom element $\bot$, but identifying carriers along subsort injections (this is captured by the equivalence relation $\sim_M$). This map $Alph$ extends to a functor from the model category to the category Set.

Given a CASL model M, we use the shorthand $M_\bot$ for the totalised version of M, i.e., carrier sets include a bottom element $M_\bot(s) = M(s) \cup \{\bot_s\}$ and the interpretation of function and predicate symbols is strictly extended. Given a sort symbol s, a CASL model M, and $x \in M_\bot(s)$ we write $\overline{x}_M^s$ to denote the alphabet element $[(s, x)]_{\sim_M}$. Further more we lift this notation to sorts, namely $\overline{s}_M = \{\overline{x}_M^s \mid x \in M_\bot(s)\} \subseteq Alph(M)$ for the set of communications that can arise from the sort s in the model M. Finally, given a set of sorts X, we write $\overline{X}_M = \bigcup_{s \in X} \overline{s}_M$. We drop the subscripts M and superscripts s when clear from the context.

3.4   Models and Satisfaction

A CSP-CASL model consists of a data (i.e., CASL) model and a collection of interpretations for processes. Concerning the interpretation of processes, let $\mathcal{D}(A)$

be a CSP domain constructed relatively to some alphabet of communications A. Examples of such domains $\mathcal{D}(A)$ include $\mathcal{T}(A)$ of the CSP traces model, $\mathcal{N}(A)$ of the CSP failures/divergences model, and $\mathcal{F}(A)$ of the CSP stable failures model, see [21] for details. Each of these domains gives rise to a different institution. Actually, $\mathcal{D}$ extends to an endofunctor on the category Set.

Given a CSP-CASL signature $\Sigma_{CC} = (\Sigma_{Data}, \Sigma_{Proc})$, a Σ_{CC}-model is a pair (M, I), where M is a CASL model for Σ_{Data} and I gives type correct interpretations of the process signature in the CSP domain $\mathcal{D}(Alph(M))$. All CSP models describe, which traces a process can execute. In the following we denote these traces with the function cTr.[3] Type correctness of (M, I) requires that the interpretation map I applied to a process name $n \in N_{\langle s_1, \ldots, s_k \rangle, comms}$ for all parameters $a_i \in \overline{s_i}, 1 \leq i \leq k$, yields an interpretation with $cTr(I(n(a_1, \ldots, a_k))) \in \mathcal{T}(\overline{comms})$. It is this type correctness condition which allows us to define the reduct functor and to prove the satisfaction condition.

Satisfaction of data sentences w.r.t. a Σ_{CC}-model is inherited from CASL. Satisfaction of a process sentence $n(x_1, \ldots, x_k) = pt$ over signature Σ_{CC} and global variable system X_G with respect to a Σ_{CC}-model (M, I) is defined as follows:

$$(M, I) \models_{\Sigma_{CC}} (n(x_1, \ldots, x_k) = pt)$$
$$\text{if and only if}$$
$$\text{for all variable valuations } \mu_G : X_G \to M_\perp :$$
$$I(n(\overline{\mu_G(x_1)}_M, \ldots, \overline{\mu_G(x_k)}_M)) = [\![pt]\!]_{(M,I),\mu_G,\emptyset}\mathcal{D}.$$

Here, $[\![pt]\!]_{(M,I),\mu_G,\mu_L}$ is the evaluation of process term pt according to CASL with respect to model (M, I) and global and local variable valuations μ_G and μ_L. $[\![pt']\!]_\mathcal{D}$ is the denotation of process term pt' in the CSP domain $\mathcal{D}$. For further details and also the definition and discussion of model morphisms see [16].

3.5 Pushouts and Amalgamation

The existence of pushouts and amalgamation properties shows that an institution has good modularity properties. Amalgamation is a major technical assumption in the study of specification semantics [6, 22]. An institution is said to be *semi-exact*, if for any pushout of signatures

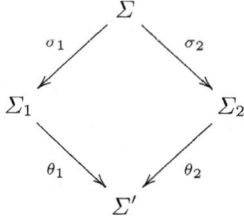

---

[3] The controlled traces are the traces as given as denotations in the traces model – in $\mathcal{F}$, they are directly given, in $\mathcal{N}$, they can be computed out of the divergences and failures.

any pair $(M_1, M_2) \in \mathbf{Mod}(\Sigma_1) \times \mathbf{Mod}(\Sigma_2)$ that is *compatible*, in the sense that M_1 and M_2 reduce to the same Σ-model, can be *amalgamated* to a unique Σ'-model M (i.e., there exists a unique $M \in \mathbf{Mod}(\Sigma')$ that reduces to M_1 and M_2, respectively), and similarly for model morphisms.

Proposition 1. CSP-CASL *signature morphisms between signatures with acyclic subsort relations are injective on sorts. Thus, CspCaslSig does not have pushouts.*

As in [14], there is a way out: Let $CspCaslSig^{plain}$ be $CspCaslSig$ with the reflection and weak non-extension restriction dropped. Then we have:

Proposition 2. $CspCaslSig^{plain}$ *has pushouts, and any such pushout of a span in $CspCaslSig$ actually is a square in $CspCaslSig$ (although not a pushout in $CspCaslSig$).*

Pushouts in $CspCaslSig^{plain}$ give us an amalgamation property:

Proposition 3. $CspCaslSig^{plain}$*-pushouts of $CspCaslSig$-morphisms have the semi-exactness property for the traces model and the stable failures model.*

In fact, this result generalises easily to multiple pushouts. Moreover, the initial (i.e., empty) signature has the terminal model category. Since all colimits can be formed by the initial object and multiple pushouts, this shows that we even have exactness (when colimits are taken in $CspCaslSig^{plain}$).

Altogether, Proposition 3 shows that CASL-style parameterisation, CASL architectural specifications and much more also work for CSP-CASL.

3.6   CSP-CASL with Channels

We often use channels in CSP-CASL. This leads to further institutions, with extended notions of signatures and sentences. Most prominently, the notion of a signature is extended by a third component C:

$$(\Sigma_{Data}, C, \Sigma_{Proc})$$

Here, C is a finite set of names typed by non-empty lists over S. We require C to be closed under the subsort relation[4] $\leq^*$ i.e., if $c : \langle s_1, \ldots, s_k \rangle \in C$ and $\langle u_1, \ldots, u_k \rangle \leq^* \langle s_1, \ldots, s_k \rangle$ then $c : \langle u_1, \ldots, u_k \rangle \in C$.

CSP-CASL with channels can be reduced to CSP-CASL (without channels) as follows: each CSP-CASL signature with a channel component is translated to a CSP-CASL theory $\Phi(\Sigma)$, where each channel is coded as a new sort (isomorphic to the sort of the channel) and each CSP-CASL Σ-sentence φ is translated to a CSP-CASL $\Phi(\Sigma)$-sentence $\alpha(\varphi)$ by reducing channel communication to ordinary communication using the new channel sorts. Models and satisfaction can then be easily borrowed from CSP-CASL by letting $\mathbf{Mod}^{CC}(\Sigma) := \mathbf{Mod}^{CC}(\Phi(\Sigma))$ and $M \models_{\Sigma}^{CC} \varphi$ iff $M \models_{\Phi(\Sigma)}^{CC} \alpha(\varphi)$. This is an instance of borrowing logical structure in the sense of [5].

In the rest of the paper we use the term CSP-CASL to represent CSP-CASL with channels.

---

[4] $\leq^*$ stands for the pointwise extension of $\leq$ to strings of sorts.

4   Refinement and a Structured Proof Calculus

Refinement allows us to develop systems in a stepwise manner. Here we discuss refinement for CSP-CASL as well as its proof calculus on structured specifications.

4.1   Refinement

CSP has a notion of refinement between individual processes, e.g., in the traces model, $pt \sqsubseteq pt'$ means that pt' has fewer traces than pt, i.e., $traces(pt') \subseteq traces(pt)$. In the context of this paper we write $pt \sqsubseteq pt'$ for $pt \sqsubseteq_\mathcal{D} pt'$ if the specific choice of $\mathcal{D} \in \{\mathcal{T}, \mathcal{N}, \mathcal{F}\}$ does not matter. Similarly, the CASL family of languages uses model class inclusion as the simplest notion of refinement [3]: $SP_1 \rightsquigarrow SP_2$ if SP_2 has fewer models than SP_1, i.e., $\mathbf{Mod}(SP_2) \subseteq \mathbf{Mod}(SP_1)$. To cater for renaming, this notion can be extended by a signature morphism σ. In this case one defines $SP_1 \rightsquigarrow^\sigma SP_2$ if the reduct of SP_2 has fewer models than SP_1, i.e., $\mathbf{Mod}(SP_2)|_\sigma \subseteq \mathbf{Mod}(SP_1)$. When combining these worlds through institution theory, one has to recognise that these two refinement notions follow different ideas: While CSP refinement talks about refinement of individual models, CASL refinement talks about refinement of model *classes*.

This should become clear with the following notion: A CSP-CASL specification SP is *single-valued*, if there is no looseness in the processes, that is, any two SP-models with the same data part coincide. Now, traditional CSP refinement is about refinement between different single-valued process specifications – reducing the amount of internal non-determinism – whereas model class inclusion mainly captures different degrees of *looseness* of specifications.

How can we reconcile these two worlds? Here, we want to capture different degrees of looseness not only for data, but also for processes! Hence we adopt the model class inclusion notion of refinement, applied to the CSP-CASL institution. However, in order to capture CSP refinement between different single-valued processes (which alone, would obviously never lead to model class inclusion), we also provide a notion of refinement closure (and here, "refinement" is meant in the CSP sense, not in the model class inclusion sense).

Given a CSP-CASL specification SP with signature $(\Sigma_{Data}, \Sigma_{Proc})$, its refinement closure $RefCl(SP)$ is defined as follows:

− the signature of $RefCl(SP)$ is that of SP,
− the model class of $RefCl(SP)$ consists of those CSP-CASL models (M', I') for which there exists a model (M, I) of SP such that
  • $M = M'$, i.e., they have the same data part,
  • for each $n \in \Sigma_{Proc}$ and all suitable data elements $a_1, \ldots, a_k$,

$$I(n(a_1, \ldots, a_k)) \sqsubseteq I'(n(a_1, \ldots, a_k))$$

  in the sense of CSP.

Alternatively, the semantics of $RefCl(SP)$ can be expressed as a structured specification

$$SP \text{ then } p_1 \sqsubseteq q_1, \ldots, p_n \sqsubseteq q_n \text{ hide } p_1, \ldots, p_n \text{ with } q_1 \mapsto p_1, \ldots, q_n \mapsto p_n$$

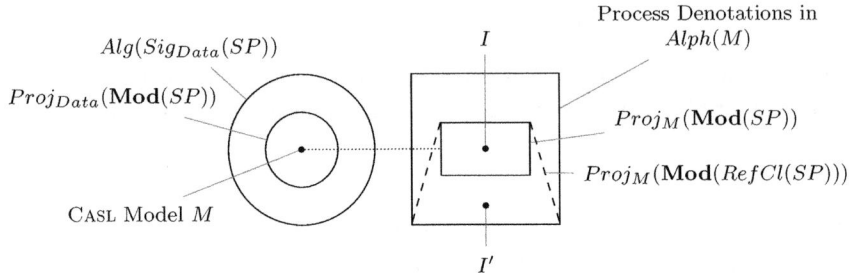

Fig. 2. A diagram showing refinement between CSP-CASL models

where $p_1, \ldots, p_n$ are the process names of SP (we assume here that all of them are unparameterised), $q_1, \ldots, q_n$ are new process names, and $p \sqsubseteq q$ stands for $p = p \sqcap q$.

Figure 2 depicts the notion of refinement closure. Given a model M of the data part of SP, we consider all of its possible "partners" relative to SP: $Proj_M(\mathbf{Mod}(SP)) = \{I \mid (M, I) \in \mathbf{Mod}(SP)\}$ – this is represented by the rectangle. The refinement closure includes all I' such that there exists some $I \in Proj_M(\mathbf{Mod}(SP))$ that refines to I'.

With this notion, we are ready to define a notion of refinement that is suitable for CSP-CASL:

$$SP_1 \leadsto^\theta_{\mathcal{D}} SP_2 \text{ iff } \mathbf{Mod}_{\mathcal{D}}(SP_2)|_\theta \subseteq \mathbf{Mod}_{\mathcal{D}}(RefCl(SP_1))$$

for $\mathcal{D} \in \{\mathcal{T}, \mathcal{N}, \mathcal{F}\}$. We omit $\mathcal{D}$ if it is clear from the context and we also omit θ if it is the identity signature morphism. We write $\_|_\theta$ to denote the CSP-CASL reduct functor. This notion reconciles CASL refinement based on model class inclusion with CSP refinement based on inclusion of trace sets, failure sets, etc. Two specifications SP_1 and SP_2 are *equivalent*, written $SP_1 \equiv SP_2$, if their signatures and model classes coincide.

Proposition 4 (Basic Refinement Properties).

1. *RefCl is monotonic, that is: if* $\mathbf{Mod}(SP_1) \subseteq \mathbf{Mod}(SP_2)$,
   then $\mathbf{Mod}(RefCl(SP_1)) \subseteq \mathbf{Mod}(RefCl(SP_2))$.
2. *RefCl is idempotent, that is* $RefCl(SP) \equiv RefCl(RefCl(SP))$.
3. $\leadsto$ *is reflexive and transitive.*
4. *If* $SP_1 \leadsto SP_2$ *and* $SP_2 \leadsto SP_1$, *then* $RefCl(SP_1) \equiv RefCl(SP_2)$.
5. *If* $SP_1 \leadsto SP_2$, $SP_2 \leadsto SP_1$, *and both are single-valued, then* $SP_1 \equiv SP_2$.

Following ideas given in [11] we obtain a decomposition theorem for basic (or unstructured) specifications. This allows us to (syntactically) decompose a basic CSP-CASL specification[5] SP into a data part (D) and a process part (P), which we shortly write as (D, P).

---

[5] Such a specification may have a structured CASL specification as the data part D.

Proposition 5 (Decomposition).

$$\frac{\mathbf{Mod}(D')|_\sigma \subseteq \mathbf{Mod}(D) \qquad (D',\theta(P)) \rightsquigarrow_{\mathcal{D}} (D',P')}{(D,P) \rightsquigarrow_{\mathcal{D}}^{\theta} (D',P')}$$

where $\theta = (\sigma,\nu)$ is a CSP-CASL *signature morphism, and $\mathcal{D} \in \{\mathcal{T},\mathcal{N},\mathcal{F}\}$.*

The above proposition allows us to decompose a CSP-CASL refinement to a CASL refinement (i.e., $\mathbf{Mod}(D')|_\sigma \subseteq \mathbf{Mod}(D)$) and a process refinement $(D',\theta(P)) \rightsquigarrow_{\mathcal{D}}$ (D',P'). Here, $\theta(P)$ is the renaming of the process part along θ. The former proof obligation can be discharged using CASL's proof tool, namely HETS [13]. The latter can be proven using the tool CSP-CASL-Prover [17].

4.2   Compositional Proof Rules along Structuring

The results of Section 3 allow us to re-use institution independent structuring operations of CASL [15, 22], which are defined in terms of signatures and models:

Presentations: For any CSP-CASL signature Σ_{CC} and finite set $\Gamma \subseteq \mathbf{Sen}(\Sigma_{CC})$ of Σ_{CC}-sentences, the presentation $\langle \Sigma_{CC},\Gamma \rangle$ is a specification with:
  $\mathbf{Sig}(\langle \Sigma_{CC},\Gamma \rangle) \quad := \Sigma_{CC}$
  $\mathbf{Mod}(\langle \Sigma_{CC},\Gamma \rangle) := \{(M,I) \in \mathbf{Mod}(\Sigma_{CC}) \mid (M,I) \models \Gamma\}$

Union: For any CSP-CASL signature Σ_{CC} and any Σ_{CC}-specifications SP_1 and SP_2, their union SP_1 **and** SP_2 is the specification with:
  $\mathbf{Sig}(SP_1 \text{ and } SP_2) \quad := \Sigma_{CC}$
  $\mathbf{Mod}(SP_1 \text{ and } SP_2) := \mathbf{Mod}(SP_1) \cap \mathbf{Mod}(SP_2)$

Translation: For any signature morphism $\theta : \Sigma_{CC} \to \Sigma'_{CC}$ and Σ_{CC}-specification SP, SP **rename** θ is the specification with:
  $\mathbf{Sig}(SP \text{ rename } \theta) \quad := \Sigma'_{CC}$
  $\mathbf{Mod}(SP \text{ rename } \theta) := \{(M',I') \in Mod(\Sigma'_{CC}) \mid (M',I')|_\theta \in \mathbf{Mod}(SP)\}$

Hiding: For any signature morphism $\theta : \Sigma_{CC} \to \Sigma'_{CC}$ and Σ'_{CC}-specification SP', SP' **hide** θ is the specification with:
  $\mathbf{Sig}(SP' \text{ hide } \theta) \quad := \Sigma_{CC}$
  $\mathbf{Mod}(SP' \text{ hide } \theta) := \{(M',I')|_\theta \mid (M',I') \in Mod(SP')\}$

As a first proof of concept, we show that the specification building operators are monotonic w.r.t. the structuring operations, cf. [3]. This requires, in our case, certain side conditions, most prominently for the structured union operation on specifications. Here, the conditions deal with the following non-monotonic situation of CSP-CASL refinement: There exist CSP-CASL specifications SP_1, SP'_1 and SP_2 with[6]

$$SP_1 \rightsquigarrow SP'_1, \mathbf{Mod}(SP_1 \text{ and } SP_2) = \emptyset, \mathbf{Mod}(SP'_1 \text{ and } SP_2) \neq \emptyset.$$

---

[6] Consider over the traces model $SP_1 = (D, P = a \to Stop)$, $SP_2 = (D, P = Stop)$, and $SP'_1 = (D, P = Stop)$ where D is a consistent CASL specification that declares a constant a. Then SP_1 **and** SP_2 is inconsistent, $SP_1 \rightsquigarrow_{\mathcal{T}} SP'_1$, and SP'_1 **and** SP_2 has models (M,I) with $I(P) = \{\langle\rangle\}$ and $M \in \mathbf{Mod}(D)$.

Definition 6. *Two* CSP-CASL *specifications* SP_1 *and* SP_2 *over the same signature are* process consistent, *written as proc-consistent(SP_1, SP_2) if for all* $M \in (Proj_{Data}(\mathbf{Mod}(SP_1)) \cap Proj_{Data}(\mathbf{Mod}(SP_2)))$, *there exists* $(M, J) \in \mathbf{Mod}(SP_1) \cap \mathbf{Mod}(SP_2)$.

Proposition 7. *The following proof rules[7] are sound over* $\mathcal{T}$, $\mathcal{N}$, *and* $\mathcal{F}$:

$$\frac{SP_1 \rightsquigarrow SP_1' \quad proc\text{-}consistent(SP_1, SP_2) \quad single\text{-}valued\ (SP_i)\ for\ i = 1 \vee i = 2}{(SP_1\ \mathbf{and}\ SP_2) \rightsquigarrow (SP_1'\ \mathbf{and}\ SP_2)}$$

$$\frac{SP_1 \rightsquigarrow SP_1' \quad SP_1 \equiv RefCl(SP_1)}{(SP_1\ \mathbf{and}\ SP_2) \rightsquigarrow (SP_1'\ \mathbf{and}\ SP_2)}$$

$$\frac{SP \rightsquigarrow SP' \quad \theta\ is\ injective\ on\ \text{process names}}{(SP\ \mathbf{rename}\ \theta) \rightsquigarrow (SP'\ \mathbf{rename}\ \theta)}$$

$$\frac{SP \rightsquigarrow SP'}{(SP\ \mathbf{hide}\ \theta) \rightsquigarrow (SP'\ \mathbf{hide}\ \theta)}$$

where $\theta : \Sigma_{CC} \rightarrow \Sigma'_{CC}$.

The rules for **and** involve rather strong preconditions, where we hope that it will be possible to obtain better results in the future.

Renaming and refinement involving the same signature morphism can be exchanged:

Proposition 8. *The following implications hold:*

1. $(SP\ \mathbf{rename}\ \theta) \rightsquigarrow SP'$ *implies* $SP \rightsquigarrow^\theta SP'$.
2. *Provided that* θ *is injective on process names, we also have:*
   $SP \rightsquigarrow^\theta SP'$ *implies* $(SP\ \mathbf{rename}\ \theta) \rightsquigarrow SP'$.

5   Compositional Verification of Deadlock Freedom

Our new version of CSP-CASL extends CSP-CASL as was presented in [11] with loose processes. However, our definitions and semantical constructions coincide for single valued specifications.

The deadlock analysis presented in [11] is practically limited to dealing with a small number of processes in parallel. It involves the construction of a so-called sequential process – which has a size that is exponential in the number of parallel components involved. Here we prove deadlock freedom in a far more elegant way.

For the rest of this section, as usual for deadlock analysis in the context of CSP, we work in the stable failures model $\mathcal{F}$ only. Furthermore we assume all processes and process terms to be divergence free.

---

[7] Note that θ being injective on process names can have restrictions on the data part of the signature morphism as data forms part of the identity of process names.

5.1   Deadlock Freedom in Structured Specifications

We first define what it means for a process term to be deadlock free in the context of a specification (be it basic or structured). We then present a collection of proof rules for deadlock freedom over the structuring operators.

Definition 9 (Deadlock freedom). *Let SP be a* CSP-CASL *specification with signature* Σ_{CC}, X_G *and* X_L *be global and local variable systems respectively over* Σ_{CC}, *and let pt be a process term over signature* Σ_{CC} *with variable systems* X_G *and* X_L. *We say: pt is* deadlock free *in specification SP, written as*

$$pt \; isDFin \; SP$$

if for all models $(M, I) \in \mathbf{Mod}(SP)$, *for all variable valuations* $\mu_G : X_G \to M_\perp$ *and* $\mu_L : X_L \to M_\perp$, *and for all traces* $s \in Alph(M)^*$ *it holds that* $(s, Alph(M)^{\checkmark}) \notin failures(\llbracket pt \rrbracket_{(M,I),\mu_G,\mu_L})$.

Deadlock freedom is compatible with the structuring operations:

Proposition 10. *The following proof rules are sound:*

$$\frac{SP \rightsquigarrow_{\mathcal{F}}^{\theta} SP' \quad pt \; isDFin \; SP}{\theta(pt) \; isDFin \; SP'} \qquad \frac{pt \; isDFin \; SP_1}{pt \; isDFin \; (SP_1 \; \mathbf{and} \; SP_2)}$$

$$\frac{pt \; isDFin \; SP}{\theta(pt) \; isDFin \; (SP \; \mathbf{rename} \; \theta)} \qquad \frac{\theta(pt) \; isDFin \; SP'}{pt \; isDFin \; (SP' \; \mathbf{hide} \; \theta)}$$

where $\theta : \Sigma_{CC} \to \Sigma'_{CC}$.

The above proof rules allow one to show deadlock freedom by decomposing structured specifications. However, it may still be a difficult task to prove deadlock freedom for complex systems involving parallel processes. We describe a technique for dealing with this situation in the following section.

5.2   Composing Networks

In order to study networks of processes, we lift a definition, originally formulated over CSP in [18], to CSP-CASL. This captures the notion of processes being responsive to one and another, i.e., not causing deadlock to occur.

Definition 11. *Assume the setting of Definition 9. Let P and Q be process terms over signature* Σ_{CC} *with variable systems* X_G *and* X_L. *Let* A_P *and* A_Q *be downward and upward closed super sets of the constituent alphabet sort set of the process terms P and Q respectively (i.e.,* $sorts(P) \subseteq A_P$, $\downarrow A_P = A_P$, *and* $\uparrow A_P = A_P$, *similar for* A_Q)[8], *and let* $J = A_P \cap A_Q$ *be the set of all shared communications sorts. Let* $J' \in J_\downarrow$ *and* $X = \overline{J'} \cup \{\checkmark\}$. *Then we define:*

$$Q :: A_Q \; ResToLive^{\checkmark} \; P :: A_P \; on \; J' \; in \; SP$$

---

[8] Upward closure is defined in the obvious way: $\uparrow X = \{y \in S \mid \exists x \in X : x \leq y\}$. The condition "upward and downward closed" is required due to CASL subsorting. It ensures that the sort set J comprises all shared communications.

if for all models $(M, I) \in \mathbf{Mod}(SP)$, *all variable valuations* $\mu_G : X_G \to M_\perp$ *and* $\mu_L : X_L \to M_\perp$, *and for all traces* $s \in Alph(M)^{\checkmark}$ *it holds that*

$$(s, X) \in failures(\llbracket P \, \| \lceil J \rceil \| \, Q \rrbracket_{(M,I),\mu_G,\mu_L}) \implies (s, X) \in failures(\llbracket P \rrbracket_{(M,I),\mu_G,\mu_L})$$

In this definition, Q can be seen as a server and P as a client. The server is responsive to the client if whenever the client needs participation from the server, the server is prepared to engage in it.

A *network* is a special way of defining a process in CSP. Formally, a network V is a finite set of pairs $\{(P_i, A_i) \, | \, i \in G\}$, where G is a nonempty, finite index set, P_i is a CSP process, and $A_i \subseteq A$ is the set of communications which P_i can engage in, for all $i \in G$. The process defined by such a network V is

$$Network(V) := \; \|_{i \in G} \, (P_i, A_i)$$

where $\|_{i \in G} \, (P_i, A_i)$ is the replicated alphabetised parallel operator of CSP. As the semantics of $\|_{i \in G}$ is independent of the order of its arguments, it is sufficient to define networks over index sets. A network $Network(\{P\})$ over a single process P is equivalent to the process P itself. Note that the process *System* in Figure 1 is defined as the network consisting of the processes *Coordinator*, *Customer*, *Warehouse*, and *PaymentSystem* with suitable communication sets. Deadlock freedom of such networks can be proven in a compositional way:

Proposition 12. *Given a* CSP-CASL *specification* SP *and process terms* P_i *for* $1 \leq i \leq k$ *and a process term* Q. *Let* A_i *and* A_Q *be downward and upwards closed supersets of the constituent alphabet of* P_i *for* $1 \leq i \leq k$ *and* Q *respectively. If*

- $A_i \cap A_j \cap A_Q = \emptyset$ *for all* i *and* j *where* $1 \leq i, j \leq k$ *and* $i \neq j$,
- $A_i \cap A_Q \neq \emptyset$ *for at least one* i *where* $1 \leq i \leq k$,
- $Network(\{(P_1, A_1), \ldots, (P_k, A_k)\})$ *isDFin* SP, *and*
- $Q :: A_Q \, ResToLive^{\checkmark} \, P_i :: A_i$ *on* $(A_i \cap A_Q)$ *in* SP *for each* i *where* $1 \leq i \leq k$ *and* $A_i \cap A_Q \neq \emptyset$

then $Network(\{(P_1, A_1), \ldots, (P_k, A_k), (Q, A_Q)\})$ *isDFin* SP.

This proposition provides an elegant proof technique: The network under consideration becomes smaller; the property "responds to live" has a characterisation in terms of refinement and thus can be proven, e.g., by CSP-CASL-Prover; the conditions concerning communication alphabets can be proven algorithmically. In order to lift this proof technique to structured specifications, we provide a proof calculus with regards to the property "responds to live":

Proposition 13. *The following proof rules are sound:*

$$\frac{Q :: A_Q \, ResToLive^{\checkmark} \, P :: A_P \text{ on } J' \text{ in } SP_1}{Q :: A_Q \, ResToLive^{\checkmark} \, P :: A_P \text{ on } J' \text{ in } (SP_1 \textbf{ and } SP_2)}$$

$$\frac{Q :: A_Q \, ResToLive^{\checkmark} \, P :: A_P \text{ on } J' \text{ in } SP}{\theta(Q) :: \sigma(A_Q) \, ResToLive^{\checkmark} \, \theta(P) :: \sigma(A_P) \text{ on } \sigma(J') \text{ in } (SP \textbf{ rename } \theta)}$$

$$\frac{\theta(Q) :: \sigma(A_Q) \; ResToLive^{\checkmark} \; \theta(P) :: \sigma(A_P) \; on \; \sigma(J') \; in \; SP'}{Q :: A_Q \; ResToLive^{\checkmark} \; P :: A_P \; on \; J' \; in \; (SP' \; \mathbf{hide} \; \theta)}$$

where $\theta = (\sigma, \nu) : \Sigma_{CC} \to \Sigma'_{CC}$.

The above propositions illustrate the successful application of techniques from CSP together with the institution independent structuring mechanisms. We expect other techniques from CSP also to lift successfully to CSP-CASL.

6   Example: Online-Shop

In this section we present a proof typical for our calculus. It concerns an online shopping system as has been studied in the literature several times.

6.1   The Specification in Detail

The online shop is a typical distributed system. It has several components, namely a customer, a warehouse, a payment system, and a coordinator. The communication structure is pointwise only: The coordinator communicates with the three other components in a star like network. The customer, warehouse and payment system only communicate with the coordinator.

The customer may ask the coordinator to perform actions such as: To login, to add an item to the basket, to remove an item from the basket, to checkout, etc. The coordinator then responds to the customer with an appropriate response message. All communication (on a channel) follows this pattern of a request message followed by a response message (except for the *Logout* message, which is more of a command). The coordinator may ask the warehouse to reserve an item, to release an item that has previously been reserved, and to dispatch the reserved items. The payment system allows the coordinator to take payments for goods.

We specify the shop on various levels of abstraction. The architectural shop (see Figure 1) describes the network layout, which remains unchanged in the development. The development is restricted to individually refining the four components. Here we present the first two levels of abstraction for our example, namely: The architectural level for describing the basic interfaces, and the abstract component level (ACL) for specifying the type system and its interplay with process behaviour. Each component contains a 'main' process as its starting point.

Within the rest of this section we use C to denote customer, Co to denote coordinator, W to denote warehouse and PS to denote payment system. We also drop the communications sets within the network construction and take them to be the declared communications sets of the process names implicitly, for instance: by $Network(\{Customer\})$, we mean $Network(\{(Customer, C\_C)\})$.

spec ACL_CUSTOMER =
    data sorts *LoginReq, Logout, GoodLoginRes, BadLoginRes,*
           AddItemReq, AddItemRes, ... < D_C
    channel *C_C : D_C*
    process *Customer : C_C ; Customer_GoodLogin : C_C ;*
           Customer_BadLogin : C_C ; Customer_AddItem : C_C ;
           Customer_Body : C_C ; Customer_Quit : C_C ; ...
           Customer = C_C ! x :: LoginReq_def →
                    (Customer_GoodLogin ; Customer_Body □
                    Customer_BadLogin ; Customer)
           Customer_GoodLogin = C_C ? x :: GoodLoginRes_def → SKIP
           Customer_BadLogin = C_C ? x :: BadLoginRes_def → SKIP
           Customer_AddItem = C_C ! x :: AddItemReq_def →
                    C_C ? y :: AddItemRes_def →
                    Customer_Body

           ...
           Customer_Quit = C_C ! x :: Logout_def → SKIP
           Customer_Body = Customer_AddItem ⊓ ...
                    ⊓ *Customer_Quit*
end

spec ACL_COORDINATOR =
    data ...
    channels *C_C : D_C; C_W : D_W; C_PS : D_PS*
    process ...
           Coordinator_AddItem = C_C ? x :: AddItemReq_def →
                    C_W ! y2 :: ReserveItemReq_def →
                    C_W ? x2 :: ReserveItemRes_def →
                    C_C ! y :: AddItemRes_def →
                    Customer_Body
           Coordinator_Body = Coordinator_AddItem □ ...
                    □ *Coordinator_Quit*
end

Fig. 3. Specification of the ACL customer and coordinator specifications

6.2   Deadlock Analysis

We illustrate how to prove deadlock freedom using the technique presented in Section 5. We discuss the core part of the proof, and explain how to scale it up for the whole system. The proof rule from Proposition 12 reduces the network of processes step by step. We start at the point where the network has been reduced to two processes only:

spec REDUCED_ARCH_SHOP [*RefCl*(ARCH_C)] [*RefCl*(ARCH_CO)] =
    process *System' : C_C ;*
           System' = Coordinator [| C_C || C_C |] Customer
end

The specification REDUCED_ARCH_SHOP instantiated with ACL components is semantically equivalent to the following specification (without parameterisation):

REDUCED_ACL_SHOP =
((($(ReflCl(\text{ARCH\_C})$ **rename** $\theta_1)$ **and**
    $(ReflCl(\text{ARCH\_CO})$ **rename** $\theta_2)$ **and** BODY
) **rename** $\theta_3)$ **and** (ACL_C **rename** $\theta_4)$ **and** (ACL_CO **rename** $\theta_5)$)

Here all signature morphisms involved are embeddings and the specification BODY is a basic specification with the signature equal to the union of the signatures of the ACL customer and coordinator along with the new process name $System'$, and where the only axiom is that of

$System' = Coordinator \parallel C\_C \parallel C\_C \parallel Customer.$

Our aim is to prove that the process term bound to $System'$ is deadlock free within the specification REDUCED_ACL_SHOP. To this end, we apply Proposition 12 and obtain:

$Network(\{Customer, Coordinator\})$ $isDFin$ REDUCED_ACL_SHOP
if (a) C $isDFin$ REDUCED_ACL_SHOP and
   (b) $Co :: C\_C$ $ResToLive^{\checkmark}$ $C :: C\_C$ on C_C in REDUCED_ACL_SHOP

To discharge obligation (a), we apply the **and** rule from Proposition 10 several times and reduce it to $(C$ $isDFin$ ACL_C **rename** $\theta_4)$. Applying the renaming rule (also from Proposition 10) results in $(C$ $isDFin$ ACL_C$)$. As ACL_C is a basic specification and the customer process does not involve any parallel operator we can easily discharge this obligation with CSP-CASL-Prover.

Concerning obligation (b), we apply the **and** rule from Proposition 13 several times and reduce it to:

$Co :: C\_C$ $ResToLive^{\checkmark}$ $C :: C\_C$ on C_C in
   ((ACL_C **rename** $\theta_4)$ **and** (ACL_CO **rename** $\theta_5))$

As ACL_C and ACL_CO are basic CSP-CASL specifications we can discharge the proof obligation by applying the flattening operation and then using CSP-CASL-Prover. This obligation holds because the coordinator allows the customer to choose the initial action (a request message) and then provides a response message to the customer for this particular type of request (possibly after further communications with other components).

The full proof of deadlock freedom has the same structure. Proposition 12 reduces $Network(\{Customer, Coordinator, PaymentSystem, Warehouse\})$ down to $Network(\{Customer\})$ by removing first $Warehouse$, then $PaymentSystem$, and – as shown above – $Customer$ from the network. The resulting obligations can then be reduced to a format where they can be discharged with CSP-CASL-Prover.

7   Conclusion and Future Work

We have presented institutions for CSP-CASL, where we added the new feature of loose process semantics. This setting allowed us to define and study structuring, parameterisation and refinement of CSP-CASL specifications. We gave several proof calculi for compositional reasoning along the structure of CSP-CASL specifications: One dedicated to refinement, the other for deadlock analysis.

Future work will include the development of further proof rules for structured operations: We intend to improve the refinement rules for **and**, and we want to develop proof rules for the structured **free** operation, with a special emphasis on connection with the CSP fixed point theory. Furthermore, we plan to apply structuring to our EP2 case study, and to implement the presented calculi within the standard proof tool for CASL, namely, HETS [13].

Acknowledgement. The authors are grateful to Erwin Catesbeiana for his structured advice on how to navigate through deadlocked situations. This work has been supported by the German Federal Ministry of Education and Research (Project 01 IW 10002 SHIP).

References

[1] eft/pos 2000 Specification, version 1.0.1. EP2 Consortium (2002)

[2] Allen, R., Garlan, D.: A formal basis for architectural connection. ACM Trans. Softw. Eng. Methodol. 6(3), 213–249 (1997)

[3] Bidoit, M., Cengarle, V.V., Hennicker, R.: Proof systems for structured specifications and their refinements. In: Astesiano, E., Kreowski, H.-J., Krieg-Brückner, B. (eds.) Algebraic Fondations of System Specification, pp. 385–434. Springer, Heidelberg (1999)

[4] Bidoit, M., Mosses, P.D. (eds.): CASL User Manual. LNCS, vol. 2900. Springer, Heidelberg (2004)

[5] Cerioli, M., Meseguer, J.: May I borrow your logic (Transporting logical structures along maps). Theoretical Computer Science 173, 311–347 (1997)

[6] Diaconescu, R., Goguen, J., Stefaneas, P.: Logical support for modularisation. In: Logical Environments, Cambridge, pp. 83–130 (1993)

[7] Fischer, C.: How to Combine Z with a Process Algebra. In: P. Bowen, J., Fett, A., Hinchey, M.G. (eds.) ZUM 1998. LNCS, vol. 1493, pp. 5–25. Springer, Heidelberg (1998)

[8] Gimblett, A., Roggenbach, M., Schlingloff, B.-H.: Towards a Formal Specification of an Electronic Payment System in CSP-CASL. In: Fiadeiro, J.L., Mosses, P.D., Yu, Y. (eds.) WADT 2004. LNCS, vol. 3423, pp. 61–78. Springer, Heidelberg (2005)

[9] Goguen, J.A., Burstall, R.M.: Institutions: Abstract model theory for specification and programming. J. ACM 39(1), 95–146 (1992)

[10] Hoare, C.A.R.: Communicating Sequential Processes. Prentice Hall (1985)

[11] Kahsai, T., Roggenbach, M.: Property Preserving Refinement for CSP-CASL. In: Corradini, A., Montanari, U. (eds.) WADT 2008. LNCS, vol. 5486, pp. 206–220. Springer, Heidelberg (2009)

[12] Mossakowski, T.: ModalCASL. Language Summary (2004), http://www.informatik.uni-bremen.de/~till/papers/Modal-Summary.pdf

[13] Mossakowski, T., Maeder, C., Lüttich, K.: The Heterogeneous Tool Set, HETS. In: Grumberg, O., Huth, M. (eds.) TACAS 2007. LNCS, vol. 4424, pp. 519–522. Springer, Heidelberg (2007)

[14] Mossakowski, T., Roggenbach, M.: Structured CSP – A Process Algebra as an Institution. In: Fiadeiro, J.L., Schobbens, P.-Y. (eds.) WADT 2006. LNCS, vol. 4409, pp. 92–110. Springer, Heidelberg (2007)

[15] Mosses, P.D. (ed.): CASL Reference Manual. LNCS, vol. 2960. Springer, Heidelberg (2004)

[16] O'Reilly, L., Kahsai, T., Mossakowski, T., Roggenbach, M.: The CSP-CASL institution. Technical Report CSR-1-2011, Swansea University (2011)

[17] O'Reilly, L., Roggenbach, M., Isobe, Y.: CSP-CASL-Prover: A generic tool for process and data refinement. ENTCS 250(2), 69–84 (2009)

[18] Reed, J.N., Sinclair, J.E., Roscoe, A.W.: Responsiveness of interoperating components. Formal Asp. Comput. 16(4), 394–411 (2004)

[19] Reggio, G., Astesiano, E., Choppy, C.: CASL-LTL. Technical Report DISI-TR-99-34, Università di Genova (2000)

[20] Roggenbach, M.: CSP-CASL: A new integration of process algebra and algebraic specification. Theoretical Computer Science 354(1), 42–71 (2006)

[21] Roscoe, A.W.: Understanding Concurrent Systems. Springer, Heidelberg (2010)

[22] Sannella, D., Tarlecki, A.: Specifications in an arbitrary institution. Information and Computation 76, 165–210 (1988)

[23] Scattergood, B.: The semantics and implementation of machine-readable CSP. PhD thesis, Oxford University (1998)

[24] Wehrheim, H.: Behavioural subtyping in object-oriented specification formalisms, Habilitation thesis, Carl-von-Ossietzky-Universität Oldenburg (2002)

[25] Zawłocki, A.: Architectural Specifications for Reactive Systems. In: Fiadeiro, J.L., Mosses, P.D., Yu, Y. (eds.) WADT 2004. LNCS, vol. 3423, pp. 252–269. Springer, Heidelberg (2005)

Proving Properties about Functions on Lists Involving Element Tests

Daniel Seidel[*] and Janis Voigtländer

Rheinische Friedrich-Wilhelms-Universität Bonn
Institut für Informatik
Römerstraße 164
53117 Bonn, Germany
{ds,jv}@iai.uni-bonn.de

Abstract. Bundy and Richardson [4] developed a method for reasoning about functions manipulating lists which is based on separating shape from content, and then exploiting a mathematically convenient representation for expressing shape-only manipulations. Later, Prince et al. [7] extended the technique to other data structures, and gave it a more formal basis via the theory of containers. All these results are restricted to fully polymorphic functions. For example, functions using equality tests on list elements are out of reach. We remedy this situation by developing new abstractions and representations for less polymorphic functions. In Haskell speak, we extend the earlier approach to be applicable in the presence of (certain) type class constraints.

1   Introduction

Abstraction is a useful strategy to get a clear view on the things that matter. Regarding proofs about program equivalences, it is beneficial to have an abstract representation of data structures and functions, holding exactly the information necessary for the intended reasoning in an easily accessible form. For lists, Bundy and Richardson [4] introduced a higher-order formulation in which a list is a pair (n, f) where n is a natural number representing the length of the list, i.e., its shape, and f is a content function taking each position in the list to its corresponding element. Bundy and Richardson's motivation was that reasoning about such representations can be easier than reasoning about standard lists. In a more precise and more general form, the idea later recurred as reasoning via *container representations* [1,7].

The usefulness of the abstraction from the actual elements stored in a list is made apparent by the fact that certain *container morphisms*, taking a list (in this case) to another one, do not inspect or alter the image of f. An example for such a container morphism is the function $reverse^c$, the container version of the usual function reversing a list. The application of this container morphism is given as follows:

$$reverse^c \ (n, f) = (n, \lambda i \to f \ (n - i - 1))$$

---

[*] This author was supported by the DFG under grant VO 1512/1-1.

T. Mossakowski and H.-J. Kreowski (Eds.): WADT 2010, LNCS 7137, pp. 270–286, 2012.

In general, a morphism shuffles positions (here by composing f with the function $\lambda i \to n - i - 1$) and can alter the length of the list, remove elements, duplicate others. It cannot modify the elements themselves or add completely new elements.

The advantage of the container representation, which led Bundy and Richardson to using that representation, is that proofs about programs expressible as the composition of container morphisms become (simple) arithmetic proofs. For example, the proof that reversing a (finite) list twice is the identity is obtained very easily as follows:

$$
\begin{aligned}
reverse^c \ (reverse^c \ (n, f)) &= reverse^c \ (n, \lambda i \to f \ (n - i - 1)) \\
&= (n, \lambda i \to f \ (n - (n - i - 1) - 1)) \\
&= (n, \lambda i \to f \ i) \\
&= (n, f)
\end{aligned}
$$

Prince et al. [7] use, from Abbott et al. [1,2], that container morphisms correspond to parametrically polymorphic functions (or, *natural transformations*). Such polymorphic functions act independently of the concrete input type and hence, necessarily, independently of concrete elements of a type. Particularly, a fully polymorphic function from lists to lists, expressed via the type $[\alpha] \to [\alpha]$, maps for every type τ input lists of type $[\tau]$ to output lists of type $[\tau]$ without using any specifics of the type τ. For example, a possible definition of *reverse* in Haskell [6] is:

$$
\begin{aligned}
reverse & \quad :: [\alpha] \to [\alpha] \\
reverse \ [] & \ = [] \\
reverse \ (x : xs) & = (reverse \ xs) \ +\!\!+ \ [x]
\end{aligned}
$$

Using category theoretic notions, Prince et al. observe that such polymorphic functions from lists to lists are isomorphic to the list container morphisms. The correspondence also generalises to other, strictly positive, data types.

What both Bundy and Richardson [4] and Prince et al. [7] fail to do is to reason about functions that are not fully polymorphic. An example, discussed in both papers, is a function *member* that checks whether a given value is an element of a given list. In Haskell:

$$
\begin{aligned}
member & \quad :: Eq \ \alpha \Rightarrow \alpha \to [\alpha] \to \mathbf{Bool} \\
member \ x \ [] & \ = \mathbf{False} \\
member \ x \ (y : ys) & = (x == y) \ || \ (member \ x \ ys)
\end{aligned}
$$

Since programmed equivalence (the binary **Bool**-valued function (==)) depends on the type at which it is used, *member* cannot be given the fully polymorphic type $\alpha \to [\alpha] \to \mathbf{Bool}$. It instead comes with the constraint "$Eq \ \alpha \Rightarrow$", using Haskell's type class mechanism [10]. In the discussions of both Bundy and Richardson [4] and Prince et al. [7], the outcome is that the proposed reasoning method is not effective for *member*. Similarly, reasoning would not work for the function *nub* that eliminates duplicates from a list. These kind of functions also fall outside the realm of *shapely operations* in the calculus of Jay [5].

While Bundy and Richardson only identified the problematic case, and Prince et al. went a step further by observing that the problem can be explained by a lack of polymorphism, we do provide a solution. In retrospect, at least the basic idea behind our solution may seem obvious: if a function is not polymorphic enough, then exploit information about to what actual extent it loses its polymorphism. In Haskell, that information is provided exactly by type class constraints like "$Eq\ \alpha \Rightarrow$". One example is the type of *member* seen above, another is that the already mentioned function *nub* will naturally be given the type $Eq\ \alpha \Rightarrow [\alpha] \to [\alpha]$. Of course, there is no reason at all to expect that the latter corresponds to an ordinary container morphism, because those were shown to be isomorphic to functions of the more general type $[\alpha] \to [\alpha]$ instead. But we can investigate refined notions of container representations and container morphisms, so that effective reasoning in the spirit of the earlier method becomes possible again. That is what we do in this paper.

In Section 2 we reconsider the connection between fully polymorphic functions and container morphisms. This sets the stage for our original development in later parts of the paper. In particular, it explains the use of *free theorems* [9], which in the guise of category-theoretic *naturality* is also at the heart of the isomorphism Prince et al. [7] use, and which in the form of free theorems for functions with type class constraints (also called ad-hoc polymorphic functions) will also pave the way to our results. In contrast to Prince et al., we do not use dependent types and therefore have slightly different formalisations of container values and container morphisms. Our reason for abstaining from using dependent types is notational convenience. Already by comparing the formulations of otherwise equivalent results and examples by Bundy and Richardson [4] and Prince et al. [7], it becomes clear that the former is lighter on notation. For the treatment of ad-hoc/type class polymorphism we found that the overhead of keeping exact dependent typing is even more cumbersome. However, there is something to lose by using less exact typing: we will not have an exact isomorphism as that employed by Prince et al. [7]. But we show in Section 2 that our setup is nevertheless sufficient for doing the kind of reasoning the overall method is aiming for. Moreover, it is perfectly possible to add all the dependent types back in, both in Section 2 and for our extensions to handle Eq-polymorphism, as presented in Section 3. Our approach is not limited to the type class Eq. In a similar way, container values, container morphisms, and the reasoning method can be extended to handle other type classes. We demonstrate this, still in Section 3, for the type class Ord, and offer some further perspective in Section 4.

2   The Earlier Results on Lists, Rephrased

In what follows, we use Haskell both as the language for writing functions about which we might want to prove properties, and as the specification language for container values and container morphisms, though for the latter use we will stretch Haskell a bit by including general math concepts. Moreover, we do not care about laziness in Haskell, or mixing strict and lazy evaluation using Haskell's *seq*-primitive. In fact, we assume a completely strict dialect of Haskell.

Let us first clarify some notations. The set of natural numbers is denoted by **Nat**. Depending on the context, a natural number n represents either the number $n \in$ **Nat** or the set of natural numbers $\{0, \ldots, n-1\}$. Furthermore, the type constructor for lists, already used in Haskell types in the introduction, is defined by

$$[\tau] = \{[x_0, \ldots, x_{n-1}] \mid n \in \textbf{Nat}, \forall i \in n.\; x_i :: \tau\}$$

Lists can alternatively be defined as container values, meaning by a shape (the length) and a content function (mapping each position to its entry). An appropriate definition (without using container terminology) was already introduced by Bundy and Richardson [4]. We restate it here by defining the set $\mathcal{C}(\tau)$ of list container values of type τ as

$$\mathcal{C}(\tau) = \{(n, f) \mid n :: \textbf{Nat},\; f :: \textbf{Nat} \to \tau\}$$

where the fs need not be totally defined, i.e., can be partial functions. But in every container value (n, f), we require f to be defined at least for all natural numbers less than n, i.e., on every position of the represented list.

In the following lemma we give a pair of functions that map back ($\square^{-1}$) and forth ($\square$) between container values and lists and nearly constitute an isomorphism. We continue to use the standard expression syntax of Haskell (while on the type level, $\mathcal{C}(\alpha)$ is "special syntax" that would not be found in actual Haskell). The operator !! takes a list and a position and returns the list entry at that position (counting from 0), and *map* is the usual function that applies its argument function to each element in its input list.

Lemma 1. *For each type τ as instantiation for α, the functions $\square$ and $\square^{-1}$ defined as*

$$\square \qquad :: \mathcal{C}(\alpha) \to [\alpha] \qquad\qquad \square^{-1} \quad :: [\alpha] \to \mathcal{C}(\alpha)$$
$$\square\;(n, f) = \mathit{map}\; f\; [0 \mathrel{..} (n-1)] \qquad \square^{-1}\; \mathit{xs} = (\mathit{length}\; \mathit{xs}, \mathit{xs}\; !!)$$

satisfy the following three properties:

1. $(\square \circ \square^{-1}) = \mathit{id}_{[\tau]}$
2. $(\square^{-1} \circ \square) \subseteq\; \equiv_{\mathcal{C}(\tau)}$, *where* $\equiv_{\mathcal{C}(\tau)} = \{((n, f), (n, f')) \mid \forall i \in n.\; f\; i = f'\; i\}$
3. $\forall (n, f), (n', f') \in \mathcal{C}(\tau).\; (n, f) \equiv_{\mathcal{C}(\tau)} (n', f')$ *iff* $\square\;(n, f) = \square\;(n', f')$

Proof. First, we show that $(\square \circ \square^{-1})\; \mathit{xs} = \mathit{xs}$ holds for every τ and $\mathit{xs} :: [\tau]$, by a straightforward induction on the length of xs. Second, we prove that $(\square^{-1} \circ \square) \subseteq\; \equiv_{\mathcal{C}(\tau)}$. We can reason as follows, for every container value:

$$\square^{-1}\;(\square\;(n, f)) = (\mathit{length}\;(\mathit{map}\; f\; [0 \mathrel{..} (n-1)]), (\mathit{map}\; f\; [0 \mathrel{..} (n-1)])\; !!)$$
$$= (n, f \circ ([0 \mathrel{..} (n-1)]\; !!))$$
$$= (n, f|_n)$$

where $g|_n$ means the restriction of a function g to the domain n. The calculation steps are all by definitions and obvious properties of *length*, *map*, and (!!). Finally, the property $((n, f), (n', f')) \in\; \equiv_{\mathcal{C}(\tau)}$ iff $\square\;(n, f) = \square\;(n', f')$ follows from the definition of $\square$.

Note that $\square$ and $\square^{-1}$ indeed only *nearly* constitute an isomorphism. For example, let $\tau = \mathbf{Char}$ and let $f_1 :: \mathbf{Nat} \to \mathbf{Char}$ be the partial function with graph $\{(0, 'a'), (1, 'b')\}$ and $f_2 :: \mathbf{Nat} \to \mathbf{Char}$ the one whose graph additionally contains $(2, 'c')$. Then $(2, f_1)$ and $(2, f_2)$ are two different elements of $\mathcal{C}(\mathbf{Char})$, but $\square \ (2, f_1) = ['a', 'b'] = \square \ (2, f_2)$.

We define a *container morphism* as a family $(s_n, P_n)_{n \in \mathbf{Nat}}$ of pairs comprising a natural number s_n, intuitively the output list length for any input list of length n, and a function $P_n :: \mathbf{Nat} \to \mathbf{Nat}$, intuitively mapping positions in the output list to positions in the input list when the latter has length n. For each container morphism and each $n \in \mathbf{Nat}$, we allow P_n to be a partial function, but require that for every $i \in s_n$, we have $(P_n \ i) \in n$. The latter guarantees that all output positions are covered and that we never map an output position to a non-existing input position. We often abbreviate $(s_n, P_n)_{n \in \mathbf{Nat}}$ as (s, P). The application of a container morphism to a container value is defined as

$$(s, P) \ (n, f) = (s_n, f \circ P_n)$$

Here are some container morphisms that intuitively correspond to well-known Haskell functions of type $[\alpha] \to [\alpha]$:

$$reverse^c = (n, \lambda i \to n - i - 1)_{n \in \mathbf{Nat}}$$
$$init^c = (n - 1, id)_{n \in \mathbf{Nat}}$$
$$tail^c = (n - 1, \lambda i \to i + 1)_{n \in \mathbf{Nat}}$$

Before we can prove a systematic connection between fully polymorphic functions (in strict Haskell) and container morphisms, and function composition in either world, we need to say a few words on free theorems. Such theorems are statements about functions only dependent on the function type, relying on a formalisation of parametricity [8] for the functional language at hand. For example, in strict Haskell, the free theorem for the type $[\alpha] \to \mathbf{Nat}$ states that for every function $f :: \tau_1 \to \tau_2$ with τ_1 and τ_2 arbitrary, every function $g :: [\alpha] \to \mathbf{Nat}$, and every list $xs :: [\tau_1]$, we have $g \ (map \ f \ xs) = g \ xs$ if f is defined for all elements of xs. The intuition is that in a purely functional language g's behaviour can clearly only depend on its input argument. Moreover, since g is fully polymorphic in the type α of elements of that input list, g cannot inspect those elements in any way. Hence, g's behaviour, and thus output, can only depend on the *structure* of its input list. Since a general property of *map* is that it does not change structure (and the output list is defined if f is defined on all input list elements), $g \ (map \ f \ xs) = g \ xs$ follows. Reasoning by parametricity/free theorems allows to derive similar statements for a wide variety of types.

Incidentally, since the function *length* itself has exactly the above mentioned polymorphic type, one of the "obvious properties" (actually two, another one for (!!)) in the proof of Lemma 1 could have been deduced without considering the concrete function *length*, just its type. But the real value of free theorems is when we really do not know what concrete function we deal with, such as when we want to prove that *every* strict-Haskell function of type $[\alpha] \to [\alpha]$ corresponds to some container morphism.

Since free theorems are available for free, i.e., can be automatically generated, we will use them as given, without considering further formal background here. Let us note, though, in preparation for Section 3, that free theorems in the presence of type class polymorphism can be established by an indirection via types (and functions) obtained through the dictionary translation method of Wadler and Blott [10].

Theorem 1. *For every function* $g :: [\alpha] \rightarrow [\alpha]$, *there exists a container morphism* (s, P) *such that* $g \circ \square = \square \circ (s, P)$.

Proof. Let $g :: [\alpha] \rightarrow [\alpha]$. Then the free theorem for g's type tells us that $g \; (map \; h \; l) = map \; h \; (g \; l)$ for every choice of types τ_1, τ_2, function $h :: \tau_1 \rightarrow \tau_2$, and list $l :: [\tau_1]$ if h is defined for all elements of l. Hence, we can reason as follows, for every container value:

$$
\begin{aligned}
g \; (\square \; (n, f)) &= g \; (map \; f \; [0 \; .. \; (n-1)]) \\
&= map \; f \; (g \; [0 \; .. \; (n-1)]) \\
&= \square \; (\square^{-1} \; (map \; f \; (g \; [0 \; .. \; (n-1)]))) \\
&= \square \; (length \; (g \; [0 \; .. \; (n-1)]), f \circ ((g \; [0 \; .. \; (n-1)]) \; !!)) \\
&= \square \; ((length \; (g \; [0 \; .. \; (n-1)]), (g \; [0 \; .. \; (n-1)]) \; !!)_{n \in \mathbf{Nat}} \; (n, f))
\end{aligned}
$$

where the second step is by the free theorem, the third by Lemma 1(1), the fourth by the definition of $\square^{-1}$ and properties of *length*, *map*, and (!!), and the last step by the definition of the application of a container morphism to a container value.

Note that there is not a unique container morphism corresponding, in the sense of Theorem 1, to a function $g :: [\alpha] \rightarrow [\alpha]$. For example, for the standard Haskell definition of $init :: [\alpha] \rightarrow [\alpha]$, both $init^c = (n-1, id)_{n \in \mathbf{Nat}}$ and $init^c = (n-1, id|_{n-1})_{n \in \mathbf{Nat}}$ satisfy $init \circ \square = \square \circ init^c$. This (direction of) non-uniqueness does no harm to our reasoning application, though. Together with Lemma 1, Theorem 1 allows the calculation with container morphisms instead of polymorphic functions. The required results are stated in the following corollary and lemma.

Corollary 1. *For every function* $g :: [\alpha] \rightarrow [\alpha]$, *there exists a container morphism* (s, P) *such that* $g = \square \circ (s, P) \circ \square^{-1}$.

Proof. By Theorem 1 and Lemma 1(1).

Lemma 2. *Let* $g, g' :: [\alpha] \rightarrow [\alpha]$. *Let* (s, P), (s', P') *be container morphisms such that* $g = \square \circ (s, P) \circ \square^{-1}$ *and* $g' = \square \circ (s', P') \circ \square^{-1}$. *Then we have* $g \circ g' = \square \circ (s, P) \circ (s', P') \circ \square^{-1}$.

Proof. By the assumptions, we have $g \circ g' = \square \circ (s, P) \circ \square^{-1} \circ \square \circ (s', P') \circ \square^{-1}$, so it would suffice to show that $\square \circ (s, P) \circ \square^{-1} \circ \square = \square \circ (s, P)$. By Lemma 1(3), this is equivalent to, for every type τ and $(n, f) \in \mathcal{C}(\tau)$,

$$
(s, P) \; (\square^{-1} \; (\square \; (n, f))) \equiv_{\mathcal{C}(\tau)} (s, P) \; (n, f)
$$

But by Lemma 1(2), we have $\square^{-1}\ (\square\ (n,f)) \equiv_{\mathcal{C}(\tau)} (n,f)$, and it is easy to show from the definitions that for every $(n,f),(n',f')$ with $(n,f) \equiv_{\mathcal{C}(\tau)} (n',f')$, it holds that $(s,P)\ (n,f) \equiv_{\mathcal{C}(\tau)} (s,P)\ (n',f')$.

Let us manifest the usefulness of our formal material by an example. Assume we want to prove that $reverse \circ tail = init \circ reverse$ holds. We have

$$reverse = \square \circ reverse^c \circ \square^{-1}$$
$$init = \square \circ init^c \circ \square^{-1}$$
$$tail = \square \circ tail^c \circ \square^{-1}$$

for standard Haskell definitions of the list functions and $reverse^c$, $init^c$, and $tail^c$ as given above Theorem 1.[1] By Lemma 2, it suffices to prove that

$$\square \circ reverse^c \circ tail^c \circ \square^{-1} = \square \circ init^c \circ reverse^c \circ \square^{-1}$$

and by Lemma 1(3) indeed to prove that for every type τ and $(n,f) \in \mathcal{C}(\tau)$,

$$(reverse^c \circ tail^c)\ (n,f) \equiv_{\mathcal{C}(\tau)} (init^c \circ reverse^c)\ (n,f)$$

We can calculate for the left-hand side

$$
\begin{aligned}
(reverse^c \circ tail^c)\ (n,f) &= reverse^c\ (n-1, \lambda i \to f\ (i+1))\\
&= (n-1, \lambda i \to f\ (((n-1)-i-1)+1))\\
&= (n-1, \lambda i \to f\ (n-1-i))
\end{aligned}
$$

and for the right-hand side

$$
\begin{aligned}
(init^c \circ reverse^c)\ (n,f) &= init^c\ (n, \lambda i \to f\ (n-i-1))\\
&= (n-1, \lambda i \to f\ (n-i-1))
\end{aligned}
$$

to see that the claim holds.

Let us contrast the above proof with an attempt at directly proving $reverse \circ tail = init \circ reverse$ using the Haskell definition of $reverse$ from the introduction as well as some suitable definitions of $tail$ and $init$. The interesting case is the one of a non-empty list: $reverse\ (tail\ (x:xs)) = init\ (reverse\ (x:xs))$, which reduces to the proof obligation $reverse\ xs = init\ ((reverse\ xs) +\!\!+ [x])$. Now an *inductive* proof using the defining equations of $init$ would be required, where first the given proof obligation would have to be *generalised* to an actually suitable induction hypothesis (since simply performing induction on xs in $reverse\ xs = init\ ((reverse\ xs) +\!\!+ [x])$ leads nowhere). In contrast, the above proof requires neither induction nor inventing a generalisation. It just performs simple arithmetics.

---

[1] Clearly, neither Theorem 1 nor Corollary 1 prove the equivalence $reverse = \square \circ reverse^c \circ \square^{-1}$ for the specific syntactic definitions of $reverse$ and $reverse^c$ given in the introduction (and likewise for $init$ and $tail$). The theorem and corollary provide, for every g, *one* suitable definition for g^c. It might not be the one we find useful for reasoning. Finding such a useful syntactic representation, like $reverse^c = (n, \lambda i \to n-i-1)_{n \in \mathbf{Nat}}$, must be done on a case-by-case basis, but is often very natural, like in all cases here.

3   Refining the Container-Related Notions

The results in the previous section can be extended in two directions. One is to consider not only functions from lists to lists, but also functions between other data structures that can be viewed as container values. That direction is already explored by Prince et al. [7]. The extension that we consider is orthogonal to that first one. In particular, while we focus on functions from lists to lists here, we are confident that our results could be easily combined with the results of Prince et al. [7] to handle functions (involving element tests like equivalence and ordering) between arbitrary container structures.

Considering functions like *nub*, removing all duplicates from a list, or *sort*, sorting a list's elements, it is clear that they are not fully polymorphic in their list element type. The functions require the availability of an equivalence test or an order defined on elements of the input list. Hence, Theorem 1 is not applicable anymore. Our aim now is to appropriately adapt the notions of container value and container morphism to get equally useful results for functions of types $Eq\ \alpha\ \Rightarrow\ [\alpha]\ \rightarrow\ [\alpha]$ and $Ord\ \alpha\ \Rightarrow\ [\alpha]\ \rightarrow\ [\alpha]$ as the earlier works provide for functions of type $[\alpha]\ \rightarrow\ [\alpha]$.

It is important to note that our view on type classes is that they really hold what they pretend to provide. In the case of *Eq*, that means that every type in *Eq* indeed carries an *equivalence* relation. In real Haskell, the implemented relations can be arbitrary (no reflexivity, transitivity, or symmetry are guaranteed or checked). In the same spirit, in Section 3.2 we expect types in the type class *Ord* to carry an actual total preorder (a reflexive, transitive, and total relation).

3.1   The Type Class *Eq*

To capture what happens if elements in a list are testable for equivalence, the container notions have to be adjusted. We use $\mathcal{E}(M)$ to denote the class of all (decidable) equivalence relations over a set M. For simplicity of notation, we freely regard an equivalence relation $\cong$ on a subset of **Nat** as the equivalence relation $\cong \cup\ id_{\mathbf{Nat}}$ on **Nat** when appropriate. For a type τ that is an instance of *Eq*, we denote by $\cong_\tau$ the corresponding fixed (in a given program) equivalence relation. Since in Haskell, it is actually accessible via the binary **Bool**-valued function (==), we set:

$$\cong_\tau\ =\ \{(x,y)\ |\ x :: \tau,\ y :: \tau,\ (x == y) = \mathbf{True}\}$$

Definition 1. *Let τ be some type that is an instance of Eq. An Eq-container value of type τ is a triple $(n, \cong, f)$ with $n :: \mathbf{Nat}$, $\cong\ \in \mathcal{E}(\mathbf{Nat})$, and $f :: \mathbf{Nat} \rightarrow \tau$ a partial function such that*

$$\forall i, j \in n.\ i \cong j \Leftrightarrow (f\ i) \cong_\tau (f\ j)$$

or, equivalently,[2]

---

[2] The kernel of a function over a relation is defined as $(\ker_\cong f) = \{(i,j)\ |\ (f\ i) \cong (f\ j)\}$.

$$(\ker_{\cong_\tau} f|_n) = (\cong \cap (n \times n)) \tag{1}$$

The set of all such container values is denoted by $\mathcal{C}^{Eq}(\tau)$.

The roles of n and f in the above definition are as before in the case of ordinary list containers $\mathcal{C}(\tau)$. Condition (1) implies the previous side condition that the function f is defined at least for all natural numbers less than n. But condition (1) is stronger than that. It involves the key new ingredient of Eq-container values, namely the second component $\cong$. The role of that equivalence relation is to capture information, in terms of list positions, about equivalence tests between elements accessible via f. For a concrete example, assume $\tau = \mathbf{Char}$ and that the equivalence relation $\cong_{\mathbf{Char}}$ were such that upper- and lowercase of the same letter were considered equivalent, while different letters were considered inequivalent. Then the list ['a', 'A', 'b'] :: [**Char**] could be represented as an Eq-container value as $(3, \cong, f)$, where $\cong = \{(0,0), (0,1), (1,0), (1,1), (2,2)\}$ and f maps 0 to 'a', 1 to 'A', and 2 to 'b'.

Some further explanations seem in order, to avoid possible misconceptions. First, from Definition 1, in particular from the presence of $\cong$ (though not $\cong_\tau$) in container value triples, it might seem that each Eq-container value somehow stores its own completely private equivalence relation so that, within the same program, two members $(n_1, \cong_1, f_1)$ and $(n_2, \cong_2, f_2)$ of the same $\mathcal{C}^{Eq}(\tau)$ can interpret equivalence between elements of type τ in two different ways. Under that perception, for example, some $(2, \cong_1, f_1), (2, \cong_2, f_2) \in \mathcal{C}^{Eq}(\mathbf{Char})$ could represent the same list ['a', 'A'] :: [**Char**] while somehow $\cong_1$ and $\cong_2$ could be chosen in such a way that in one case when ['a', 'A'] is passed to some function g :: $Eq\ \alpha \Rightarrow [\alpha] \rightarrow [\alpha]$ (or its "container version") the two list elements are considered equivalent, while in the other case they are not.

But that is *not* the case! Actually, by condition (1) we have that n, f, and τ (through $\cong_\tau$) uniquely determine $\cong$ (or at least its relevant part, on $n \times n$). So why, then, do we include the $\cong$ in $(n, \cong, f) \in \mathcal{C}^{Eq}(\tau)$ at all? The point is that we will be able (in Definitions 2 and 3 below) to describe the behaviour of (a container analogue of) a function g :: $Eq\ \alpha \Rightarrow [\alpha] \rightarrow [\alpha]$ on $(n, \cong, f)$ solely by relying on n and $\cong$, rather than looking into f. That is the key abstraction/enabler for exploiting the type class polymorphism when reasoning about such functions: that the behaviour of such a function g can be understood by just considering relative equivalences between list elements (as captured via $\cong$), rather than the concrete list elements themselves (as still accessible via f, but deliberately not used in determining g's behaviour). So explicitly representing and (while preserving the invariant (1); see Lemma 4) manipulating $\cong$ is crucial to effectively "let the symbols do the work".

Hopefully having accepted $\cong$ as an explicit component of Eq-container values, note further that we use f as a function from list positions into τ. We could have been tempted to instead define f as a function from *equivalence classes* of positions, with respect to $\cong$, into τ, rather than from the positions themselves. While these choices may appear to be interchangeable, there is actually a crucial difference. With our choice we can distinguish elements that are equivalent with respect to $\cong_\tau$, but not equal. For example, consider the list $['a','A'] :: [\mathbf{Char}]$ and assume that the equivalence relation $\cong_{\mathbf{Char}}$ were again the one mentioned in the paragraph directly following Definition 1. Then a container representation working with a function from equivalence classes of positions would, at length two, only be able to represent lists with two equal elements $(['a','a'], ['b','b'], ['A','A'], \ldots)$ and lists with different letters $(['a','b'], ['b','a'], ['a','B'], \ldots)$, but not the list $['a','A']$ as distinguishable from $['a','a']$ and $['A','A']$. One might be willing to accept this limited expressiveness, as indeed when the equivalence provided by the type class instance for **Char** is "same letter"-ness, then all of $['a','A']$, $['a','a']$, and $['A','a']$ ought to be considered equivalent with respect to the inferred type class instance for $[\mathbf{Char}]$. But after all, equivalent with respect to a type class instance is not the same as semantically equal, and we want to keep that distinction in our reasoning. For example, we want to still be able to observe that applying (the container morphism corresponding to) *reverse* to $['a','A']$ gives $['A','a']$, and not $['a','a']$ or $['A','A']$.

After having made and justified these important decisions, we can set up a pair of functions between *Eq*-container values and lists satisfying similar properties as the pair of functions $\Box$ and $\Box^{-1}$ defined in Lemma 1.

Lemma 3. *For each type τ that is an instance of Eq, the instantiations of the functions $\Box^{Eq}$ and $(\Box^{Eq})^{-1}$ defined as*

$$
\begin{aligned}
&\Box^{Eq} &&:: Eq\ \alpha \Rightarrow \mathcal{C}^{Eq}(\alpha) \to [\alpha] \\
&\Box^{Eq}\ (n, \cong, f) = map\ f\ [0\ ..\ (n-1)] \\
&(\Box^{Eq})^{-1} &&:: Eq\ \alpha \Rightarrow [\alpha] \to \mathcal{C}^{Eq}(\alpha) \\
&(\Box^{Eq})^{-1}\ xs &&= (length\ xs, \ker_{\cong_\alpha}\ (xs\ !!), xs\ !!)
\end{aligned}
$$

satisfy the following three properties:

1. $(\Box^{Eq} \circ (\Box^{Eq})^{-1}) = id_{[\tau]}$
2. $((\Box^{Eq})^{-1} \circ \Box^{Eq}) \subseteq \equiv_{\mathcal{C}^{Eq}(\tau)},$
   where $\equiv_{\mathcal{C}^{Eq}(\tau)} = \{((n, \cong, f), (n, \cong', f')) \mid \forall i \in n.\ f\ i = f'\ i\}$
3. $\forall (n, \cong, f), (n', \cong', f') \in \mathcal{C}^{Eq}(\tau).$
   $(n, \cong, f) \equiv_{\mathcal{C}^{Eq}(\tau)} (n', \cong', f')$ *iff* $\Box^{Eq}\ (n, \cong, f) = \Box^{Eq}\ (n', \cong', f')$

Proof. The proofs of properties (1)–(3) are similar to the proof of Lemma 1. An important aspect to show is that indeed $((\Box^{Eq})^{-1}\ xs) \in \mathcal{C}^{Eq}(\tau)$ for every $xs :: [\tau]$, particularly so for condition (1) from Definition 1. But the required statement is obtained relatively directly from the definition of $(\Box^{Eq})^{-1}$.

Now, appropriate morphisms between *Eq*-container values, and their application, are defined as follows.

Definition 2. *An Eq-container morphism (s, P) is a family of pairs $(s_n^{\cong}, P_n^{\cong})$ $_{n \in \mathbf{Nat}, \cong \in \mathcal{E}(\mathbf{Nat})}$ such that $s_n^{\cong} :: \mathbf{Nat}$ and $P_n^{\cong} :: \mathbf{Nat} \to \mathbf{Nat}$ a partial function with $(P_n^{\cong}\ i) \in n$ for every $i \in s_n^{\cong}$.*

The intuitions for $s_n^{\cong}$ and $P_n^{\cong}$ are as for ordinary container morphisms before, except that now both can depend on the new parameter $\cong$ in addition to n. After all, we need to be prepared for the fact that the behaviour (i.e., determining the length of the output list and the distribution of elements in it) of a function involving equivalence tests cannot anymore be described by just inspecting the input list length. In addition, information about such equivalence tests may have to be accessed.

Definition 3. *Let* $(n, \cong, f)$ *be an Eq-container value and* (s, P) *an Eq-container morphism. The application of* (s, P) *to* $(n, \cong, f)$ *is defined as*

$$(s, P) \; (n, \cong, f) = (s_n^{\cong}, \ker_{\cong} \; P_n^{\cong}, f \circ P_n^{\cong})$$

The first and last components of the output triple are analogous to the ordinary case without element tests. For the middle component, we capture the position-wise equivalence of output list elements in terms of the mapping to input positions and what we know about, again position-wise, equivalence of input list elements.

The following lemma states the well-behavedness of the notions defined above.

Lemma 4. *Let* τ *be a type that is an instance of Eq. Let* $c \in \mathcal{C}^{Eq}(\tau)$ *and let* m *be an Eq-container morphism. Then we have* $(m \; c) \in \mathcal{C}^{Eq}(\tau)$.

Proof. The critical point to prove is the condition (1) from Definition 1 on the content function of the resulting container value. Let $c = (n, \cong, f)$ and $m = (s, P)$. We have $(m \; c) = (s_n^{\cong}, \ker_{\cong} \; P_n^{\cong}, f \circ P_n^{\cong})$ and hence need to show that

$$(\ker_{\cong_\tau} (f \circ P_n^{\cong})|_{s_n^{\cong}}) = ((\ker_{\cong} \; P_n^{\cong}) \cap (s_n^{\cong} \times s_n^{\cong}))$$

is satisfied. But that is an easy exercise, using $(\ker_{\cong_\tau} f|_n) = (\cong \cap (n \times n))$.

Comparing the definitions of morphisms on ordinary container values and on Eq-container values, we can easily translate the former ones into the latter ones.

Note 1. Every (ordinary) container morphism $(s_n, P_n)_{n \in \mathbf{Nat}}$ can be viewed as the Eq-container morphism $(s_n, P_n)_{n \in \mathbf{Nat}, \cong \in \mathcal{E}(\mathbf{Nat})}$.

To verify that our definitions of Eq-container values and Eq-container morphisms are useful when reasoning about strict-Haskell functions of type $Eq \; \alpha \Rightarrow [\alpha] \to [\alpha]$, we need results similar to Theorem 1, Corollary 1, and Lemma 2. Indeed, such results are possible and given below.

Theorem 2. *For every function* $g :: Eq \; \alpha \Rightarrow [\alpha] \to [\alpha]$, *there exists an Eq-container morphism* (s, P) *such that* $g \circ \Box^{Eq} = \Box^{Eq} \circ (s, P)$.

Proof. Let $g :: Eq \; \alpha \Rightarrow [\alpha] \to [\alpha]$. Then the free theorem for g's type tells us that $g \; (map \; h \; l) = map \; h \; (g \; l)$ for every choice of types τ_1, τ_2 that are instances of Eq, function $h :: \tau_1 \to \tau_2$, and list $l :: [\tau_1]$, provided that $(\ker_{\cong_{\tau_2}} h) = \cong_{\tau_1}$ and that h is defined for all elements of l. Now, let $(n, \cong, f) \in \mathcal{C}^{Eq}(\tau)$. By the definition of Eq-container values, we know that the function f satisfies $(\ker_{\cong_\tau} f|_n) = (\cong \cap (n \times n))$. So for $h = f|_n$, $\tau_1 = n$, $\cong_{\tau_1} = (\cong \cap (n \times n))$, and $\tau_2 = \tau$ we can apply the free theorem above and obtain $g \ (map \ f|_n \ l) = map \ f|_n \ (g \ l)$ for every list $l :: [n]$. Hence, we can reason similarly to the proof of Theorem 1 as follows:[3]

$$
\begin{aligned}
g_{\cong_\tau} \ (\square^{Eq} \ (n, \cong, f)) &= g_{\cong_\tau} \ (map \ f \ [0 \ .. \ (n-1)]) \\
&= g_{\cong_\tau} \ (map \ f|_n \ [0 \ .. \ (n-1)]) \\
&= map \ f|_n \ (g_{\cong \cap (n \times n)} \ [0 \ .. \ (n-1)]) \\
&= \square^{Eq} \ ((\square^{Eq})^{-1} \ (map \ f|_n \ (g_{\cong \cap (n \times n)} \ [0 \ .. \ (n-1)]))) \\
&= \square^{Eq} \ (length \ (g_{\cong \cap (n \times n)} \ [0 \ .. \ (n-1)]), \\
&\qquad \ker_{\cong_\tau} (f|_n \circ ((g_{\cong \cap (n \times n)} \ [0 \ .. \ (n-1)]) \ !!)), \\
&\qquad f|_n \circ ((g_{\cong \cap (n \times n)} \ [0 \ .. \ (n-1)]) \ !!)) \\
&= \square^{Eq} \ ((s, P) \ (n, \cong, f))
\end{aligned}
$$

where we set

$$(s, P) = (length \ (g_{\cong \cap (n \times n)} \ [0 \ .. \ (n-1)]), (g_{\cong \cap (n \times n)} \ [0 \ .. \ (n-1)]) \ !!)_{n \in \mathbf{Nat}, \cong \in \mathcal{E}(\mathbf{Nat})}$$

and use

$$f \circ ((g_{\cong \cap (n \times n)} \ [0 \ .. \ (n-1)]) \ !!) = f|_n \circ ((g_{\cong \cap (n \times n)} \ [0 \ .. \ (n-1)]) \ !!)$$

as well as

$$\ker_{\cong} ((g_{\cong \cap (n \times n)} \ [0 \ .. \ (n-1)]) \ !!) = \ker_{\cong_\tau} (f|_n \circ ((g_{\cong \cap (n \times n)} \ [0 \ .. \ (n-1)]) \ !!))$$

These two statements used here are true since $g_{\cong \cap (n \times n)} \ [0 \ .. \ (n-1)] :: [n]$ contains only elements from 0 to $n-1$ and since, for the second statement, we know that $(\ker_{\cong_\tau} \ f|_n) = (\cong \cap (n \times n))$.

Corollary 2. *For every function* $g :: Eq \ \alpha \Rightarrow [\alpha] \to [\alpha]$, *there exists an Eq-container morphism* (s, P) *such that* $g = \square^{Eq} \circ (s, P) \circ (\square^{Eq})^{-1}$.

Proof. By Theorem 2 and Lemma 3(1).

Lemma 5. *Let* $g, g' :: Eq \Rightarrow [\alpha] \to [\alpha]$. *Let* $(s, P), (s', P')$ *be Eq-container morphisms such that* $g = \square^{Eq} \circ (s, P) \circ (\square^{Eq})^{-1}$ *and* $g' = \square^{Eq} \circ (s', P') \circ (\square^{Eq})^{-1}$. *Then we have* $g \circ g' = \square^{Eq} \circ (s, P) \circ (s', P') \circ (\square^{Eq})^{-1}$.

Proof. Similarly to the proof of Lemma 2.

We have now established all the formal setup that is required for reasoning about functions of type $Eq \Rightarrow [\alpha] \to [\alpha]$ by instead reasoning about Eq-container morphisms. To manifest this with some examples, consider first the following container morphism versions of *nub* and *rmSingles*, where the first of these functions removes duplicates from a list and the second one throws away each element that appears only once in a given list (in both cases, ultimately with respect to an equivalence relation provided via a type class instance for *Eq*, of course):

---

[3] To highlight the changes of the equivalence relation that g uses, we have throughout subscripted each instance of g with the equivalence relation it actually works with.

$$nub^c = (s_n^\cong, P_n^\cong)_{n\in\mathbf{Nat},\cong\in\mathcal{E}(\mathbf{Nat})}$$
$$\text{with } s_n^\cong = |n_{/\cong}| \text{ and}$$
$$P_n^\cong = \lambda i \to \min\{j : |\{[k]_\cong : k \leq j\}| = i+1\}$$

$$rmSingles^c = (s_n^\cong, P_n^\cong)_{n\in\mathbf{Nat},\cong\in\mathcal{E}(\mathbf{Nat})}$$
$$\text{with } s_n^\cong = \sum_{e\in n_{/\cong},|e|>1} |e| \text{ and}$$
$$P_n^\cong = \lambda i \to \min\{j : |\{j' \in \bigcup_{e\in n_{/\cong},|e|>1} e : j' \leq j\}| = i+1\}$$

Note that we use standard notations $n_{/\cong}$ for factorisation with respect to an equivalence relation and $[k]_\cong$ for building equivalence classes.

As already noticed, we can view "ordinary" container morphisms as Eq-container morphisms as well. For an example, we give the application of $init^c$ to an Eq-container value. As we use them in the following examples of proofs, we show the result of applying nub^c and $rmSingles^c$, in general, as well.

$$init^c\ (n,\cong,f) = (n-1,\cong,f)$$
$$nub^c\ (n,\cong,f) = (|n_{/\cong}|, id, \lambda i \to f\ (\min\{j : |\{[k]_\cong : k \leq j\}| = i+1\}))$$
$$rmSingles^c\ (n,\cong,f) = (\sum_{e\in n_{/\cong},|e|>1} |e|,$$
$$\ker_\cong\ (\lambda i \to \min\{j : |\{j' \in \bigcup_{e\in n_{/\cong},|e|>1} e : j' \leq j\}|$$
$$= i+1\}),$$
$$\lambda i \to f\ (\min\{j : |\{j' \in \bigcup_{e\in n_{/\cong},|e|>1} e : j' \leq j\}|$$
$$= i+1\}))$$

Note that we used algebraic simplifications like that ($\ker_\cong\ id$) is $\cong$ and that the kernel of an (up to the relevant $\cong$) injective function is the identity.

Let us now demonstrate the usefulness of reasoning with our extended container notions, based on three examples.

An example proof with Eq-container morphisms. We wish to show that $nub \circ init$ always returns a prefix of the result of just nub. Using our new setup, we can do this by showing that for every Eq-container value c,

$$\mathbf{prefix}\ ((nub^c \circ init^c)\ c)\ (nub^c\ c) \tag{2}$$

holds, where **prefix** is defined by

$$\mathbf{prefix}\ (n_1,\cong_1,f_1)\ (n_2,\cong_2,f_2) \Leftrightarrow n_1 \leq n_2 \wedge \forall i \in n_1.\ f_1\ i = f_2\ i$$

To prove the desired statement, we take an arbitrary Eq-container value $c = (n,\cong,f)$ and first calculate both arguments to **prefix** in (2) above. We get

$$nub^c\ (init^c\ (n,\cong,f))$$
$$= nub^c\ (n-1,\cong,f)$$
$$= (|(n-1)_{/\cong}|, id, \lambda i \to f\ (\min\{j : |\{[k]_\cong : k \leq j\}| = i+1\}))$$

and

$$nub^c\ (n,\cong,f) = (|n_{/\cong}|, id, \lambda i \to f\ (\min\{j : |\{[k]_\cong : k \leq j\}| = i+1\}))$$

To verify the **prefix** property, we then have to establish the following statements:

1. $|(n-1)_{/\cong}| \le |n_{/\cong}|$
2. $\forall i \in |(n-1)_{/\cong}|.\ f\ (\min\{j\ :\ |\{[k]_\cong\ :\ k \le j\}| = i+1\}) = f\ (\min\{j\ :\ |\{[k]_\cong\ :\ k \le j\}| = i+1\})$

of which the first is a simple property of factorisation (of a subset, with respect to the same equivalence relation), and of which the second is a syntactic tautology.

Another example proof. We wish to show that $rmSingles \circ nub$ always returns an empty list. Using our new setup, we can do this by showing that for every Eq-container value c, the container value $(rmSingles^c \circ nub^c)\ c$ has 0 in its length component. So let $c = (n, \cong, f)$ be an Eq-container value. Then:

$$rmSingles^c\ (nub^c\ (n, \cong, f))$$
$$= rmSingles^c\ (|n_{/\cong}|, id, \lambda i \to f\ (\min\{j\ :\ |\{[k]_\cong\ :\ k \le j\}| = i+1\}))$$
$$= (\sum_{e \in |n_{/\cong}|_{/id}, |e| > 1} |e|, \cdots, \cdots)$$
$$= (0, \cdots, \cdots)$$

And yet another example proof. We wish to show that nub is idempotent, i.e., $nub \circ nub = nub$. Using our new setup, we can do this by showing that $nub^c \circ nub^c = nub^c$. So let $c = (n, \cong, f)$ be an Eq-container value. Then:

$$nub^c\ (nub^c\ (n, \cong, f))$$
$$= nub^c\ (|n_{/\cong}|, id, h)$$
$$\qquad \text{with } h = \lambda i \to f\ (\min\{j\ :\ |\{[k]_\cong\ :\ k \le j\}| = i+1\})$$
$$= (||n_{/\cong}|_{/id}|, id, \lambda i \to h\ (\min\{j\ :\ |\{[k]_{id}\ :\ k \le j\}| = i+1\}))$$
$$= (|n_{/\cong}|, id, h)$$
$$= nub^c\ (n, \cong, f)$$

where except for the next-to-last one all steps are simply by applying definitions. That one interesting step is valid by $|m_{/id}| = m$ for every $m \in \mathbf{Nat}$,[4] and by the fact that for every $i \in \mathbf{Nat}$,

$$\min\{j\ :\ |\{[k]_{id}\ :\ k \le j\}| = i+1\} = \min\{j\ :\ |\{\{k\}\ :\ k \le j\}| = i+1\}$$
$$= \min\{j\ :\ |\{\{0\}, \{1\}, \ldots, \{j\}\}| = i+1\}$$
$$= i$$

3.2   The Type Class *Ord*

As a second example for the adjustment of the container notions to type classes, we consider the type class *Ord*. Similarly to the adjustment for the type class

---

[4] Note that our notation overloading is at work here, according to which $m \in \mathbf{Nat}$ can represent the actual number m in one context and the set of numbers $\{0, \ldots, m-1\}$ in another context.

Eq, the container values and container morphisms have to be aware of the operation(s) now available on the formerly completely polymorphic list content. We use an approach analogous to that in Section 3.1, but replace equivalence relations by total preorders. We use $\mathcal{O}(M)$ to denote the class of all total preorders over a set M. For simplicity of notation, we freely regard a total preorder $\preceq$ on a subset n of **Nat** as the total preorder $\preceq \cup \{(i,j) \mid (i \in n \wedge n \leq j) \vee (n \leq i \leq j)\}$ on **Nat** when appropriate.

As the remaining definitions, results, and proofs are in close analogy to the ones for the type class Eq, we will give them in condensed form only, and omit all proofs. For a type τ that is an instance of Ord, we denote by $\preceq_\tau$ the corresponding fixed (in a given program) total preorder:

$$\preceq_\tau = \{(x,y) \mid x :: \tau,\, y :: \tau,\, (x <= y) = \textbf{True}\}$$

Definition 4. *Let τ be some type that is an instance of Ord. An Ord-container value of type τ is a triple $(n, \preceq, f)$ with $n :: \textbf{Nat}$, $\preceq \in \mathcal{O}(\textbf{Nat})$, and $f :: \textbf{Nat} \to \tau$ a partial function such that $(\ker_{\preceq_\tau} f|_n) = (\preceq \cap (n \times n))$.*[5] *The set of all such container values is denoted by $\mathcal{C}^{Ord}(\tau)$.*

Definition 5. *We define functions $\square^{Ord}$ and $(\square^{Ord})^{-1}$ as*

$$\square^{Ord} \qquad\quad :: Ord\ \alpha \Rightarrow \mathcal{C}^{Ord}(\alpha) \to [\alpha]$$
$$\square^{Ord}\ (n, \preceq, f) = map\ f\ [0\ ..\ (n-1)]$$

$$(\square^{Ord})^{-1} \qquad :: Ord\ \alpha \Rightarrow [\alpha] \to \mathcal{C}^{Ord}(\alpha)$$
$$(\square^{Ord})^{-1}\ xs\ = (length\ xs, \ker_{\preceq_\alpha} (xs\ !!), xs\ !!)$$

Definition 6. *An Ord-container morphism (s, P) is a family of pairs $(s_n^{\preceq}, P_n^{\preceq})$ $_{n \in \textbf{Nat}, \preceq \in \mathcal{O}(\textbf{Nat})}$ such that $s_n^{\preceq} :: \textbf{Nat}$ and $P_n^{\preceq} :: \textbf{Nat} \to \textbf{Nat}$ a partial function with $(P_n^{\preceq}\ i) \in n$ for every $i \in s_n^{\cong}$.*

Definition 7. *Let $(n, \preceq, f)$ be an Ord-container value and (s, P) an Ord-container morphism. The application of (s, P) to $(n, \preceq, f)$ is defined as*

$$(s, P)\ (n, \preceq, f) = (s_n^{\preceq}, \ker_{\preceq} P_n^{\preceq}, f \circ P_n^{\preceq})$$

Theorem 3. *For every function $g :: Ord\ \alpha \Rightarrow [\alpha] \to [\alpha]$, there exists an Ord-container morphism (s, P) such that $g = \square^{Ord} \circ (s, P) \circ (\square^{Ord})^{-1}$.*

Lemma 6. *Let $g, g' :: Ord \Rightarrow [\alpha] \to [\alpha]$. Let $(s, P), (s', P')$ be Ord-container morphisms such that $g = \square^{Ord} \circ (s, P) \circ (\square^{Ord})^{-1}$ and $g' = \square^{Ord} \circ (s', P') \circ (\square^{Ord})^{-1}$. Then we have $g \circ g' = \square^{Ord} \circ (s, P) \circ (s', P') \circ (\square^{Ord})^{-1}$.*

Comparing the definitions of morphisms on ordinary container values and on Ord-container values, we can again easily translate the former ones into the latter ones, analogously to Note 1. Moreover, every Eq-container morphism can be viewed as an Ord-container morphism as well.

---

[5] Note that the kernel of a function can not only be taken over an equivalence relation.

Note 2. Every *Eq*-container morphism $(s_n^{\cong}, P_n^{\cong})_{n \in \mathbf{Nat}, \cong \in \mathcal{E}(\mathbf{Nat})}$ can be viewed as the *Ord*-container morphism $(s_n^{\preceq \cap \succeq}, P_n^{\preceq \cap \succeq})_{n \in \mathbf{Nat}, \preceq \in \mathcal{O}(\mathbf{Nat})}$.

Clearly, there are also *Ord*-container morphisms that are no ordinary container morphisms and no *Eq*-container morphisms. They correspond exactly to the functions of type $Ord \Rightarrow [\alpha] \to [\alpha]$ that are not of type $Eq\ \alpha \Rightarrow [\alpha] \to [\alpha]$ (or even of type $[\alpha] \to [\alpha]$). For example, the Haskell function

```
least         :: Ord α ⇒ [α] → [α]
least []      = []
least (x : xs) = [go x xs]
          where  go x []      = x
                 go x (y : ys) = go (if x <= y then x else y) ys
```

corresponds to:

$$least^c = (s_n^{\preceq}, P_n^{\preceq})_{n \in \mathbf{Nat}, \preceq \in \mathcal{O}(\mathbf{Nat})}$$
$$\text{with}\quad s_n^{\preceq} = \min\{n, 1\} \text{ and}$$
$$P_n^{\preceq} = \lambda i \to \min\{j\ :\ \forall k \in (n \setminus j).\ j \preceq k\}$$

4 Conclusion and Future Work

We have extended the ellipsis [4] or container [7] technique for reasoning about functions on lists to the case of the presence of element tests. The key insight was to use, in the proofs of Theorems 2 and 3, an extension of free theorems [8,9] to ad-hoc polymorphism à la type classes [10]. An obvious goal for future work is to see what needs to be done to make reasoning with our refined container-related notions, as we have performed on examples by hand, more effective and mechanisable. Just as the techniques of Bundy and Richardson [4] and Prince et al. [7] have to rely on good proof tactics for arithmetics, our method will have to rely on tactics that additionally take properties of equivalence relations and total preorders into account, and that can exploit algebraic notions like the kernel of a function over a relation, etc.

Another issue is that of transforming function definitions we want to reason about into suitable container morphism representations in the first place. As we have seen with examples like $rmSingles^c$, describing a structure change as a result of element tests can be somewhat involved to express by a mathematical formula. Only more practical experience will be able to tell how problematic that really is. Note, though, that container morphism representations need not necessarily be provided by the "customer" of a proof system. Indeed, in Bundy and Richardson's setup the container versions of list functions were used internally only, not exposed to the user.

How about further extensions? We have already mentioned that moving from lists to a broader range of data structures is largely orthogonal to taking element tests into account. A more challenging extension is to treat other type classes than *Eq* and *Ord*. The framework of free theorems is readily available for other type classes as well. However, finding the right abstractions and morphism notions may appear to require new insights for each new class. For example, while both *Eq* and *Ord* mathematically correspond to relations, or to ways of *observing* elements of an unspecified type, what about type classes that provide ways of *constructing* elements via some operations, say class *Monoid*? Interestingly, recent work by Bernardy et al. [3] could shed some light here. For the purpose of testing (not verification), they essentially characterise polymorphic functions in terms of monomorphic inputs, such as characterising a function of type $[\alpha] \to [\alpha]$ in terms of its action on integer lists of the form $[1 .. n]$. For more complicated types, in particular higher-order ones, they work from a classification of function arguments (typically themselves functions) into observers and constructors, and describe a methodology for finding fixed types and monomorphic inputs that completely determine a function's behaviour. Via the dictionary translation method, type class constraints lead to precisely such different kinds of function arguments, so there is a good chance for leverage here.

Acknowledgements. We thank the anonymous reviewers for their comments and suggestions for improving the paper.

References

1. Abbott, M., Altenkirch, T., Ghani, N.: Categories of Containers. In: Gordon, A.D. (ed.) FOSSACS 2003. LNCS, vol. 2620, pp. 23–38. Springer, Heidelberg (2003)
2. Abbott, M., Altenkirch, T., Ghani, N.: Containers: Constructing strictly positive types. Theoretical Computer Science 342(1), 3–27 (2005)
3. Bernardy, J.-P., Jansson, P., Claessen, K.: Testing Polymorphic Properties. In: Gordon, A.D. (ed.) ESOP 2010. LNCS, vol. 6012, pp. 125–144. Springer, Heidelberg (2010)
4. Bundy, A., Richardson, J.: Proofs about Lists Using Ellipsis. In: Ganzinger, H., McAllester, D., Voronkov, A. (eds.) LPAR 1999. LNCS, vol. 1705, pp. 1–12. Springer, Heidelberg (1999)
5. Jay, C.B.: A semantics for shape. Science of Computer Programming 25(2–3), 251–283 (1995)
6. Peyton Jones, S.L. (ed.): Haskell 98 Language and Libraries: The Revised Report. Cambridge University Press (2003)
7. Prince, R., Ghani, N., McBride, C.: Proving Properties about Lists Using Containers. In: Garrigue, J., Hermenegildo, M.V. (eds.) FLOPS 2008. LNCS, vol. 4989, pp. 97–112. Springer, Heidelberg (2008)
8. Reynolds, J.C.: Types, abstraction and parametric polymorphism. In: Proc. of Information Processing, pp. 513–523. Elsevier (1983)
9. Wadler, P.: Theorems for free! In: Proc. of FPCA 1989, pp. 347–359. ACM (1989)
10. Wadler, P., Blott, S.: How to make ad-hoc polymorphism less ad hoc. In: Proc. of POPL 1989, pp. 60–76. ACM (1989)

Test-Case Generation for Maude Functional Modules[*]

Adrián Riesco

Facultad de Informática, Universidad Complutense de Madrid, Spain
ariesco@fdi.ucm.es

Abstract. Testing takes much of the time of the software development process, so several efforts have been devoted to automate it. We present here a tool that is able to generate test cases for Maude functional modules, and check their correctness with respect to a given specification or select a subset of these test cases to be checked by the user by using different strategies. Since these processes are very expensive we also present different trusting techniques to ease them.

Keywords: Test cases, Maude, black-box testing, white-box testing, code coverage.

1 Introduction

Testing takes much of the time of the software development process, so several efforts have been devoted to automate it. Although initially much progress was done in testing for imperative languages [17,14,12], during the last years several efforts have been devoted to develop test-case generators for declarative languages [9,10,4,2,5], being specially notable the development of Quickcheck [4], a very powerful test-case generator developed for Haskell (and coded in Haskell itself) that has been adapted to imperative languages as Java[1] or C++,[2] thus filling the gap between testing strategies for imperative and declarative languages. To perform testing we use *test cases*, whose definition depends on the programming language being tested, that the programmer uses to examine his program by checking the correctness of these test cases against an oracle, which usually is a specification of the system or the programmer himself.

These test cases are generated following two different strategies: black-box and white-box testing. The former uses a specification language, usually with a formal semantics, to generate the test cases that are later translated to test cases in the implementation language; a semantical relation must be established between the test cases in both languages to determine the correctness of the implementation. Examples of black-box testing are the translation to Java and C++ of the test cases generated by Quickcheck presented above and the language Congu,[3] a framework to create algebraic specifications to test Java programs. On the other hand, white-box testing (also known as glass-box testing) uses the

---

[*] Research supported by MICINN Spanish project *DESAFIOS10* (TIN2009-14599-C03-01) and Comunidad de Madrid program *PROMETIDOS* (S2009/TIC-1465).

[1] https://quickcheck.dev.java.net/

[2] http://software.legiasoft.com/quickcheck/

[3] http://gloss.di.fc.ul.pt/congu/

T. Mossakowski and H.-J. Kreowski (Eds.): WADT 2010, LNCS 7137, pp. 287–301, 2012.

current implementation of the system to select the most appropriate test cases. Both approaches have been followed in imperative and declarative contexts; black-box testing has been studied in imperative languages [12,13], in declarative languages [4,15,2], and in general contexts [1,11], while white-box testing has been investigated in [17,14] for imperative programming and in [10,9] for declarative programming.

Maude [6] is a high-level language and high-performance system supporting both equational and rewriting logic computation for a wide range of applications. Maude modules correspond to specifications in rewriting logic [16], a simple and expressive logic which allows the representation of many models of concurrent and distributed systems. This logic is an extension of equational logic; in particular, Maude functional modules correspond to specifications in membership equational logic [3], which, in addition to equations, allows the statement of membership axioms characterizing the elements of a sort. Rewriting logic extends membership equational logic by adding rewrite rules, that represent transitions in a concurrent system. Maude system modules are used to define specifications in this logic.

Although the initial aim of the Maude system was to be used as a specification language, the last releases of the system introduce new features such as TCP/IP sockets [6, Chapter 11] and unification [7] that encourage to use Maude as a programming language. Thus, Maude specifications grow in size and complexity, growing consequently the difficulty to debug and analyze them. As part of an ongoing project to debug Maude specifications, we have already implemented a declarative debugger for Maude [19] that allows to debug both wrong and missing answers (incorrect and incomplete results, respectively). Following this line, this paper presents a methodology to test Maude functional modules by using both black-box testing, where the specification language is Maude itself, and white-box testing, where we adapt some strategies already developed for declarative languages and, in addition, present a new strategy to test sort inferences. These techniques have been implemented in Maude and integrated with the the declarative debugger, which allows the user to debug the erroneous test cases at once.

Exploiting the fact that rewriting logic is *reflective* [8], a key distinguishing feature of Maude is its systematic and efficient use of reflection through its predefined META-LEVEL module [6, Chap. 14], that allows access to metalevel entities such as specifications or computations as usual data. Therefore, we are able to generate and check the test cases in Maude itself. The system provides another module, Full Maude [6, Chap. 18], that includes features for parsing, evaluating, and pretty-printing terms, improving the input/output interaction. By extending Full Maude our test-case generator, including its user interactions, is implemented in Maude itself.

Although the Maude metalevel allows an efficient implementation of black-box testing by providing mechanisms to test the correctness of a Maude module against another one and, consequently, its performance should be comparable to Quickcheck's [4], the main drawback of our approach is the term generation: while Quickcheck uses narrowing to obtain the test cases, the Maude's machinery for narrowing is still under development[4] and cannot be used with general Maude theories, so we incrementally

---

[4] Currently, the narrowing command only supports some theories, does not allows the user to introduce a condition the returned terms are assumed to fulfill, and does not allow incremental searches, that is, we cannot obtain new results without computing again the previous ones.

generate terms and then check whether they are appropriate for testing. Although our black-box testing is less efficient that the one in Quickcheck, we overcome it by providing white-box testing. This testing is based on [10], which adopted some techniques from imperative languages to select a set of terms fulfilling a giving coverage, that is, a number of statements that must be executed to consider the specification tested. We improve these coverage techniques by providing coverage for membership inferences, which takes into account both positive (statements used) and negative (statements that could not be used) information.

The rest of the paper is organized as follows. After briefly introducing Maude functional modules in Section 2, we describe how the terms are generated in Section 3. Our methodology to test Maude functional modules is described in Section 4, while Section 5 outlines the implementation of the tool. Section 6 concludes and outlines some future work.

More information about the test-case generator, related papers, examples, and its source code can be found at http://maude.sip.ucm.es/testing/.

2 Maude

Maude [6] is a declarative language based on both equational and rewriting logic for the specification and implementation of a whole range of models and systems. Functional modules define data types and operations on them by means of *membership equational logic* theories [3] that support multiple sorts, subsort relations, equations, and assertions of membership in a sort. In this way, Maude makes possible the faithful specification of data types (like sorted lists or search trees) whose data are not only defined by means of constructors, but also by the satisfaction of additional properties. It is important to note that in membership equational logic sorts are grouped into equivalence classes called *kinds*. For this purpose, two sorts are grouped together in the same equivalence class if and only if they belong to the same connected component.

For our purposes in this work we take advantage of the fact that membership equational logic theories are assumed to be terminating, confluent, and sort decreasing [6]. In this way, we can use a calculus that modifies the usual one shown in [3] by considering that equations are only applied from left to right, which allows us to infer judgments of the form $t \rightarrow_n t'$ and $t :_{ls} s$, introduced in [18] and which indicate, respectively, that the normal form of t is t' and that the least sort of t is s. Models of these judgments, given a signature Σ and a set of equations and membership axioms E, are Σ-term models $\mathcal{T}_{\Sigma/E}$ [16]; see [18] for details in the relation between models and judgments.

Below we present the basics of Maude functional modules and present an example that will be used throughout the rest of the paper.

2.1 Maude Functional Modules

Maude functional modules [6, Chapter 4], introduced with syntax fmod ... endfm, are executable membership equational specifications and their semantics is given by the corresponding initial membership algebra in the class of algebras satisfying the specification.

In a functional module we can declare sorts (by means of keyword sort(s)); subsort relations between sorts (subsort); operators (op) for building values of these sorts, giving the sorts of their arguments and result, and which may have attributes such as being associative (assoc) or commutative (comm), for example;[5] memberships (mb) asserting that a term has a sort; and equations (eq) identifying terms. Both memberships and equations can be conditional (cmb and ceq). In Maude the user can specify each operator with its own syntax, which can be prefix, postfix, infix, or any "mixfix" combination. This is done by indicating with underscores the places where the arguments appear in the mixfix syntax. Another interesting feature for our tool is that Maude allows both equations and membership axioms to be identified with a label, which is introduced after either the keyword eq or ceq (mb or cmb for memberships).[6]

Maude does automatic kind inference from the sorts declared by the user and their subsort relations. Kinds are *not* declared explicitly, and correspond to the connected components of the subsort relation. The kind corresponding to a sort s is denoted [s].

For example, we show how to specify lists of natural numbers in the module LIST below. We declare the sort List for these lists, while the subsort declaration indicates that a single natural number is also a list:

```
fmod LIST is
  pr NAT .

  sort List .
  subsort Nat < List .
```

Lists are built with the operator nil for empty lists and the juxtaposition operator _ _, which is associative and has nil as identity, for bigger lists:

```
  op nil : -> List [ctor] .
  op _ _ : List List -> List [ctor assoc id: nil] .
```

Finally, we define a function reverse to reverse a list. Note that this function is buggy: the equation labeled with rev1 should return nil instead of 0:

```
  var N : Nat .
  var L : List .

  op reverse : List -> List .
  eq [rev1] : reverse(nil) = 0 .
  eq [rev2] : reverse(N L) = reverse(L) N .
endfm
```

3   Term Generation

The tool is able to generate terms by using the constructor information provided by the user. As a first approach, we computed them in a recursive fashion: starting with constants, in each step the new terms were computed from the ones previously obtained.

[5] It is important to note that the equational theory works modulo these axioms.
[6] It is also possible to write this label at the end of the statement as an attribute, although we will always use labels in the way described above.

We can also understand this approach as a grammar, where sorts are non-terminals, constants are terminals, and operators are production rules. After each step, membership axioms were applied to ensure the terms were assigned the appropriate sort. For example, the terms in the LIST example above (assuming that the predefined natural numbers have constructors 0 and s_ for zero and successor) were generated as follows:

1. The constants 0 and nil for natural numbers and lists respectively are built.
2. Sort inference is applied. Thus, the term 0 is also considered a term of sort List.
3. Nontrivial constructors are applied. The term s(0) (pretty printed as 1 by Maude) is built for natural numbers, while the term 0 0 is generated for lists.
4. Steps 2 and 3 are applied until enough terms have been generated.

However, although this method builds up to several thousand terms very quickly, it presents a major drawback: most of the terms are very similar and thus they find the same bugs, while some other problems, that would be found with more complex terms, cannot be found due to the quick growth in the number of terms, that prevents the system from computing more terms once a few steps have been performed (although the user can select the number of steps that are applied in function of the complexity of the constructors, the amount of time required for big bounds greatly limits this option).

To palliate this problem we tried to use the narrowing features available in Maude, using the constructors to distinguish between the different kinds of terms and then trying to fulfill the conditions imposed by the equations and membership axioms. However, these narrowing features do not support general theories and some combinations of attributes cannot be used. Another major problem is that the narrowing command returns the first n solutions but, since it does not receive the condition to be fulfilled but only the lefthand side of the statement to be matched, it is possible to obtain terms that finally cannot be used as term cases, and thus more terms are needed but, with the current format, the system has to recalculate the n previous solutions. We expect this command to be improved soon and thus incorporate this feature to our test generator.

Since narrowing did not improve the tool as expected, for the time being we decided to randomly remove some terms in each iteration of the previous algorithm in order to reduce the number of combinations in the next levels and thus be able to generate bigger terms.[7] Once these terms are computed, we can start the testing process.

4 Testing Maude Functional Modules

We define a test case in Maude as a judgment $t \rightarrow_n t'$ or $t :_{ls} s$, where t and t' are terms and s is a sort. We describe in this section how, starting with the terms generated in the previous section, test cases of this form are generated in Maude and used for testing. First, we show in Section 4.1 how they can be checked against a correct specification; then Section 4.2 describes how to select a set of terms to be inspected by the user depending on different strategies. Finally, Section 4.3 explains how to improve the testing process by allowing the user to select some statements as trusted, preventing the tool from taking them into account when creating this set of terms.

---

[7] Although this technique does not guarantee that the terms are more suitable for testing, we have checked that it works better in practice.

4.1 Black-Box Testing

Usually, a good approach to testing consists in checking the correctness of several test cases against a specification of the system [4,12]. In our case this relation can be easily established because both the correct specification and the program under test are Maude specifications: assuming that $\mathcal{T}$ is the model of the correct specification and $\mathcal{T}'$ the model of the specification under test, then a test case j fulfills the specification when $\mathcal{T} \models j \iff \mathcal{T}' \models j$. This technique can be efficiently adopted in our prototype thanks to the reflective capabilities of Maude, that allow us to use modules as data. Thus, the tool compares the results obtained from the current specification with respect to the correct one and extracts several pieces of information: the results are different (either they have different constructors or the terms are equal but the inferred sorts are different), the term is not in normal form, or the results are incomparable. Note that it is not necessary to have a correct module with the same functions used in the tested module: if a property over the function to be tested can be defined, it is enough to define this property in a correct module as a constant function that always returns true:

```
fmod MY-SPEC is        fmod PROP is              fmod CORRECT is
   ...                     pr MY-SPEC .              pr MY-SPEC .
endfm                     op prop : ... -> Bool .   op prop : ... -> Bool .
                          eq prop(...) = ... .      eq prop(...) = true .
                          ...                     endfm
                       endfm
```

More specifically, we can define the property revProp for our lists specification, stating that the reverse of a composition of lists is equal to the composition of the reverses of the lists in inverse order, as follows:

```
fmod REV_LIST is
 pr LIST .
 vars L1 L2 : List .
 var N : Nat .

 op revProp : List List -> Bool .
 eq [prop] : revProp(L1, L2) = reverse(L2) reverse(L1) == reverse(L1 L2) .
endfm
```

Now, we create a new module CORRECT_LIST where a function with the same name and profile is defined as the constant true, that is, our specification indicates that this property is true:

```
fmod CORRECT_LIST is
  pr LIST .
  vars L1 L2 : List .

  op revProp : List List -> Bool .
  eq revProp(L1, L2) = true .
endfm
```

Now, we can use our tool to check the property. First, we identify which is the correct module, and then we start the testing process with the `test` command:

```
Maude> (correct test module CORRECT_LIST .)
CORRECT_LIST selected as correct module for testing.

Maude> (test in REV_LIST : revProp .)
8464 test cases were generated.
8464 test cases are incorrect with respect to the correct module.
```

Notice that the property never holds. We can ask the tool to show some of the incorrect test cases found, and use the debugger to fix the specification:

```
Maude> (show 1 incorrect .)
The following test cases are incorrect with respect to the correct module:
1. The term test(0,0) has been reduced to false

Maude> (invoke debugger with incorrect test case 1 .)
Declarative debugging of wrong answers started.
...
The buggy node is:
reverse(nil) -> 0
with the associated equation: rev1
```

Complete explanations of this example and the ones in the following sections, including the debugging sessions, are available at http://maude.sip.ucm.es/testing/.

4.2   White-Box Testing

Since Maude is a specification language itself, the user does not always have another specification (or is able to define a property) to check the results with. In this case the correctness of the test cases depends on the intended semantics given by the user, and hence a strategy that selects a subset of the generated terms, called *code coverage*, is needed in order to be easily checked by humans. We assume that this intended interpretation is a Σ-term model I corresponding to the model that the user had in mind while writing the specification, and thus we require that, given a test case j and the initial model $\mathcal{T}$ of the specification, $I \models j \iff \mathcal{T} \models j$.

Covering Equations. In [10] some strategies for selecting a coverage in functional languages are described: *global branch coverage* and *function coverage*. The former selects a set of terms such that they cover all branches (both direct and indirect) of the function being tested; the latter tries that, in addition to all branches of the original call to the function, also all branches of all recursive calls to that function have to be considered. Although function coverage is more difficult to apply, it detects more bugs in general than global branch coverage.

In the Maude case, these strategies select a subset of the equations and membership axioms in the specification and then looks for a set of test cases whose inference requires the application of the statements previously selected:

– Global branch coverage tries to find terms that use all the statements potentially used by the function under test (which, of course, also includes the functions in the conditions). That is, the coverage of a function symbol f using this strategy includes all the equations whose lefthand side matches the term $f(x_1, \ldots, x_n)$, where $x_1, \ldots, x_n$ are variables on the kinds specified by the program, and, for each equation $l = r$ if $\bigwedge_{i=1}^{n} t_i = t_i' \wedge \bigwedge_{j=1}^{m} t_j'' : s_j$ added to the coverage we must also add all the membership axioms for each sort s_j and the coverage for all the function symbols in the equation, $funs(r) \cup \bigcup_{i=1}^{n} funs(t_i) \cup funs(t_i') \cup \bigcup_{j=1}^{m} funs(t_j'')$, where

$$
\begin{aligned}
funs(f(t_1, \ldots, t_n)) &= \{f\} \cup funs(t_1) \cup \ldots \cup funs(t_n) \\
funs(a) &= \{a\} \\
funs(X) &= \emptyset
\end{aligned}
$$

For example, if we want to test the function `revProp` from our lists specifications, we should cover the equations `prop`, `rev1`, and `rev2`. We can use the tool to test it with the commands:

```
Maude> (global coverage .)
Global Branch Coverage selected

Maude> (test in REV_LIST : revProp .)
1 test cases have to be checked by the user:
    1. The term revProp(nil,0) has been reduced to false

All the statements were covered.

Maude> (invoke debugger with user test case 1 .)
...
```

Actually, reducing this term we cover `prop` (it is the only equation that can be initially used), and `rev1` and `rev2` (by reducing `reverse(nil)` and `reverse(0)` once the first equation has been applied). Once again, we can invoke the debugger to fix the specification by using this term.

– Function coverage checks that all the statements that can be applied for a given function are applied by all the recursive calls (including all those calls in the conditions) in the program. That is, if we try to compute the coverage of a function symbol f with respect to the recursive calls to a function r, then we must find all the appearances of r traversing the specification in the same way we explained for global branch coverage. Once all the reachable calls to r from f have been found, the coverage requires each of them to execute all the equations whose lefthand side matches $r(x_1, \ldots, x_n)$, with $x_1, \ldots, x_n$ variables of the appropriate kind.

In our lists example, if we want to test `revProp` taking into account the calls to `reverse` we have to distinguish between the four different calls to this function: the first one in `rev2` and three more in `prop`. Each one of these calls must execute both `rev1` and `rev2`. We can use our test-case generator to look for a coverage with the commands:

```
Maude> (function coverage .)
Function Coverage selected

Maude> (test in REV_LIST : revProp wrt reverse .)

2 test cases have to be checked by the user:
    1. The term revProp(0,0) has been reduced to false
    2. The term revProp(nil,nil) has been reduced to false

All calls were covered.

Maude> (invoke debugger with user test case 2 .)
...
```

In that case it is impossible to complete the coverage with only one term, because the calls in prop can only execute one of the equations for reverse with each test case: with the first test case all these calls execute rev2, while with the second one they execute rev1. Regarding the recursive call in rev2, it executes both equations when reducing reverse(0 0) from the first test case. Finally, note that both test cases detect the error and can be used to debug the specification.

Testing memberships. Maude functional modules contain not only equations; as said in the introduction, they also allow the user to define membership axioms and, although initially one could think that the strategies described above can be straightforwardly adapted to work in this case, we soon notice that to apply the axioms (and thus computing an erroneous sort) is as important as *not to apply* them (and thus obtaining a least sort bigger than expected). This problem does not arise with equations, because when a term is not reduced the test generator indicates it is not in normal form by using the constructors, while in this case the system cannot state whether the inferred sort is the least one or just one possible sort of the term.

For this reason, a new coverage strategy that takes into account this information (that we call negative) has been developed: some of the terms in the coverage have to apply all reachable statements *but also* some other terms have to fail, in a special way we will explain below, when trying to apply them. However, some constraints have to be applied to this negative information in order to obtain a realistic coverage strategy:

– It should not consider as negative information trivial failures, which in fact usually occurs when matching the current term with the lefthand side of a membership axiom. For example, assume we are defining the sort OList for ordered lists[8] and we state the following axiom:

```
cmb E E' L : OList if E <= E' /\ E' L : OList .
```

Of course, this membership cannot be applied to the test cases nil or 0, but this information is probably unimportant to the user, since even the number of subterms are different.

---

[8] We prefer "ordered lists" over "sorted lists" because "sort" is already used to refer to types in this context.

– However, asking the term to match the lefthand side of the axiom can also be too
restrictive, since the lefthand side can contain information about the sorts of the
terms. For example, we could replace the previous membership axiom for our ordered lists specification with the following one:

```
cmb E OL : OList if E' OL' := OL /\ E <= E' .
```

where the variables OL and OL' have sort OList. In that case, if we only consider
terms matching the lefthand side as negative information we are discarding important terms: those that cannot be applied because the membership for OList is
wrong and thus prevents the term from matching.

– To solve these problems we have decided to consider as valid test cases those that
match the lefthand side of the membership axiom *at the kind level*. That is, we
consider the variables in the lefthand side as declared in their corresponding kind
and then we add the matching at the sort level as the new first condition of the
membership axiom.

Besides this problem, another question arises when taking into account the negative
information: is it necessary to check that each condition fails? Although in general
this approach would detect more errors, with medium examples the computation of
the coverage takes too much time to be useful. For this reason we have decided to
consider that a membership axiom provides enough negative information when any of
its conditions (including the ad hoc condition indicating that the sorts of the terms are
correct) fails.

Following the ideas presented previously, assume that we specify ordered lists of
natural numbers with:

```
(fmod OLIST is
 pr NAT .

 sorts List OList .
 subsort OList < List .

 op nil : -> OList [ctor] .
 op _:_ : Nat List -> List [ctor] .
 cmb [oll] : (N : N' : L) : OList if N <= N' /\ N' : L : OList .
endfm)
```

That is, the membership axiom stating that singleton lists are ordered lists is missing.
We can look for test cases for this specification with the command:

```
Maude> (test sort in OLIST : OList .)

1 test cases have to be checked by the user:
    1. The term 0 : 0 : nil has least sort List

The following statements were not checked with the given test cases:
oll
```

```
All the negative information was covered.

Maude> (invoke debugger with user test case 1 .)
...
The buggy node is:
The least sort of 0 : nil is List
Either the operator _:_ needs more membership axioms or the conditions
of the current axioms are not written in the intended way.
```

The tool is not able to apply o11 (actually, it cannot be applied without the membership axiom for singleton lists) but it informs the user that it has found a term that, although it matches the lefthand side of one of the memberships, it cannot be finally applied. In fact, the term should have as least sort OList instead of List, and thus it reveals the failure in our specification.

4.3 Enhancing the Performance

While developing the Maude declarative debugger several buggy specifications, describing all possible errors, were developed. The tool has successfully generated test cases for all the functional examples. However, the main drawback of the tool is its poor performance when facing large specifications, specially when computing the code coverage.

More specifically, although the term generator is able to build up to ten thousand test cases, only the testing with respect to a correct module can use all these test cases, while when computing the coverage it is recommended to select a lower bound for the number of test cases to be checked. The coverage is computed quite slowly (it works with less than one thousand cases), due both to the fact that it performs several operations at the metalevel (see Section 5 for details) and that it computes the minimum coverage, which has exponential complexity.

To improve the performance, a trusting mechanism that hastens the computation of the coverage has been developed: some statements can be pointed out as correct, and thus the tool will omit them when computing the required coverage. The tool offers several options to trust the statements: only labeled statements are taken into account when generating the coverage, specific statements can be trusted, and even complete modules can be selected as correct.

The previous examples are very simple and thus the trusting mechanisms cannot be applied with all their power. We could trust the equation rev2 with the commands:

```
Maude> (set test select on .)
Debug select is on for test generation.

Maude> (test include REV_LIST .)
Labels prop rev1 rev2 have been added to the coverage.

Maude> (test deselect rev2 .)
Labels rev2 have been excluded from the coverage.
```

The first command initializes the trusting mode, the second one introduces all the labels in the (flattened) module REV_LIST as suspicious, and the third one trusts the equation rev2. We can use now function coverage with our initial example:

```
Maude> (test in REV_LIST : revProp wrt reverse .)

1 test cases have to be checked by the user:
    1. The term revProp(nil,nil) has been reduced to false

All calls were covered.
```

Note that now one test case is enough to cover all the (non-trusted) equations for reverse.

Finally, the tool also allows to trust a specific kind of statement of different modules with the command:

```
(test include/exclude eqs/mbs MODULES .)
```

where MODULES is a list of module names separated by spaces.

5   Implementation

We present in this section how the ideas shown in the previous sections have been implemented. This implementation makes extensive use of Maude metalevel [6, Chapter 3], which allows metalevel entities such as terms and modules be used as usual data. Moreover, the test-case generator, as well as the declarative debugger, is implemented on top of Full Maude [6, Chap. 18], which improves the input/output loop provided by the LOOP-MODE [6, Chapter 17] with several parsing features. In this way, we are able to generate the term cases, compute the coverage, check the correctness of the test cases against a correct module, and implement the user interface in Maude itself.

The first phase in the implementation of the tool is the term generator. To build the terms the tool traverses all the operators in the specification looking for those with the ctor attribute indicating that they are constructors of the given sort. As explained in Section 3, it first selects the constant constructors (those whose arity is nil) and then the rest of operators are used, using as arguments the terms obtained in the previous steps. However, when creating these new terms we must be careful with the operator attributes, that can identify terms that at first sight are different. To take into account these attributes we use the predefined function metaNormalize, that computes the normal form of the term with respect to the equational theory consisting of these equational attributes. Finally, after each step we use the predefined function leastSort to obtain the least sort of the term and then add it to the set of all its supersorts.

Black-box testing is implemented in a straightforward way; we use the function metaReduce in both the correct module and the module under test, and then we check that both the term and the sort correspond. White-box testing is more complicated: starting from the function to be tested, we check all the possible paths in order to keep the reachable statements, in the case of global branch coverage, or the reachable recursive calls, in the case of function coverage. Once the needed coverage has been computed we execute the test cases obtained in the previous step; however, the usual way of executing a term in a functional module is just obtaining the result, while in our case we need to examine each term to keep the coverage thus far. To do this we use the function metaMatch to check whether the current term matches the lefthand side of an equation and fulfills the conditions and, in case the matching succeeds, we apply the obtained substitution to the righthand side, which generates the next term to be examined.

Regarding the interaction with the user, we have extended the internal state of the loop shown in [19] with attributes to keep the type of coverage selected, the trusting information, the test cases, and the type of error detected by each test case (in case we are using black-box testing). With these attributes and the new commands described in this paper we are able to combine the declarative debugger with the test-case generator, which shows the scalability of the system.

6   Concluding Remarks and Ongoing Work

This work is the first step toward developing a test-case generator for Maude specifications. Currently, the tool allows the user to debug functional modules following two different strategies: black box and white box. While the former compares the results obtained in the module under test with those obtained in a correct specification, the latter selects a set of terms in such a way that they fulfill a so called code coverage. In addition to known coverage strategies like global branch and function coverage, that have been adapted to the Maude case, we have designed a membership coverage that takes into account not only the statements applied, but also the memberships that were not applied.

Regarding scalability, we distinguish between the scalability with respect to the complexity of the constructors and with respect to the number of statements. In the first case the tool only scales well for medium-sized specifications, because the number of terms generated for a given sort in each step of the term-generation process depends on the number of terms built for the sorts used as arguments and thus, if several levels are needed to build the sort (i.e., if the sort is complex) then each step is very expensive and only a few can be taken before the system collapses. In the second case, the tool works even for large specifications, since the complexity does not depend on the size of the specification but on the complexity of the function being tested (number of statements/recursive calls); moreover, the trusting mechanisms work better for large (and structured) specifications, since we expect the user to test the imported modules before using them, and thus they can be trusted.

For the reasons sketched above, most of the ongoing work is devoted to improve the performance of the tool. We are now working on the term generator. The narrowing command working on the Maude metalevel is being enhanced to allow consecutive searches in an efficient way (currently, it recomputes the previous results). Using this command we can generate terms, check whether these terms fulfill the conditions of any of the statements under test, and then continue generating terms until the required number of terms have been generated. It will also be required an extension of narrowing to more theories than the currently supported, especially taking membership axioms into account.

The prototype can be improved, first, with new strategies (both new coverage strategies and black-box testing) and, second, by enhancing its performance by providing new trusting mechanisms. We also intend to improve the current coverage strategies: currently, the smallest set of terms fulfilling the selected strategy are presented; however, it could be easier for the user to check a big set of simple terms than a small set of very complex terms. Thus, we are developing different strategies to allow the user to select the most appropriate set of test cases depending on his expertise. We also plan to allow the user to fix some complex values (e.g. tables and arrays which do not change the behavior of the function) in the functions to be tested, so the test-case generator can focus on the rest of parameters. We intend to improve the performance of all these tasks by using a distributed architecture, where each processor is in charge of a specific task while another processor gathers and handles all the information.

Since Maude is a specification language, it would be interesting to use Maude to specify a system and another language to implement it. Currently, this approach is been followed to teach data structures at the Universidad Complutense: the data structures are first specified in Maude and then implemented in C++. To test them a translation from Maude to C++, written by hand for each data structure, is required. The results obtained from this experience will be used to develop translations to other languages.

An extension to system modules is also outlined; since these modules are not required to be either terminating or confluent, the test cases must take into account different information. Probably, a coverage strategy that checks which terms cannot be further rewritten (i.e., provides negative information) will be useful. Finally, the graphical user interface is being updated to connect the test-case generator with the Maude declarative debugger.

Acknowledgments. I thank Sebastian Fischer for his kind explanations of his coverage strategies, Fernando Orejas for his help in preliminary versions of the paper, Ricardo Peña for his useful comments on previous versions of the tool, and Markus Roggenbach for his help with the final version of the paper.

References

1. Bernot, G.: Testing Against Formal Specifications: A Theoretical View. In: Abramsky, S., Maibaum, T.S.E. (eds.) TAPSOFT 1991, CCPSD 1991, and ADC-Talks 1991. LNCS, vol. 494, pp. 99–119. Springer, Heidelberg (1991)
2. Borba, P., Cavalcanti, A., Sampaio, A., Woodcook, J. (eds.): PSSE 2007. LNCS, vol. 6153. Springer, Heidelberg (2010)
3. Bouhoula, A., Jouannaud, J.-P., Meseguer, J.: Specification and proof in membership equational logic. Theoretical Computer Science 236, 35–132 (2000)
4. Claessen, K., Hughes, J.: Quickcheck: A lightweight tool for random testing of Haskell programs. In: ACM SIGPLAN Notices, pp. 268–279. ACM Press (2000)
5. Claessen, K., Smallbone, N., Hughes, J.: QUICKSPEC: Guessing Formal Specifications Using Testing. In: Fraser, G., Gargantini, A. (eds.) TAP 2010. LNCS, vol. 6143, pp. 6–21. Springer, Heidelberg (2010)
6. Clavel, M., Durán, F., Eker, S., Lincoln, P., Martí-Oliet, N., Bevilacqua, V., Talcott, C.: All About Maude - A High-Performance Logical Framework. LNCS, vol. 4350. Springer, Heidelberg (2007)

7. Clavel, M., Durán, F., Eker, S., Lincoln, P., Martí-Oliet, N., Meseguer, J., Talcott, C.: Maude Manual (Version 2.5) (June 2010), http://maude.cs.uiuc.edu/maude2-manual

8. Clavel, M., Meseguer, J., Palomino, M.: Reflection in membership equational logic, many-sorted equational logic, Horn logic with equality, and rewriting logic. Theoretical Computer Science 373(1-2), 70–91 (2007)

9. Degrave, F., Schrijvers, T., Vanhoof, W.: Automatic generation of test inputs for Mercury. In: 18th International Symposium on Logic-Based Program Synthesis and Transformation (LOPSTR 2008), Valencia, Spain, July 17-18, 2008, Revised Selected Papers, pp. 71–86. Springer, Heidelberg (2009)

10. Fischer, S., Kuchen, H.: Systematic generation of glass-box test cases for functional logic programs. In: Proceedings of the 9th ACM SIGPLAN International Conference on Principles and Practice of Declarative Programming, PPDP 2007, pp. 63–74. ACM Press, New York (2007)

11. Gaudel, M.-C., Le Gall, P.: Testing Data Types Implementations from Algebraic Specifications. In: Hierons, R.M., Bowen, J.P., Harman, M. (eds.) FORTEST. LNCS, vol. 4949, pp. 209–239. Springer, Heidelberg (2008)

12. Hierons, R.M., Bogdanov, K., Bowen, J.P., Rance Cleaveland, J.D., Dick, J., Gheorghe, M., Harman, M., Kapoor, K., Krause, P., Lüttgen, G., Simons, A.J.H., Vilkomir, S., Woodward, M.R., Zedan, H.: Using formal specifications to support testing. ACM Computing Surveys 41(2), 1–76 (2009)

13. Koopman, P., Alimarine, A., Tretmans, J., Plasmeijer, R.: GAST: Generic Automated Software Testing. In: Peña, R., Arts, T. (eds.) IFL 2002. LNCS, vol. 2670, pp. 84–100. Springer, Heidelberg (2003)

14. Lembeck, C., Caballero, R., Müller, R.A., Kuchen, H.: Constraint solving for generating glass-box test cases. In: Kuchen, H. (ed.) Proceedings of International Workshop on Functional and (Constraint) Logic Programming (WFLP 2004), pp. 19–32 (2004)

15. Machado, P.D.L.: On Oracles for Interpreting Test Results against Algebraic Specifications. In: Haeberer, A.M. (ed.) AMAST 1998. LNCS, vol. 1548, pp. 502–518. Springer, Heidelberg (1998)

16. Meseguer, J.: Conditional rewriting logic as a unified model of concurrency. Theoretical Computer Science 96(1), 73–155 (1992)

17. Müller, R.A., Lembeck, C., Kuchen, H.: A symbolic Java virtual machine for test case generation. In: IASTED Conf. on Software Engineering, pp. 365–371 (2004)

18. Riesco, A., Verdejo, A., Martí-Oliet, N.: Enhancing the Debugging of Maude Specifications. In: Ölveczky, P.C. (ed.) WRLA 2010. LNCS, vol. 6381, pp. 226–242. Springer, Heidelberg (2010)

19. Riesco, A., Verdejo, A., Martí-Oliet, N., Caballero, R.: Declarative debugging of rewriting logic specifications. Technical Report SIC-02-10, Dpto. Sistemas Informáticos y Computación, Universidad Complutense de Madrid (2010), http://maude.sip.ucm.es/debugging

Author Index